Karl Frauendorfer · Hans Glavitsch
Rainer Bacher (Eds.)

Optimization in Planning and Operation of Electric Power Systems

Lecture Notes of the SVOR/ASRO Tutorial
Thun, Switzerland, October 14-16, 1992

With 57 Figures

Springer-Verlag Berlin Heidelberg GmbH

PD Dr. Karl Frauendorfer
Institute for Operations Research
University of St. Gallen
for Business Administration,
Economics, Law and Social Sciences
Bodanstr. 6
CH-9000 St. Gallen
Switzerland

Prof. Dr. Hans Glavitsch
Dr. Rainer Bacher
Institute of Electric Power Transmission
and High Voltage Technology
ETH-Zentrum
CH-8092 Zurich
Switzerland

ISBN 978-3-7908-0718-9 ISBN 978-3-662-12646-2 (eBook)
DOI 10.1007/978-3-662-12646-2

88/7130-543210 - Printed on acid-free paper

PREFACE

Permanently increasing requirements in power supply necessitate efficient control of electric power systems. An emerging subject of importance is optimization which has been the challenging principal theme of a tutorial on *"Optimization in Planning and Operation of Electric Power Systems"* held in Thun (Switzerland) in October 1992. This tutorial was organized by the Swiss Association of Operations Research (SVOR) in collaboration with the Power Engineering Society (PES) as member of the Swiss Institute of Electrical Engineers (SEV). The objective of the tutorial has not been only to present the state-of-the-art of mathematical programming and power system control but also to promote the exchange of experience between the specialist of the respective disciplines.

Lectures on modelling aspects of unit commitment and optimal power flow have provided the introduction to power systems control and to its associated problem statement. Due to the nature of the underlying optimization problems recent developments in advanced and well-established mathematical programming methodologies have been presented, illustrating in which way dynamic, separable, continuous and stochastic features might be exploited. In completing the various methodologies a number of presentations have stated experiences with optimization packages currently used for unit commitment and optimal power flow calculations.

Due to the success of this tutorial and the many fruitful communications between operations research experts, analysts on the application's side and users in the power industry, the organizers have decided to edit the submitted revised lecture notes on *mathematical programming methodologies*, on *unit commitment* and on *optimal power flow* for publication as proceedings. This work is thought to represent a state-of-the-art of the presented topics and their applications in power system control. We would like to thank Dr. H. Schiltknecht, president of the SVOR, for his support and for the generous freedom, we received during the organization of the tutorial. We have also greatly appreciated the financial support by the Swiss Academy of Engineering Societies (SATW).

K. Frauendorfer, H. Glavitsch, R. Bacher
(editors)

St.Gallen, Zurich, May 1993

CONTENTS

I Mathematical Programming Methodology 1

Basic Issues in Lagrangian Optimization 3
R.T. Rockafellar

1. Formulation of optimization problems 3
2. Optimality conditions 8
3. Extended problem models 17

Dynamic Programming - Basic Concepts and Applications 31
K. Neumann

1. Deterministic dynamic programming 31
2. Stochastic dynamic programming 45

Interior Point Methodology for Linear Programming: Duality, Sensitivity Analysis and Computational Aspects 57
B. Jansen, C. Roos, T. Terlaky, J.-Ph. Vial

1. Introduction 58
2. A new approach to the theory of linear programming 60
3. Parametric analysis 81
4. A primal-dual interior point algorithm 99
5. Implementation 110
6. Conclusions 118

Approaches to Stochastic Programming with Application to Electric Power Systems 125
G.B. Dantzig, G. Infanger

1. Introduction 125
2. Two-stage stochastic linear programs 126
3. Multi-stage stochastic linear programs 127
4. Multidimensional integration 129
5. Importance sampling 130
6. Benders decomposition 133
7. Implementation 134
8. Conclusion 137

II Unit Commitment 141

Unit Commitment and Thermal Optimization - Problem Statement 143
H. Braun

1. Introduction 144
2. The planning tasks in the electric energy supply 148
3. Structure of the power systems 152
4. Long-term planning 152
5. Short term operational planning 159
6. Summary and outlook 168

Experiences with Optimization Packages for Unit Commitment 173
H. Sanders, K. Linke

1. Introduction 173
2. Load modelling 174
3. Generation modelling 176
4. VEW optimization levels 179
5. Annual optimization 181
6. Medium term planning 188
7. Conclusions and consequences 193

Modelling in Hydro-Thermal Optimization 199
A. Schadler, E. Steinbauer

1. Problem definition 199
2. Modelling 205
3. Optimization 209
4. Computational experience 211

III Optimal Power Flow 215

Power System Models, Objectives and Constraints in Optimal Power Flow Calculations 217
R. Bacher

1. Introduction 218
2. The role of the optimal power flow (OPF) computation within the overall power system control 220
3. The power flow model as equality constraint set of the OPF 223
4. Mathematical formulation of operational constraints 233
5. OPF objectives and objective functions 247
6. The complete OPF formulation 252

Use of Linear and Quadratic Programming Techniques in Exploiting the Nonlinear Features of the Optimal Power Flow 265
H. Glavitsch

1. Introduction 266
2. Characteristics of the nonlinearities 267
3. Incremental forms of system relations 271
4. Solution concepts and considerations on the choice of a method 277
5. Economic dispatch - a sample problem 281
6. Linear programming as a solution method 283
7. Use of a standard quadratic programming method 288
8. Dual method of quadratic programming 292
9. Loss minimization by reactive optimization based on LP 298
10. The Newton optimal power flow 299
11. Concluding remarks 299

Optimal Power Flow Packages - Requirements and Experiences 309
A. Papalexopoulos

1. Introduction 310
2. Overview of the OPF problem 312
3. Operational requirements for an on-line OPF implementation 317
4. Experience with OPF packages in a practical environment 338
5. Conclusions 344

Cost/Benefits Analysis of the Optimal Power Flow 349
K. Kato

1. Introduction 350
2. Approximate analysis 350
3. Accurate analysis 351
4. Conclusions 364

Chapter I

MATHEMATICAL PROGRAMMING METHODOLOGY

BASIC ISSUES IN LAGRANGIAN OPTIMIZATION

R. Tyrrell Rockafellar

Dept. of Applied Mathematics
University of Washington FS-20
Seattle, WA 98195, USA

Abstract. These lecture notes review the basic properties of Lagrange multipliers and constraints in problems of optimization from the perspective of how they influence the setting up of a mathematical model and the solution technique that may be chosen. Conventional problem formulations with equality and inequality constraints are discussed first, and Lagrangian optimality conditions are presented in a general form which accommodates range constraints on the variables without the need for introducing constraint functions for such constraints. Particular attention is paid to the distinction between convex and nonconvex problems and how convexity can be recognized and taken advantage of.

Extended problem statements are then developed in which penalty expressions can be utilized as an alternative to black-and-white constraints. Lagrangian characterizations of optimality for such problems closely resemble the ones for conventional problems and in the presence of convexity take a saddle point form which offers additional computational potential. Extended linear-quadratic programming is explained as a special case.

1. Formulation of optimization problems

Everywhere in applied mathematics the question of how to choose an appropriate mathematical model has to be answered by art as much as by science. The model must be rich enough to provide useful qualitative insights as well as numerical answers that don't mislead. But it can't be too complicated or it will become intractable for analysis or demand data inputs that can't be supplied. In short, the model has to reflect the right balance between the practical issues to be addressed and the mathematical approaches that might be followed.

This means, of course, that to do a good job of formulating a problem a modeler needs to be aware of the pros and cons of various problem statements that might serve as templates, such as standard linear programming, quadratic programming, and the like. Knowledge of which features are advantageous, versus which are potentially troublesome, is essential. In optimization the difficulties can be all the greater because the key ideas are often different from the ones central to the rest of applied mathematics. For instance, in many subjects the crucial division is between linear and nonlinear models, but in optimization it is between convex and nonconvex. Yet convexity is not a topic much treated in a general mathematical education.

Problems of optimization always focus on the maximization or minimization of some function over some set, but the way the function and set are specified can have a great impact. One distinction is whether the decision variables involved are "discrete" or "continuous." Discrete variables with integer values, in particular logical variables which can only have the values 0 or 1, are appropriate in circumstances where a decision has to be made whether to build a new facility, or to start up a process with fixed initial costs. But the introduction of such variables in a model is a very serious step; the problem may become much harder to solve or even to analyze. Here we'll concentrate on continuous variables.

The conventional way to think about an optimization problem in finitely many continuous variables is that a function $f_0(x)$ is to be minimized over all the points $x = (x_1, \ldots, x_n)$ in some subset C of the finite-dimensional real vector space $\mathbb{R}^n$. (Maximization is equivalent to minimization through multiplication by -1.) The set C is considered to be specified by a number of side conditions on x which are called constraints, the most common form being equality constraints $f_i(x) = 0$ and inequality constraints $f_i(x) \leq 0$. As a catch-all for anything else, there may be an "abstract constraint" $x \in X$ for some subset $X \subset \mathbb{R}^n$. For instance, X can be thought of as indicating nonnegativity conditions, or upper and lower bounds, on some of the variables x_j appearing as components of x. Such conditions could be translated one by one into the form $f_i(x) \leq 0$ for additional functions f_i, but this may not be convenient.

The conventional statement of a general problem of optimization from this point of view is

$$(\mathcal{P}) \qquad \begin{aligned} &\text{minimize } f_0(x) \text{ over all } x \in X \\ &\text{such that } f_i(x) \begin{cases} \leq 0 & \text{for } i = 1, \ldots, s, \\ = 0 & \text{for } i = s+1, \ldots, m. \end{cases} \end{aligned}$$

The points x satisfying the constraints in $(\mathcal{P})$ are called the *feasible solutions* (i.e., candidates for solutions) to the problem. They form a certain set $C \subset \mathbb{R}^n$, and it is over this that the function f_0 is to be minimized. A point $\bar{x} \in C$ is a (globally) *optimal solution* to $(\mathcal{P})$ if $f_0(\bar{x}) \leq f_0(x)$ for all $x \in C$. It is a *locally optimal solution* if there is a neighborhood V of $\bar{x}$ such that $f_0(\bar{x}) \leq f_0(x)$ for all $x \in C \cap V$. The *optimal value* in $(\mathcal{P})$ is the minimum value of the objective function f_0 over C, as distinguished from the point or points where it's attained, if any.

In dealing with a problem in the format of $(\mathcal{P})$, people usually take for granted that the functions $f_0, f_1, \ldots, f_m$ are second-order smooth (i.e., have continuous second partial derivatives). We'll do that too, but there are important modeling issues here that shouldn't be swept under the rug. We'll return to them in Section 3 in discussing how penalty expressions may in some situations be a preferable substitute for "exact" equality or inequality constraints of the sort in $(\mathcal{P})$.

Concerning the set X, we'll assume here for simplicity that it's *polyhedral*, or in other words, definable in terms of a finite system of linear constraints, these being conditions that *could*, if we so wished, be written in the form $f_i(x) \leq 0$ or $f_i(x) = 0$ for additional functions f_i that are *affine* (linear-plus constant). The main example we have in mind is the case where X is a *box*, $X = X_1 \times \cdots \times X_n$ with X_j a closed (nonempty but not necessarily bounded) interval in $\mathbb{R}$. Then, of course, the condition $x \in X$ reduces to $x_j \in X_j$ for $j = 1, \ldots, n$. If $X_j = [0, \infty)$ the condition $x_j \in X_j$ requires x_j to be nonnegative. If $X_j = [a_j, b_j]$, it requires x_j to lie between the bounds a_j and b_j. If $X_j = (-\infty, \infty)$ it places no restriction on x_j. The latter case is a reminder that even the general condition $x \in X$ in $(\mathcal{P})$ doesn't necessarily restrict x, because we can always take X to be all of $\mathbb{R}^n$ when we want to deal in effect with constraints of type $f_i(x) \leq 0$ or $f_i(x) = 0$ only. The whole space $\mathbb{R}^n$ is considered to be a polyhedral subset of $\mathbb{R}^n$, as is the empty set $\emptyset$; singleton sets (consisting of exactly one point) are polyhedral as well.

A technical point that shouldn't be overlooked in setting up a model is the existence of a solution. If that isn't guaranteed by the formulation, something's wrong; note that the issue isn't whether the "application" has a solution in some sense (e.g. the existence in principle of a best mode of operating a given system), but whether the mathematical description of the problem is adequate. Under the assumptions we have given for $(\mathcal{P})$ a simple condition guaranteeing the existence of at least one optimal solution, provided there is at least one feasible solution, is the *boundedness* of the set X. (Boundedness of X means that for each coordinate x_j of x, there is an upper bound to x_j as x ranges over X and also a lower bound.) A more flexible criterion would be the boundedness, for each $\mu > 0$, of the set of all $x \in X$ satisfying $f_i(x) \leq \mu$ for $i = 0, 1, \ldots, s$ and $|f_i(x)| \leq \mu$ for $i = s+1, \ldots, m$.

Convexity has already been mentioned as a critical property in optimization which needs to be recognized and taken advantage of as far as possible when it is present. A set $C \subset \mathbb{R}^n$ is said to be *convex* if it contains along with any two different points the line segment joining those points:

$$x \in C,\ x' \in C,\ 0 < t < 1 \quad \Longrightarrow \quad (1-t)x + tx' \in C. \tag{1.1}$$

(In particular, the empty set is convex, as are sets consisting of a single point.) A real-valued function f on $\mathbb{R}^n$ is called *convex* if it satisfies the inequality

$$f((1-t)x + tx') \leq (1-t)f(x) + tf(x') \text{ for any } x \text{ and } x' \text{ when } 0 < t < 1. \tag{1.2}$$

It's *concave* if the opposite inequality always holds, and *affine* under equality; the affine functions $f : \mathbb{R}^n \to \mathbb{R}$ have the form $f(x) = v{\cdot}x + \text{const}$. Finally, f is *strictly* convex if "$\leq$" can be replaced by "$<$" in (1.2); it's strictly concave in the case of "$>$."

Convexity is a large subject which can barely be touched on here; a book with many details is [1]. The importance of convexity in optimization comes from the following crucial properties.

Theorem 1.1.

(a) *In minimizing a convex function f_0 over a convex set C, every locally optimal solution $\bar{x}$ (if there is one) is globally optimal.*

(b) *In minimizing a strictly convex function f_0 over a convex set C, there can be no more than one optimal solution.*

In contrast to these properties of "convex optimization," two major difficulties with "nonconvex optimization" stand out. First, there is *virtually no way* to arrive for sure at a globally optimal solution. There are some global optimization techniques, more or less amounting in practice to forms of random search, but even with these one generally has to be content merely with a statistical prospect of probably locating the true minimum eventually through persistence. In practice when applying an optimization package, for instance, one should be skeptical about any claims an optimal solution has been found, in the absence of convexity. Just because a sequence of points generated by a method seems to settle down and converge to something, that doesn't necessarily mean that an optimal solution is being approximated. This is a central issue in the analysis of algorithms. At best, with well designed methods that are soundly based on theoretical principles, one can hope that a locally optimal solution has been located, but that would still leave open the possibility that some other locally optimal solution—a better one—exists nearby.

Second, there is *virtually no way* to know that a problem has a *unique* optimal solution, apart from the strict convexity criterion just offered. This is even true in convex optimization. It's not wise therefore to speak of "the" solution to a problem of general type. Of course, a problem may well turn out to have a unique solution; the trouble is that we can't know that in advance, nor in the nonconvex case can we even hope to check whether an optimal solution already found (if that were possible) is unique.

These observations about what can go wrong without convexity might be regarded as raising false issues, in a sense. For some practitioners, it may be enough just to use optimization methodology to achieve improvements. Achieving the "ultimate" doesn't really matter. That's true to a degree, but only in the background of a method that provides a sequence of feasible points that get better and better. Most computational methods for problems with nonlinear constraints only approach feasibility in the limit, and that can open the door to various dangers. Another thing to remember is that optimization methods often entail the repeated solution of certain subproblems, such as in determining a good direction in which to search for improvement. One has to be careful that if such subproblems aren't solved to full optimality the method is still valid.

How do the properties in Theorem 1.1 connect with problem $(\mathcal{P})$? We'll refer to the *convex case* of $(\mathcal{P})$ when the objective f_0 and inequality constraint functions $f_1, \dots, f_s$ are convex and the equality constraint functions $f_{s+1}, \dots, f_m$ are affine.

Theorem 1.2. *In the convex case of $(\mathcal{P})$ the feasible set C is convex, so the property in Theorem 1.1(a) holds. If f_0 is not just convex but strictly convex, the property in Theorem 1.1(b) holds also.*

The next results review some criteria for a function to be convex. We denote by $\nabla f(x)$ the gradient of f at x, which is the vector of first partial derivatives. Similarly, we let $\nabla^2 f(x)$ stand for the square matrix of second partial derivatives, called the Hessian matrix of f at x. Recall that a matrix $H \in \mathbb{R}^{n \times n}$ is *positive semidefinite* when $w \cdot Hw \geq 0$ or all $w \in \mathbb{R}^n$. It is *positive definite* when $w \cdot Hw > 0$ for all $w \in \mathbb{R}^n$, except $w = 0$.

Proposition 1.3. *Let f be a function on $\mathbb{R}^n$ with continuous second derivatives.*

(a) *If f is convex, then $\nabla^2 f(x)$ is positive semidefinite for all x.*

(b) *If $\nabla^2 f(x)$ is positive semidefinite for all x, then f is convex.*

(c) *If $\nabla^2 f(x)$ is positive definite for all x, then f is strictly convex.*

Proposition 1.4.

(a) *If f_1 and f_2 are convex, then $f_1 + f_2$ is convex. If in addition either f_1 or f_2 is strictly convex, then $f_1 + f_2$ is strictly convex.*

(b) *If f is convex and $\lambda \geq 0$, then λf is convex. If f is strictly convex and $\lambda > 0$, then λf is strictly convex.*

(c) *If $f(x) = \phi(g(x))$ with g convex on $\mathbb{R}^n$ and ϕ is convex and nondecreasing on $\mathbb{R}$, then f is convex on $\mathbb{R}^n$. If in addition g is strictly convex and ϕ is increasing, then f is strictly convex.*

(d) *If $f(x) = g(Ax + b)$ for a convex function g on $\mathbb{R}^m$ and a matrix $A \in \mathbb{R}^{m \times n}$ and a vector $b \in \mathbb{R}^n$, then f is convex on $\mathbb{R}^n$. If g is strictly convex and A has rank n, then f is strictly convex.*

(e) *If $f(x) = \sup_{s \in S} g_s(x)$ for a finite or infinite collection $\{g_s\}_{s \in S}$ of convex functions on $\mathbb{R}^n$, then f is convex on $\mathbb{R}^n$.*

Later we will use these criteria in verifying the convexity of expressions defined with penalty terms that aren't differentiable.

2. Optimality conditions

First-order optimality conditions for problem $(\mathcal{P})$ will now be stated in terms of Lagrange multipliers. In order to get a simple form of expression that will later be extendible to problems with penalty functions, we use the concept and notation of normal vectors. A more general exposition of the material in this section, complete with proofs, is available in the expository article [2].

Definition 2.1. *The (outward) normal vectors to the polyhedral set X at a point $\bar{x} \in X$ are the vectors v such that*

$$v \cdot (x - \bar{x}) \leq 0 \text{ for all } x \in X.$$

The set of all these vectors is called the normal cone to X at $\bar{x}$ and is denoted by $N_X(\bar{x})$.

The term "cone" refers to the fact that for any $v \in N_X(\bar{x})$ and $\lambda \geq 0$, then $\lambda v \in N_X(\bar{x})$. In other words, $N_X(\bar{x})$ is a bundle of rays emanating from the origin—unless $\bar{x}$ is an interior point of X, in which case $N_X(\bar{x})$ consists of the zero vector alone. In the case where X is a box, the normal cone condition is

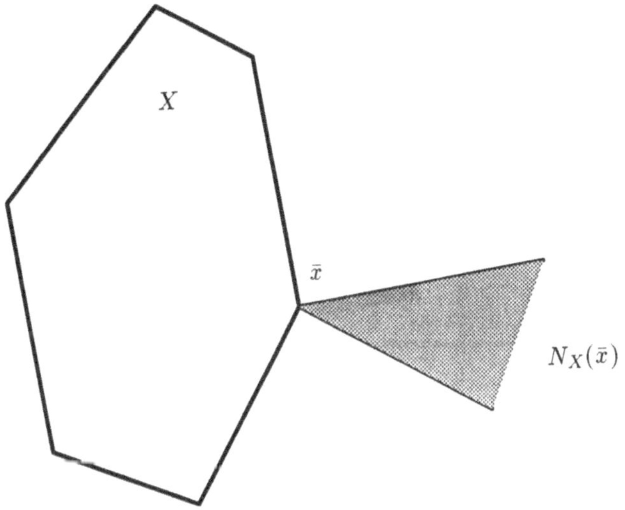

Figure 1: The Normal Cone to a convex polyhedral set at a vertex

especially easy to understand: in terms of $v = (v_1, \dots, v_n) \in \mathbb{R}^n$ we have

$$\begin{cases} \text{if } \bar{x} \in X = X_1 \times \cdots \times X_n,\ \bar{x} = (\bar{x}_1, \dots, \bar{x}_n), \text{ then} \\ v \in N_X(\bar{x}) \iff v_j \in N_{X_j}(\bar{x}_j) \text{ for } j = 1, \dots, n. \end{cases} \tag{2.1}$$

When X_j is closed interval with lower bound a_j and upper bound b_j (these bounds possibly being infinite), we get that

$$v_j \in N_{X_j}(\bar{x}_j) \text{ means } \begin{cases} v_j \geq 0 & \text{if } a_j < \bar{x}_j = b_j, \\ v_j \leq 0 & \text{if } a_i = \bar{x}_j < b_i, \\ v_j = 0 & \text{if } a_i < \bar{x}_j < b_i, \\ v_j \text{ unrestricted} & \text{if } a_i = \bar{x}_j = b_i. \end{cases} \tag{2.2}$$

In order to state the main result about first-order optimality conditions in problem $(\mathcal{P})$, we'll need a condition on the constraints. This condition will involve normal vectors to the set

$$D = \{\, u = (u_1, \dots, u_m) \,|\, u_i \leq 0 \text{ for } i \in [1, s],\ u_i = 0 \text{ for } i \in [s+1, m] \,\}. \tag{2.3}$$

The constraints in $(\mathcal{P})$ can be written as

$$x \in X,\ F(x) \in D, \text{ where } F(x) = (f_1(x), \dots, f_m(x)). \tag{2.4}$$

Note that D is another polyhedral set, actually a box:

$$D = D_1 \times \cdots \times D_m \text{ with } D_i = \begin{cases} (-\infty, 0] & \text{for } i \in [1, s], \\ [0, 0] & \text{for } i \in [s+1, m]. \end{cases} \tag{2.5}$$

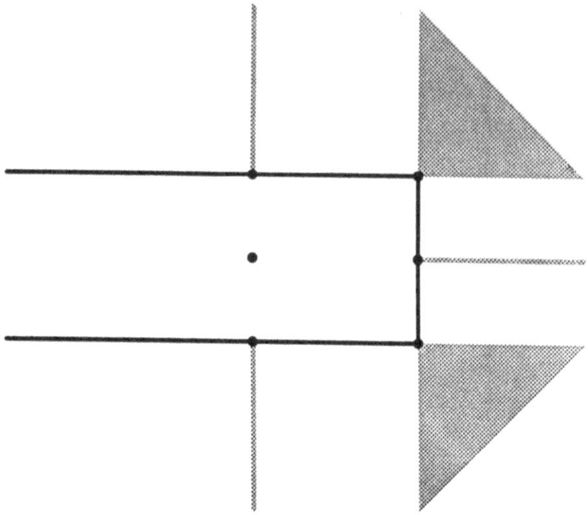

Figure 2: The Normal Cones to a Box

Definition 2.2. *The basic constraint qualification at a feasible solution* $\bar{x}$ *to problem* $(\mathcal{P})$ *is the condition:*

$$(\mathcal{Q}) \quad \begin{cases} \text{there is no vector } \bar{y} = (\bar{y}_1, \ldots, \bar{y}_m) \text{ other than } \bar{y} = 0 \text{ such that} \\ \bar{y} \in N_D(F(\bar{x})), \qquad -[\bar{y}_1 \nabla f_1(\bar{x}) + \cdots + \bar{y}_m \nabla f_m(\bar{x})] \in N_X(\bar{x}). \end{cases}$$

This condition is needed to rule out situations where the constraints fail to give a robust representation of the feasible set C around $\bar{x}$. From the product form of D in (2.5), it's clear that

$$\bar{y} \in N_D(F(\bar{x})) \iff \begin{cases} \bar{y}_i = 0 & \text{for } i \in [1, s] \text{ with } f_i(\bar{x}) < 0, \\ \bar{y}_i \geq 0 & \text{for } i \in [1, s] \text{ with } f_i(\bar{x}) = 0, \\ \bar{y}_i \text{ unrestricted} & \text{for } i \in [s+1, m]. \end{cases} \qquad (2.6)$$

Writing these sign conditions as $\bar{y} \in N_D(F(\bar{x}))$ is not only convenient but leads the way to a statement of optimality conditions that can be extended later to problems incorporating penalty expressions. Observe that if $\bar{x}$ belongs to the interior of X (as is certainly true when $X = \mathbb{R}^n$), the gradient condition in $(\mathcal{Q})$ reduces to $\sum_{i=1}^m \bar{y}_i \nabla f_i(\bar{x}) = 0$. Then $(\mathcal{Q})$ becomes the John constraint qualification [3], which is the dual form of the Mangasarian-Fromovitz constraint qualification [4].

Optimality conditions for $(\mathcal{P})$ involve the *Lagrangian function*

$$L(x, y) = f_0(x) + y_1 f_1(x) + \cdots + y_m f_m(x) \text{ for } x \in X \text{ and } y \in Y, \qquad (2.7)$$

where

$$Y = \mathbb{R}^s_+ \times \mathbb{R}^{m-s} = \{ y = (y_1, \ldots, y_m) \mid y_i \geq 0 \text{ for } i \in [1, s] \}. \tag{2.8}$$

Observe that Y too is a box, and

$$\bar{y} \in N_D(F(\bar{x})) \quad \Longleftrightarrow \quad F(\bar{x}) \in N_Y(\bar{y}). \tag{2.9}$$

This equivalence is clear from (2.6) and the fact that

$$u \in N_Y(\bar{y}) \quad \Longleftrightarrow \quad \begin{cases} u_i \geq 0 & \text{for } i \in [1, s] \text{ with } \bar{y}_i = 0, \\ u_i = 0 & \text{for } i \in [1, s] \text{ with } \bar{y}_i > 0, \\ & \quad \text{and for } i \in [s+1, m]. \end{cases} \tag{2.10}$$

Theorem 2.3. *If $\bar{x} \in X$ is a locally optimal solution to $(\mathcal{P})$ at which the basic constraint qualification $(\mathcal{Q})$ is satisfied, there must exist a vector $\bar{y} \in Y$ such that*

$$(\mathcal{L}) \qquad -\nabla_x L(\bar{x}, \bar{y}) \in N_X(\bar{x}), \qquad \nabla_y L(\bar{x}, \bar{y}) \in N_Y(\bar{y}).$$

Condition $(\mathcal{L})$ is the *Lagrange multiplier rule* in general form. Since $\nabla_y L(\bar{x}, \bar{y}) = F(\bar{x})$, the second part of $(\mathcal{L})$ is simply another statement of the sign conditions in (2.6) on the multipliers $\bar{y}_i$, but again one which will lead to extensions. The first part of $(\mathcal{L})$ becomes the equation $\nabla f_0(\bar{x}) + \sum_{i=1}^m \bar{y}_i \nabla f_i(\bar{x}) = 0$ when $\bar{x}$ is an interior point of X.

Although the normal cone notation here is a recent development, and the incorporation of the abstract constraint $x \in X$ a novel feature, the first-order optimality conditions in Theorem 2.3 are basically the ones found in every textbook on optimization. They are commonly called the *Kuhn-Tucker conditions* because of the 1951 paper of Kuhn and Tucker [5], but it's now known that the same conditions were derived in the 1939 master's thesis of Karush [6], which however was never published. For this reason they are also referred to as the *Karush-Kuhn-Tucker conditions.*

In many applications linear constraints are very important, and then the following variant of Theorem 2.3 is useful.

Theorem 2.4. *The assertion of Theorem 2.3 remains valid when the basic constraint qualification $(\mathcal{Q})$ at $\bar{x}$ is replaced by*

$$(\mathcal{Q}') \qquad \begin{cases} \text{the nonzero vectors } \bar{y} \in Y \text{ satisfying} \\ \bar{y} \in N_D(F(\bar{x})), \qquad -[\bar{y}_1 \nabla f_1(\bar{x}) + \cdots + \bar{y}_m \nabla f_m(\bar{x})] \in N_X(\bar{x}), \\ \text{if any, have } \bar{y}_i = 0 \text{ for each index } i \text{ such that } f_i \text{ is not affine.} \end{cases}$$

On the basis of this theorem, for instance, the multiplier rule $(\mathcal{L})$ is always necessary for optimality in problems having only linear constraints.

What dangers are there in applying Lagrangian optimization methodology in the absence of being able to verify, when nonlinear constraints are present, that the constraint qualification $(\mathcal{Q})$ or $(\mathcal{Q}')$ is definitely satisfied at an optimal solution? This depends on the solution technique involved, but the main difficulty is that if a technique can at best identify points singled out as candidates by the Lagrange multiplier rule, but the desired solution is not such a point, then the technique has no hope of finding it. The technique in combination with some kind of global search could seem indicate a particular point as the best, but only because it is blind to the real solution. Another possibility is that numerical instabilities may be experienced. However, there is good theoretical support for the notion that the Lagrange multiplier rule "usually" is necessary for optimality.

Lagrange multipliers have special properties under convexity which lead to another level of practical usage.

Theorem 2.5. *In the convex case of problem $(\mathcal{P})$, the Lagrangian $L(x,y)$ is convex in x and concave (actually affine) in y. The multiplier rule $(\mathcal{L})$ in Theorem 2.3 is equivalent then to the condition that*

$$\begin{cases} \text{the minimum of } L(x,\bar{y}) \text{ in } x \in X \text{ is attained at } \bar{x}, \\ \text{the maximum of } L(\bar{x},y) \text{ in } y \in Y \text{ is attained at } \bar{y}. \end{cases} \tag{2.11}$$

This theorem supports—up to a certain degree–a popular approach called *Lagrangian relaxation*, which is especially attractive in connection with ideas of decomposing a large-scale problem by introducing appropriate "prices" to achieve a decentralization of the decision process. Under this approach, a vector $\hat{y}$ is selected, and then a vector $\hat{x}$ is obtained by minimizing $L(x,\hat{y})$ subject to $x \in X$. It is hoped that through a good choice of $\hat{y}$ a nearly optimal solution $\hat{x}$ to $(\mathcal{P})$ itself will generated. But is this hope justified?

According to Theorem 2.5, if an optimal solution $\bar{x}$ *exists and satisfies the Lagrange multiplier rule $(\mathcal{L})$ along with some vector $\bar{y}$* (the latter being true when $(\mathcal{Q})$ or $(\mathcal{Q}')$ holds at $\bar{x}$), and if one is dealing with a *convex* case of $(\mathcal{P})$, then $\bar{x}$ will be *among* the vectors $\hat{x}$ obtainable under the Lagrangian relaxation approach if, through luck or design, $\hat{y}$ can be chosen equal to $\bar{y}$. With strict convexity of the objective function f_0, $\hat{x}$ will have to be $\bar{x}$, the unique optimal solution to $(\mathcal{P})$, in these circumstances. But without strict convexity of f_0, even with everything else holding, $\hat{x}$ *might not even satisfy the constraints* of $(\mathcal{P})$.

In the nonconvex case of $(\mathcal{P})$, unfortunately, just about everything can go wrong in Lagrangian relaxation. A vector $\hat{x}$ obtained in this manner, even from some "ideal" choice of $\hat{y}$, need have no relation to optimality. All that can be said then is that the minimizing value of $L(x, \hat{y})$ as x ranges over X will be a *lower bound* for the optimal value (number) associated with $(\mathcal{P})$—provided that this minimizing value is *global*, which as noted earlier is very hard to guarantee without convexity. The vector $\hat{x}$ offers nothing.

Incidentally, it's interesting to note that this negative conclusion from theory doesn't stop economists, especially in today's political climate, from flirting with the idea that if only the right markets and prices could be introduced, decisions could effectively be decentralized and society could function more efficiently. Theory provides no backing for this concept in situations where convexity is absent, such as characterize much of the real world. (It's known that in the case of a very large number of small agents, such in classical free markets, a kind of convexification is approached, but this is far from the actual economies of developed countries.)

Lagrangian relaxation can be understood further in connection with saddle points and dual problems. A pair of elements $\bar{x}$ and $\bar{y}$ is said to give a *saddle point* of L on $X \times Y$ when (2.11) holds; this can also be written as

$$L(x, \bar{y}) \geq L(\bar{x}, \bar{y}) \geq L(\bar{x}, y) \text{ for all } x \in X,\ y \in Y \quad (\text{where } \bar{x} \in X,\ \bar{y} \in Y). \tag{2.12}$$

This relation has a life of its own as an equilibrium condition for certain "games," and it leads to further properties of Lagrange multipliers which are of prime importance for many applications. In particular it gives rise to the notion of *duality* in optimization. To appreciate the meaning of duality, let's first note that problem $(\mathcal{P})$ can be viewed as the problem of minimizing over all $x \in X$ the function f defined by

$$f(x) = \begin{cases} f_0(x) & \text{if } x \in C, \\ \infty & \text{if } x \notin C, \end{cases} \tag{2.13}$$

where C is the set of feasible solutions to $(\mathcal{P})$, and that f has the Lagrangian representation

$$f(x) = \sup_{y \in Y} L(x, y) = \sup_{y \in Y} \left\{ f_0(x) + y_1 f_1(x) + \ldots + y_m f_m(x) \right\} \text{ for } x \in X, \tag{2.14}$$

where the restriction of y to Y in taking the "sup" means that the coefficients y_i can be chosen arbitrarily for the terms indexed by $i = s+1, \ldots, m$, but must be nonnegative for $i = 1, \ldots, s$. By analogy in reversing the roles of x and y, we can state the problem:

$$(\mathcal{D}) \qquad \text{maximize } g(y) = \inf_{x \in X} \left\{ f_0(x) + y_1 f_1(x) + \ldots + y_m f_m(x) \right\} \text{ over } y \in Y.$$

This is the optimization problem *dual* to problem $(\mathcal{P})$ in the Lagrangian sense.

Observe that for each vector $\hat{y}$ the subproblem solved to get the value $g(\hat{y})$ of the essential objective function g in $(\mathcal{D})$ is precisely the one indicated in the Lagrangian relaxation approach. In general g might, like f in (2.13), be extended-real-valued. To learn more about the nature of the dual problem $(\mathcal{D})$ in a given case, with particular structure assigned to X and the f_i's, we would have to identify the set of points y where $g(y) > -\infty$ and regard that as the feasible set in $(\mathcal{D})$. Examples will be considered below, but we first record the main facts relating problems $(\mathcal{P})$ and $(\mathcal{D})$.

Theorem 2.6. *In the convex case of $(\mathcal{P})$, the existence for $\bar{x}$ of a multiplier vector $\bar{y}$ satisfying the Lagrange multiplier rule $(\mathcal{L})$ is sufficient for $\bar{x}$ to be a globally optimal solution to $(\mathcal{P})$. The vectors $\bar{y}$ that appear in this condition along with $\bar{x}$ are then precisely the optimal solutions to the dual problem $(\mathcal{D})$, and the optimal values in the two problems agree: one has*

$$\min(\mathcal{P}) = \max(\mathcal{D}).$$

The final equation in Theorem 2.6 confirms that, for any $\hat{y}$, the value $g(\hat{y})$ is a lower bound for the optimal value $\min(\mathcal{P})$, and under the right circumstances of convexity, this lower bound can be elevated to the degree that it actually equals the desired optimal value. Furthermore, Theorem 2.6 give the theoretical prescription to be used in designing an algorithm to product a multiplier vector $\bar{y}$ for which the equality holds. Once again, though, without the convexity the equation between $\min(\mathcal{P})$ and $\max(\mathcal{D})$ could very well become a strict inequality $>$. Then no amount of fiddling with the values of Lagrange multipliers could be expected to produce approximate optimal solutions fo $(\mathcal{P})$ through Lagrangian relaxation.

The best known and most highly successful example of duality in optimization occurs in *linear programming*, which is the case of problem $(\mathcal{P})$ where the objective function is linear, all the constraints are linear, and

$$X = \mathbb{R}^r_+ \times \mathbb{R}^{n-r} = \left\{ x = (x_1, \ldots, x_n) \,\middle|\, x_j \geq 0 \text{ for } j \in [1, r] \right\}. \tag{2.15}$$

Adopting the notation

$$f_0(x) = c_1 x_1 + \cdots + c_n x_n,$$
$$f_i(x) = b_i - a_{i1} x_1 - \cdots - a_{in} x_n \text{ for } i = 1, \ldots, m,$$

we can express the problem in this special case as

$$(\mathcal{P}_{\text{lin}})\qquad \begin{aligned} &\text{minimize } c_1x_1 + \cdots + c_nx_n \text{ subject to } x_j \geq 0 \text{ for } j = 1,\ldots,r,\\ &a_{i1}x_1 + \cdots + a_{in}x_n \begin{cases} \geq b_i & \text{for } i = 1,\ldots,s,\\ = b_i & \text{for } i = s+1,\ldots,m. \end{cases} \end{aligned}$$

The Lagrangian function is

$$L(x,y) = \sum\nolimits_{j=1}^{n} c_jx_j + \sum\nolimits_{i=1}^{m} y_ib_i - \sum\nolimits_{i=1,j=1}^{m,n} y_ia_{ij}x_j, \tag{2.16}$$

which exhibits the same kind of symmetry between the x and y arguments as appears in the choice of X and Y. To obtain the problem dual to this, we must determine the function g defined in $(\mathcal{D})$ for this Lagrangian and see where it is finite or infinite. Elementary calculations show that $g(y) = \sum_{i=1}^{m} y_ib_i$ if $c_j - \sum_{i=1}^{m} y_ia_{ij} \geq 0$ for $j = 1,\ldots,r$ and $c_j - \sum_{i=1}^{m} y_ia_{ij} = 0$ for $j = r+1,\ldots,n$, whereas $g(y) = -\infty$ if y does not satisfy these constraints. The dual problem therefore comes out as

$$(\mathcal{D}_{\text{lin}})\qquad \begin{aligned} &\text{maximize } y_1b_1 + \cdots + y_mb_m \text{ subject to } y_i \geq 0 \text{ for } i = 1,\ldots,s,\\ &y_1a_{1j} + \cdots + y_ma_{mj} \begin{cases} \leq c_j & \text{for } j = 1,\ldots,r,\\ = c_j & \text{for } j = r+1,\ldots,n. \end{cases} \end{aligned}$$

From all this symmetry it emerges that not only do the Lagrange multiplier vectors associated with an optimal solution to $(\mathcal{P}_{\text{lin}})$ have an interpretation as optimal solutions $\bar{y}$ to $(\mathcal{D}_{\text{lin}})$, but by the same token, the Lagrange multiplier vectors associated with an optimal solution to $(\mathcal{D}_{\text{lin}})$ have an interpretation as optimal solutions $\bar{x}$ to $(\mathcal{P}_{\text{lin}})$. Each of these problems furnishes the multipliers for the other.

Corollary 2.7 (Gale-Kuhn-Tucker Theorem [7]). *If either of the linear programming problems $(\mathcal{P}_{\text{lin}})$ or $(\mathcal{D}_{\text{lin}})$ has an optimal solution, then so does the other, and*

$$\min(\mathcal{P}_{\text{lin}}) = \max(\mathcal{D}_{\text{lin}}).$$

The pairs $(\bar{x},\bar{y})$ such that $\bar{x}$ solves $(\mathcal{P}_{\text{lin}})$ and $\bar{y}$ solves $(\mathcal{D}_{\text{lin}})$ are precisely the ones that, for the choice of L, X and Y corresponding to these problems, satisfy the Lagrange multiplier rule $(\mathcal{L})$, or equivalently, give a saddle point of L on $X \times Y$.

Even for the nonconvex case of $(\mathcal{P})$, the dual problem $(\mathcal{D})$ has significance.

Proposition 2.8. *Regardless of whether* $(\mathcal{P})$ *is of convex type or not, the function* g *being maximized over the polyhedral set* Y *in* $(\mathcal{D})$ *is concave. For each* $y \in Y$ *the value* $g(y)$ *is a lower bound to the value* $\min(\mathcal{P})$. *The greatest of such lower bounds obtainable this way is* $\max(\mathcal{D})$.

In other words, by selecting any $y \in Y$ and then minimizing $L(x,y)$ over $x \in X$, one obtains a number denoted by $g(y)$ with the property that $g(y) \leq f_0(x)$ for every feasible solution x to problem $(\mathcal{P})$. This number may be useful in estimating how far a particular point $\hat{x}$ already calculated in $(\mathcal{P})$, and satisfying the constraints of $(\mathcal{P})$, may be from optimality. One will have

$$0 \leq f_0(\hat{x}) - \min(\mathcal{P}) \leq f_0(\hat{x}) - g(y), \tag{2.17}$$

so that if $f_0(\hat{x}) - g(y)$ is less than some threshold value ε, the decision can be made that $\hat{x}$ is good enough, and further computations aren't worth the effort. By applying an optimization technique to $(\mathcal{D})$, it may be possible to get better estimates of such sort. The best would be a dual optimal solution $\bar{y}$, for which $g(\bar{y}) = \max(\mathcal{D})$; then the estimate would take the form

$$0 \leq f_0(\hat{x}) - \min(\mathcal{P}) \leq f_0(\hat{x}) - \max(\mathcal{D}). \tag{2.18}$$

But in nonconvex problems where $\min(\mathcal{P}) > \max(\mathcal{D})$, the bound on the right can't be brought to 0 no matter how much effort is expended. The technique is therefore limited in its ability to estimate optimality of $\hat{x}$. Another pitfall is that the estimates only make sense if the exact value of $g(y)$ can be computed for a given $y \in Y$, or at least a lower estimate c for $g(y)$ (then one gets $f_0(\hat{x}) - c$ as an upper bound to substitute for the right side in (2.17)). But in the nonconvex case of $(\mathcal{P})$ the expression $L(x,y)$ being minimized over $x \in X$ to calculate the value $g(y)$ may be nonconvex in x, yet the minimization must be *global*. Then, as already explained in Section 1, it may be difficult or impossible to know when the global minimum has been attained.

For more on duality in convex optimization, see [1], [7], [8]. For the theory of the *augmented Lagrangian* function for $(\mathcal{P})$, which makes saddle point characterizations of optimality possible even without convexity, see [2]. Second-order optimality conditions are discussed in [2] also.

3. Extended problem models

The conventional problem statement $(\mathcal{P})$ doesn't fully convey the range of possibilities available in setting up a mathematical model in optimization. First, it gives the impression that as modelers we won't have trouble distinguishing between objectives and constraints. We're supposed to know what should be minimized and be able to express it by a *smooth* function. All other features of the situation being addressed must be formulated as "black-and-white" constraints—side conditions that have to be satisfied exactly, or we'll be infinitely unhappy. No gray areas are allowed.

The real modeling context is often very different. There may well be some conditions that the variables must satisfy exactly, because otherwise the model doesn't make sense. For instance, a nonnegativity condition $x_j \geq 0$ may fall in this category: we wouldn't know how to interpret a negative value of x_j physically and aren't in the least interested in relaxing the constraint $x_j \geq 0$ to $x_j \geq -\varepsilon$, say. Other examples of such black-and-white constraints are defining relationships between variables. A condition like $x_3 - x_1 - x_2^2 = 1$ could simply indicate the definition of x_3 in terms of x_1 and x_2, and we wouldn't want to consider relaxing it. But many of the constraints may have a "soft" character. We might write $4.3x_1 + 2.7x_2 + x_3 \leq 5.6$ as a constraint because we desire the expression on the left not to exceed 5.6, but a sort of guesswork is involved. We could be quite content when the expression on the left had the value 5.9 if that resulted in substantial benefits in other respects. Another source of fuzziness might be that coefficients like 4.3 are just estimates, or worse. Then it seems foolish to insist on the inequality being satisfied without error.

In fact, a fair description of the difficulty often faced in reality may be that there are several expressions $f_0(x), f_1(x), \ldots, f_m(x)$ of interest to the modeler, who is seeking a sort of "ideal combination" subject to the trade-offs that may be involved. Somewhat arbitrarily, one of these expressions is selected as the one to optimize while the others are held in fixed ranges, but after the optimization has been carried out, there may be second thoughts inspired by knowledge generated during the optimization process, and a modified optimization formulation may then be tested out. Besides choosing one of the functions as the objective and putting constraint bounds on the others, it's possible of course to form some combination. Examples to consider might be the minimization, subject to the underlying hard constraints on x, of a weighted sum $f(x) = f_0(x) + c_1 f_1(x) + \cdots + c_m f_m(x)$ or a weighted max

$$f(x) = f_0(x) + \max\{c_1 f_1(x), \ldots, c_m f_m(x)\}. \tag{3.1}$$

Or, taking as reference the minimization of "cost," with $f_0(x)$ expressing certain costs directly, we could consider for each other function f_i a nonlinear rescaling function ρ_i that converts the value $f_i(x)$ into an associated cost $\rho_i(f_i(x))$. Then we would want to minimize

$$f(x) = f_0(x) + \rho_1(f_1(x)) + \cdots + \rho_m(f_m(x)). \tag{3.2}$$

Although the given functions $f_0, f_1, \ldots, f_m$ may be smooth, the function f obtained in such a manner may be nonsmooth. Problems of minimizing such a function aren't well covered by the standard theory for $(\mathcal{P})$.

To get around this difficulty and enhance the possibilities for optimization modeling, we direct our attention to the following *extended* problem formulation, which was introduced in [2]:

$$(\overline{\mathcal{P}}) \qquad \text{minimize } f(x) = f_0(x) + \rho(F(x)) \text{ over } x \in X,$$

where $F(x) = (f_1(x), \ldots, f_m(x))$. In this the functions $f_0, f_1, \ldots, f_m$ will still be assumed to be smooth, and the set X to be closed, but the function ρ need not be smooth and can even take on the value ∞.

For a sense of what $(\overline{\mathcal{P}})$ covers, let's consider first the cases where ρ is *separable*, i.e.,

$$\rho(u) = \rho(u_1, \ldots, u_m) = \rho_1(u_1) + \cdots + \rho_m(u_m), \tag{3.3}$$

so that $(\overline{\mathcal{P}})$ takes the form of minimizing an expression of the form (3.2) over X. Right away we can observe that $(\overline{\mathcal{P}})$ contains $(\mathcal{P})$ as corresponding to the choice:

$$\begin{aligned} &\text{for } i = 1, \ldots, s: && \rho_i(u_i) = \begin{cases} 0 & \text{if } u_i \le 0, \\ \infty & \text{if } u_i > 0, \end{cases} \\ &\text{for } i = s+1, \ldots, m: && \rho_i(u_i) = \begin{cases} 0 & \text{if } u_i = 0, \\ \infty & \text{if } u_i \ne 0. \end{cases} \end{aligned} \tag{3.4}$$

This gives for $f(x)$ in (3.2) the value $f_0(x)$ when the point $x \in X$ is feasible in $(\mathcal{P})$, but the value ∞ if x is not feasible. As we saw earlier, the minimization of this "essential objective" function f over X is equivalent to the minimization of $f_0(x)$ subject to x being feasible.

This example may create some discomfort with its use of ∞, but it also serves as a reminder of the true nature of the modeling represented by the

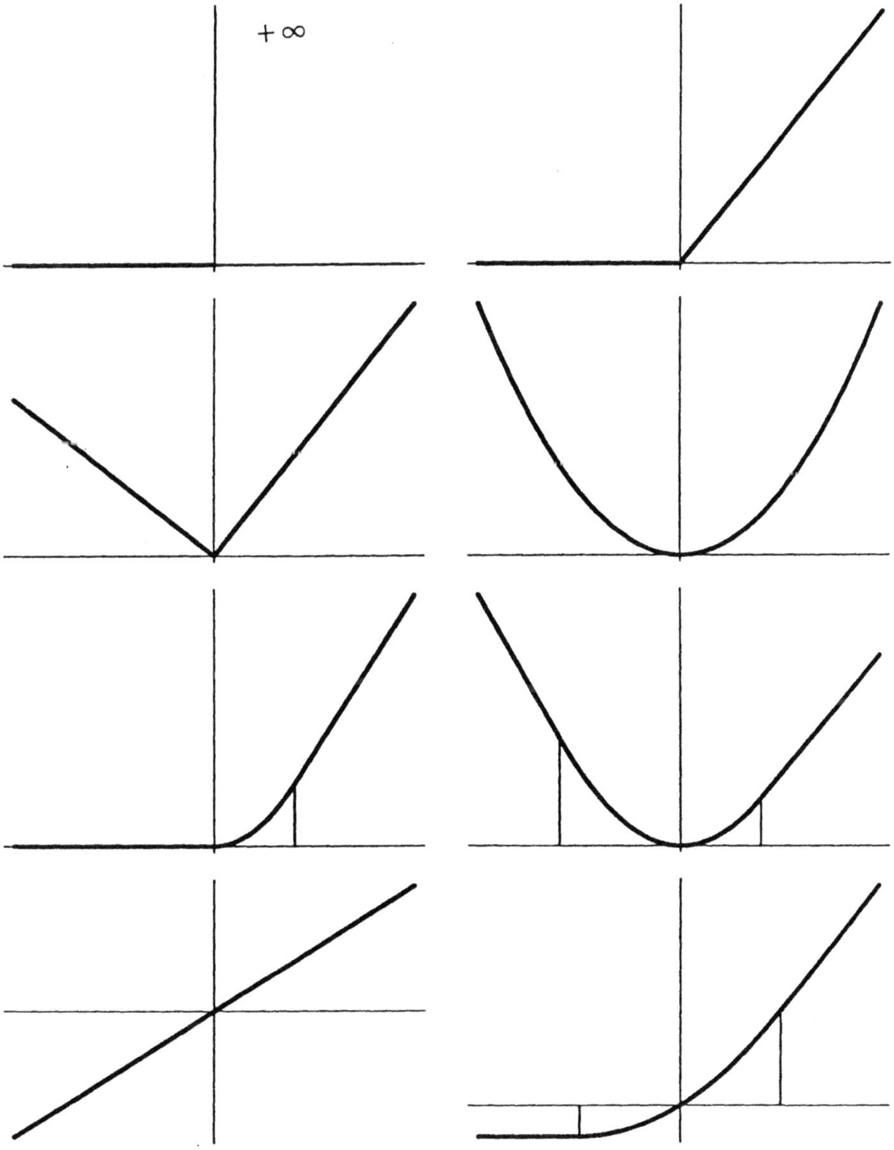

Figure 3: A collection of ρ functions: the first represents the indicator function in case of inequalities; the second penalizes the violation of inequalities linearly; the third and fourth penalize violated inequalities linearly and quadratically; the fifth and sixth penalize the violation of inequalities and equalities in a linear-quadratic sense (see below); the seventh represents Lagrangian relaxation and the eighth is a hybrid version.

conventional problem $(\mathcal{P})$. There is an infinite penalty if the stated conditions are violated, but no gray area allowing for "approximate" satisfaction.

Obviously, the function f in this infinite penalty case of $(\overline{\mathcal{P}})$ corresponding to $(\mathcal{P})$ is far from smooth and even is discontinuous, but even a finite penalty approach may be incompatible with constructing a *smooth* function to minimize. For example, in the pure *linear penalty case* of $(\overline{\mathcal{P}})$ the choice is

$$\begin{aligned} &\text{for } i = 1,\dots,s: && \rho_i(u_i) = \begin{cases} 0 & \text{if } u_i \le 0, \\ d_i u_i & \text{if } u_i > 0, \end{cases} \\ &\text{for } i = s+1,\dots,m: && \rho_i(u_i) = \begin{cases} 0 & \text{if } u_i = 0, \\ d_i |u_i| & \text{if } u_i \neq 0, \end{cases} \end{aligned} \tag{3.5}$$

with positive constants d_i. These functions have "kinks" at the origin which prevent f from being smooth. The pure *quadratic penalty case* of $(\overline{\mathcal{P}})$ instead takes

$$\begin{aligned} &\text{for } i = 1,\dots,s: && \rho_i(u_i) = \begin{cases} 0 & \text{if } u_i \le 0, \\ \frac{1}{2} d_i u_i^2 & \text{if } u_i > 0, \end{cases} \\ &\text{for } i = s+1,\dots,m: && \rho_i(u_i) = \begin{cases} 0 & \text{if } u_i = 0, \\ \frac{1}{2} d_i u_i^2 & \text{if } u_i \neq 0, \end{cases} \end{aligned} \tag{3.6}$$

with coefficients $d_i > 0$. Penalty functions of this type are first-order smooth, yet discontinuous in their second derivatives.

To illustrate the modeling considerations, consider a situation where the demand for a certain commodity is $d > 0$, and this is to be met by producing amounts $x_j \ge 0$ at plants $j = 1,\dots,n$, the costs being $\phi_j(x_j)$. One formulation as a problem of optimization would be to minimize $f_0(x) = \phi_1(x_1) + \cdots + \phi_n(x_n)$ over all vectors $x \in X = \mathbb{R}^n_+$ satisfying $d - x_1 - \cdots - x_n = 0$. But this could be a fragile approach, since it takes the target to be exact and makes no provision for not meeting it precisely. A better model could be to minimize $f_0(x) + \rho(d - x_1 - \cdots - x_n)$, where $\rho(u) = ru$ when $u \ge 0$ and $\rho(u) = q|u|$ when $u < 0$, where the parameter values r and q are positive. This would correspond to a penalty rate of r per unit of overproduction, but a penalty rate of q per unit of underproduction. The function ρ in this case is finite but has a kink at the origin (Figure 4). An alternative might be a formula for ρ that maintains the positive slope r for significantly positive u and the negative slope $-q$ for significantly negative u, but introduces a quadratic rounding between the two linear segments of the graph in order to do away with the kink.

A wide and flexible class of functions ρ_i, which aren't necessarily just penalty functions in the traditional sense, has been proposed by Rockafellar and Wets [10], [11], for modeling purposes in dynamic and stochastic programming. These are functions describable with four parameters β_i, $\hat{y}_i$, $\hat{y}_i^+$,

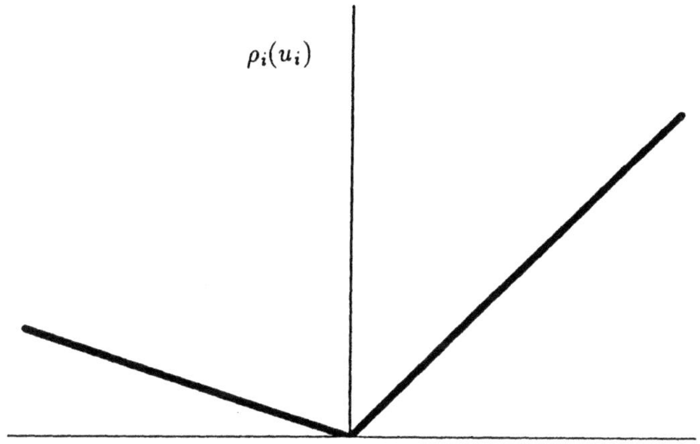

Figure 4: Linear Penalization

$\hat{y}_i^-$, where

$$0 \le \beta_i < \infty, \qquad -\infty < \hat{y}_i < \infty, \qquad -\infty \le \hat{y}_i^- \le \hat{y}_i \le \hat{y}_i^+ \le \infty.$$

The formula falls into three pieces and is best described by first introducing the auxiliary function $\hat{\rho}_i(u_i) = \hat{y}_i u_i + (1/2\beta_i)u_i^2$, this being the unique quadratic function with the property that $\hat{\rho}_i(0) = 0$, $\hat{\rho}_i'(0) = \hat{y}_i$, and $\hat{\rho}_i''(0) = 1/\beta_i$. Let $\hat{u}_i^+$ be the unique value such that $\hat{\rho}_i'(\hat{u}_i^+) = \hat{y}_i^+$, and similarly let $\hat{u}_i^-$ be the unique value such that $\hat{\rho}_i'(\hat{u}_i^-) = \hat{y}_i^-$. Then

$$\rho_i(u_i) = \begin{cases} \hat{\rho}_i(\hat{u}_i^+) + \hat{y}_i^+(u_i - \hat{u}_i^+) & \text{when } u_i \ge \hat{u}_i^+, \\ \hat{\rho}_i(u_i) & \text{when } \hat{u}_i^- \le u_i \le \hat{u}_i^+, \\ \hat{\rho}_i(\hat{u}_i^-) + \hat{y}_i^-(u_i - \hat{u}_i^-) & \text{when } u_i \le \hat{u}_i^-. \end{cases} \tag{3.7}$$

In other words, ρ_i agrees with the quadratic function $\hat{\rho}_i$, except that it extrapolates linearly to the right from the point where the slope of $\hat{\rho}_i$ is the specified value $\hat{y}_i^+$, and linearly to the left from the point where the slope is $\hat{y}_i^-$ (Figure 5). If $\hat{y}_i^+ = \infty$, this is taken to mean that the quadratic graph is followed forever to the right without switching over to a linear expression; the interpretation for $\hat{y}_i^- = -\infty$ is analogous. The case of $\beta_i = 0$ is taken to mean that there is no quadratic middle piece at all: the function is given by $\hat{y}^+ u_i$ when $u_i > 0$ and by $\hat{y}^- u_i$ when $u_i < 0$.

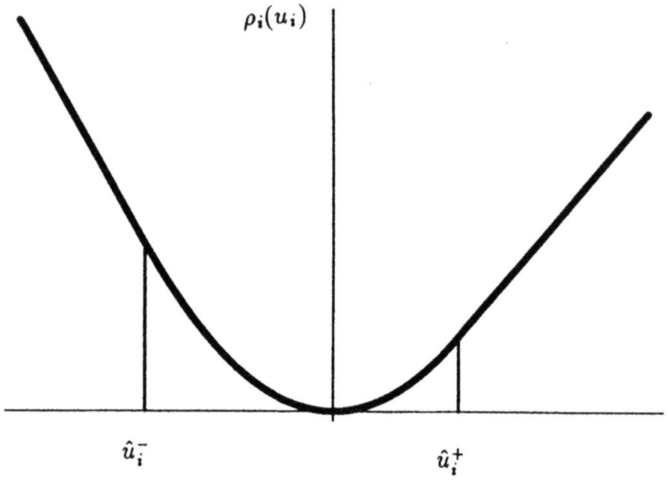

Figure 5: Linear-Quadratic Penalization

The functions in (3.5) and (3.6), and even (3.4), can be interpreted as special cases of (3.7). So too can the ρ function in the small modeling illustration. (The initial version of ρ in the illustration would correspond to $\hat{y}^+ = r$, $\hat{y}^- = -q$, $\hat{y} = 0$, and $\beta = 0$; the rounded version would differ only in having $\beta = \varepsilon > 0$.) This form also covers expressions that arise in augmented Lagrangian theory [2]. An example not of this kind, and not having the separable structure in (3.3), is

$$\rho(u) = \rho(u_1, \ldots, u_m) = \max\{u_1, \ldots, u_m\}. \tag{3.8}$$

This corresponds in $(\overline{\mathcal{P}})$ to the minimization of the expression $f(x)$ in (3.1).

Of course, a problem can also be formulated with a mixture of expressions like these. For each f_i one can decide whether to incorporate it into the model with an exact equality or inequality constraint, in effect by choosing the corresponding ρ_i as in (3.4), or one can associate it with a ρ_i conforming to the prescription in (3.5), (3.6), or more generally (3.7). Certain functions can be lumped together by a "max" expression as in (3.7), and so forth.

It may seem that the level of generality being suggested is too complicated to be usable in practice. But things are simpler than might first be imagined. All these examples show an underlying pattern which we capture by the following condition.

Definition 3.1. *The function ρ on $\mathbb{R}^m$ will be said to have an elementary dual representation if it can be expressed alternatively by*

$$\rho(u) = \sup_{y \in Y} \left\{ y{\cdot}u - k(y) \right\}, \tag{3.9}$$

where Y is some nonempty polyhedral set in $\mathbb{R}^m$ and k is some linear-quadratic convex function on $\mathbb{R}^m$. (Possibly $Y = \mathbb{R}^m$, or $k \equiv 0$. "Linear-quadratic" refers to a polynomial expression with no terms of degree higher than 2.) In the separable case (3.3) this comes down to whether each function ρ_i on $\mathbb{R}$ can be expressed alternatively by

$$\rho_i(u_i) = \sup_{y_i \in Y_i} \left\{ y_i u_i - k_i(y_i) \right\}, \tag{3.10}$$

where Y_i is some nonempty closed interval in $\mathbb{R}$ and k is some linear-quadratic convex function on $\mathbb{R}$.

Let's verify that the examples given do fit this. The case where $(\overline{\mathcal{P}})$ reduces to $(\mathcal{P})$ corresponds to $Y = \mathbb{R}^s_+ \times \mathbb{R}^{m-s}$ and $k \equiv 0$; in other words, the functions ρ_i in (3.4) achieve the representation (3.10) through

$$\begin{array}{lll} \text{for } i = 1,\ldots,s: & k_i(y_i) \equiv 0, & Y_i = [0,\infty), \\ \text{for } i = s+1,\ldots,m: & k_i(y_i) \equiv 0, & Y_i = (-\infty,\infty). \end{array} \tag{3.11}$$

The pure linear penalty case (3.5) corresponds instead to

$$\begin{array}{lll} \text{for } i = 1,\ldots,s: & k_i(y_i) \equiv 0, & Y_i = [0,d_i], \\ \text{for } i = s+1,\ldots,m: & k_i(y_i) \equiv 0, & Y_i = [-d_i,d_i]. \end{array} \tag{3.12}$$

The pure quadratic penalty case (3.6) is represented by

$$\begin{array}{lll} \text{for } i = 1,\ldots,s: & k_i(y_i) = (1/2d_i)y_i^2, & Y_i = [0,\infty), \\ \text{for } i = s+1,\ldots,m: & k_i(y_i) = (1/2d_i)y_i^2, & Y_i = (-\infty,\infty). \end{array} \tag{3.13}$$

The more general kind of ρ_i function in (3.6) corresponds to

$$k_i(y_i) = (\beta_i/2)|y_i - \hat{y}_i|^2, \qquad Y_i = [\hat{y}_i^-, \hat{y}_i^+]. \tag{3.14}$$

Finally, the max function case in (3.1) and (3.8)—which is not separable—arises from

$$k(y) \equiv 0, \qquad Y = \left\{ y \,\middle|\, y_i \geq 0,\ y_1 + \cdots + y_m = 1 \right\}. \tag{3.15}$$

Definition 3.2. *By the extended Lagrangian function corresponding to the extended problem* $(\overline{\mathcal{P}})$ *in the case where* ρ *has an elementary dual representation in the sense of Definition 3.1 for a set* Y *and function* k*, we shall mean the function*

$$\overline{L}(x,y) = f_0(x) + y_1 f_1(x) + \cdots + y_m f_m(x) - k(y) \text{ on } X \times Y.$$

Optimality conditions generalizing the ones for $(\mathcal{P})$ will be stated for $(\overline{\mathcal{P}})$ in terms of X, Y, and $\overline{L}$. For this we need to develop the correct analog of the constraint qualification $(\mathcal{Q})$ that was used for $(\mathcal{P})$.

Proposition 3.3. *When the function* ρ *has an elementary dual representation as in Definition 3.1, the set*

$$D = \{\, u = (u_1, \ldots, u_m) \mid \rho(u) < \infty \,\}$$

is nonempty and polyhedral in $\mathbb{R}^m$. *On* D, ρ *is a finite convex function, in fact* ρ *is continuous and piecewise linear-quadratic on* D. *The feasible set in* $(\overline{\mathcal{P}})$*, defined to be the set of point* $x \in X$ *where* $f(x) < \infty$*, is given by*

$$C = \{\, x \in X \mid F(x) \in D \,\}.$$

These properties are easy to see except for the polyhedral nature of D and the piecewise linear-quadratic nature of ρ on D, which are proved in [12].

This view of feasibility in problem $(\overline{\mathcal{P}})$ makes it possible to state the constraint qualification for the extended problem in the same manner as for the original problem.

Definition 3.4. *The basic constraint qualification at a feasible solution* $\bar{x}$ *to problem* $(\overline{\mathcal{P}})$*, when* ρ *has an elementary dual representation, is the condition:*

$$(\overline{\mathcal{Q}}) \quad \begin{cases} \text{there is no vector } \bar{y} = (\bar{y}_1, \ldots, \bar{y}_m) \text{ other than } \bar{y} = 0 \text{ such that} \\ \bar{y} \in N_D(F(\bar{x})), \qquad -[\bar{y}_1 \nabla f_1(\bar{x}) + \cdots + \bar{y}_m \nabla f_m(\bar{x})] \in N_X(\bar{x}), \end{cases}$$

where D *is the polyhedral set in Proposition 3.3.*

The main result about first-order optimality conditions in problem $(\overline{\mathcal{P}})$ can now be given. For details, see Rockafellar [2].

Theorem 3.5. *Suppose in* $(\overline{\mathcal{P}})$ *that* ρ *has an elementary dual representation in the sense of Definition 3.1 for a certain set* Y *and function* k. *If* $\bar{x} \in X$ *is a locally optimal solution to* $(\overline{\mathcal{P}})$ *at which the basic constraint qualification* $(\overline{\mathcal{Q}})$ *is satisfied, there must exist a vector* $\bar{y} \in Y$ *such that*

$$(\overline{\mathcal{L}}) \qquad -\nabla_x \overline{L}(\bar{x},\bar{y}) \in N_X(\bar{x}), \qquad \nabla_y \overline{L}(\bar{x},\bar{y}) \in N_Y(\bar{y}).$$

Convexity is important in $(\overline{\mathcal{P}})$ just as it was in $(\mathcal{P})$. We'll speak of the *convex case* of $(\overline{\mathcal{P}})$ when the extended Lagrangian $\overline{L}(x,y)$ is convex with respect to $x \in X$ for each $y \in Y$. (It's always concave in $y \in Y$ for each $x \in X$ by its definition.)

Theorem 3.6. *In the convex case of problem $(\overline{\mathcal{P}})$, the multiplier rule $(\overline{\mathcal{L}})$ in Theorem 3.5 is equivalent then to the saddle point condition*

$$\begin{cases} \text{the minimum of } \overline{L}(x,\bar{y}) \text{ in } x \in X \text{ is attained at } \bar{x}, \\ \text{the maximum of } \overline{L}(\bar{x},y) \text{ in } y \in Y \text{ is attained at } \bar{y}. \end{cases} \tag{3.16}$$

This saddle point condition leads to a dual problem. The extended Lagrangian has been introduced in just such a way that the function f being minimized over X in $(\overline{\mathcal{P}})$ has the representation

$$f(x) = \sup_{y \in Y} \overline{L}(x,y) = \sup_{y \in Y} \Big\{ f_0(x) + y_1 f_1(x) + \ldots + y_m f_m(x) - k(y) \Big\} \text{ for } x \in X. \tag{3.17}$$

We therefore introduce as the *extended dual problem* associated with $(\overline{\mathcal{P}})$ (when ρ has an elementary dual representation) the problem

$(\overline{\mathcal{D}})$
$$\begin{aligned} &\text{maximize } \bar{g}(y) \text{ over } y \in Y, \text{ where} \\ &\bar{g}(y) = \inf_{x \in X} \overline{L}(x,y) = \inf_{x \in X} \Big\{ f_0(x) + y_1 f_1(x) + \ldots + y_m f_m(x) - k(y) \Big\}. \end{aligned}$$

The results for $(\mathcal{P})$ and $(\mathcal{D})$ carry over to this more general pair of primal and dual problems.

Theorem 3.7. *In the convex case of $(\overline{\mathcal{P}})$, the existence for $\bar{x}$ of a multiplier vector $\bar{y}$ satisfying the extended Lagrange multiplier rule $(\overline{\mathcal{L}})$ is sufficient for $\bar{x}$ to be a globally optimal solution to $(\overline{\mathcal{P}})$. The vectors $\bar{y}$ that appear in this condition along with $\bar{x}$ are then precisely the optimal solutions to the dual problem $(\overline{\mathcal{D}})$, and the optimal values in the two problems agree: one has*

$$\min(\overline{\mathcal{P}}) = \max(\overline{\mathcal{D}}).$$

As a special case of this duality, of course, we have the earlier duality between $(\mathcal{P})$ and $(\mathcal{D})$, which corresponds to taking the function ρ to be given by the exact penalty expressions in (3.4). But the example we want to emphasize now is *extended linear-quadratic programming*, which will generalize the linear programming duality in Section 2. We take this term as referring to the case where the extended Lagrangian has the form

$$\overline{L}(x,y) = c{\cdot}x + \tfrac{1}{2}x{\cdot}Cx + b{\cdot}y - \tfrac{1}{2}y{\cdot}By - y{\cdot}Ax \tag{3.18}$$

where the matrices $C \in \mathbb{R}^{n\times n}$ and $B \in \mathbb{R}^{m\times m}$ are symmetric and positive *semi-*definite (possibly 0). To give a general expression to the two problems in this case, we use the notation

$$\rho_{YB}(u) = \sup_{y\in Y}\left\{y{\cdot}u - \tfrac{1}{2}y{\cdot}By\right\}, \qquad \rho_{XC}(v) = \sup_{x\in X}\left\{v{\cdot}x - \tfrac{1}{2}x{\cdot}Cx\right\}. \tag{3.19}$$

These functions can be made more specific according to the particular choices made of the polyhedral sets X and Y along with the matrices C and B, in accordance with the examples we have been discussing. Especially to be noted is the case where X and Y are *boxes* and C and B are *diagonal*, because then the expressions in (3.19) break down component by component.

The primal and dual problems of extended linear-quadratic programming come out in this notation as:

$(\mathcal{P}_{\text{elq}})$ minimize $c{\cdot}x + \tfrac{1}{2}x{\cdot}Cx + \rho_{YB}(b - Ax)$ over $x \in X$,

$(\mathcal{D}_{\text{elq}})$ maximize $b{\cdot}y - \tfrac{1}{2}y{\cdot}By - \rho_{XC}(A^*y - c)$ over $y \in Y$,

where A^* denotes the transpose of the matrix A. The linear programming problems $(\mathcal{P}_{\text{lin}})$ and $(\mathcal{D}_{\text{lin}})$ correspond to

$$X = \mathbb{R}^r_+ \times \mathbb{R}^{n-r}, \qquad Y = \mathbb{R}^s_+ \times \mathbb{R}^{m-s}, \qquad C = 0, \qquad B = 0.$$

Theorem 3.8. *If either of the extended linear-quadratic programming problems $(\mathcal{P}_{\text{elq}})$ or $(\mathcal{D}_{\text{elq}})$ has an optimal solution, then so does the other, and*

$$\min(\mathcal{P}_{\text{elq}}) = \max(\mathcal{D}_{\text{elq}}).$$

The pairs $(\bar{x}, \bar{y})$ such that $\bar{x}$ solves $(\mathcal{P}_{\text{elq}})$ and $\bar{y}$ solves $(\mathcal{D}_{\text{elq}})$ are precisely the ones that, for the choice of $\overline{L}$, X and Y corresponding to these problems, satisfy the extended Lagrange multiplier rule $(\overline{\mathcal{L}})$, or equivalently, give a saddle point of $\overline{L}$ on $X \times Y$.

This theorem was proved in [10]. The subject is elaborated and applied to dynamic modeling in [12]. Extended linear-quadratic programming models in multistage stochastic programming are described in [13].

Numerical approaches to extended linear-quadratic programming have been developed in Rockafellar and Wets [10], Rockafellar [14], Zhu and Rockafellar [15], Zhu [16], [17], and Chen and Rockafellar [16] for various purposes.

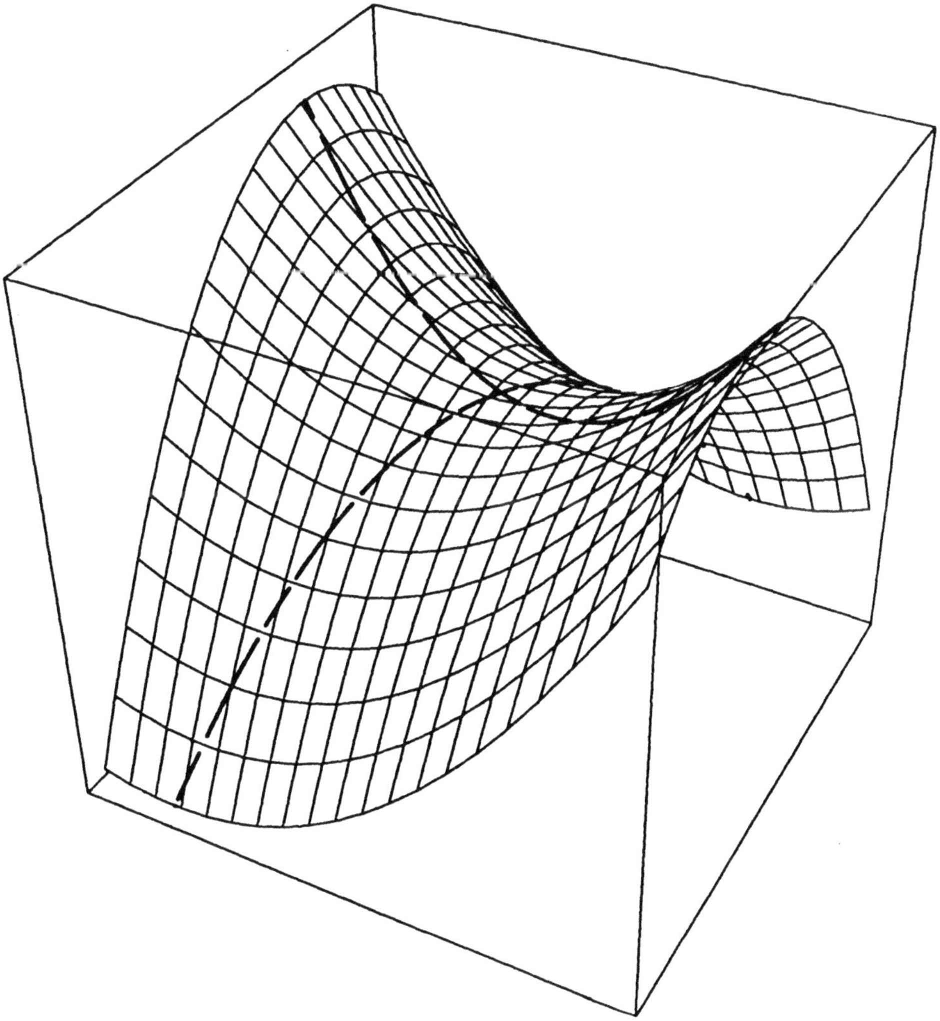

Figure 6: Biquadratic Lagrangian with associated (convex) primal and (concave) dual functional.

No doubt much more could be done, so the brief review of this work that follows should be seen as merely suggestive of some of the possibilities.

Paper [10] explains how, as a fallback option, any problem of extended linear-quadratic programming can be reformulated as one of ordinary quadratic programming—through the introduction of many extra variables. That technique, while offering reassurance that exotic new codes are not necessarily needed to get numerical answers, suffers however from two drawbacks. It greatly increases the dimension of the problem to be solved and at the same time may disrupt special structure in the objective and constraints. Ideally, such structure should instead be put to use in computation, at least if the aim is to cope with the huge optimization models that can arise from dynamics and stochastics. But for problems of modest size the reduction technique may be adequate.

Closely related is the technique of rewriting the optimality condition $(\overline{\mathcal{L}})$ as a linear "variational inequality" as described in [14]. This can in turn be translated into a complementarity relation to which algorithms for linear complementarity problems can be applied. The symmetry between the primal and dual problems is thereby preserved, although dimensionality is again increased. No write-ups are yet available on this, but numerical experiments conducted by S. J. Wright at Argonne National Laboratories near Chicago on solving extended linear-quadratic programming problems through interior-point methods for linear complementarity problems appear very promising.

Most of the algorithmic development has been undertaken in the *strictly quadratic* case, i.e., with the assumption that both of the matrices B and C in $(\mathcal{P}_{\rm elq})$ are positive definite. While this assumption apparently excludes linear programming and even the standard form of quadratic programming, it's not as severe as it first may seem. A number of approaches to solving large-scale problems introduce "proximal terms" in the objective. These are regularizing terms in the form of a strictly quadratic (although possibly small) penalty for deviation from a current estimate of the solution. They are moved and updated as computations proceed. Each subproblem with such a term does have, in effect, a positive definite matrix B. It turns out that proximal terms can be added iteratively in the dual variables of the Lagrangian as well as the primal variables, and in that way a sequence of regularized subproblems is generated in which the associated C is positive definite too. By solving the subproblems, one obtains sequences of primal and dual vectors which, in the limit, solve $(\mathcal{P}_{\rm elq})$ and $(\mathcal{D}_{\rm elq})$.

From this perspective, the solution of strictly quadratic problems is the key to the solution of more general problems. Such an approach with proximal terms has, for instance, been explored in some detail in the context of two-stage stochastic programming in [10].

In [14], a novel class of algorithms for solving strictly quadratic problems $(\mathcal{P}_{\mathrm{elq}})$ and $(\mathcal{D}_{\mathrm{elq}})$ has been developed in terms of "envelope representations" of the essential objective functions f and g. The partial approximation of f and g by an "envelope representation" is somewhat kin to using a pointwise maximum of affine functions to represent a convex function from below (which can be seen as a cutting-plane idea), but the approximations are piecewise linear-quadratic rather than just piecewise affine. The envelope representations are generated by iterative application of steps in which the Lagrangian $L(x, y)$ is minimized in $x \in X$ for fixed y, or maximized in $y \in Y$ for fixed x. For many large-scale problems arising in applications, such steps are easy to carry out, because the models can be set up in such a way that $L(x, y)$ is separable in x and y—separately, cf. [13].

Results of numerical experiments using envelope methods to solve extended linear-quadratic programming problems are reported in [15]. In particular, that paper develops special methods called primal-dual projected gradient algorithms. These methods are characterized by having two procedures go on at once—one in the primal problem and one in the dual problem—with a kind of information feedback between them. The feedback is the source of dramatic improvements in the rate of convergence. Besides being effective for moderately sized problems, the algorithms have successfully been used to solve problems in as many as 100,000 primal and 100,000 dual variables in a stable manner. This line of research has been carried further in [16] and [17].

While envelope methods take advantage of possible decomposability of large-scale problem structure through separate separability of the Lagrangian in the primal and dual variables, another form of decomposition is exploited by the Lagrangian "splitting methods" introduced in [17]. These are aimed at problems in which the Lagrangian is a kind of sum of independent sub-Lagrangians coming from prospective subproblems and a bilinear linking expression. Examples are furnished in [13], but they also arise in finite-element models for partial differential equations and associated variational inequalities. Iterations proceed with an alternation between "backward steps" which can be calculated by assigning each subproblem to a separate processor, and "forward steps" which are analogous to integrating dynamics, or calculating conditional expectations.

References

1. R. T. Rockafellar: *Convex Analysis*; Princeton University Press, Princeton, NJ, 1970.
2. R. T. Rockafellar: Lagrange multipliers and optimality; SIAM Review, 1993.
3. F. John: Extremum problems with inequalities as subsidiary conditions; in *Studies and Essays, Courant Anniversary Volume*, Interscience, New York, 1948.
4. O. L. Mangasarian and S. Fromovitz: The Fritz John conditions in the presence of equality and inequality constraints; J. Math. Anal. Appl. 17 (1967), 73–74.
5. H. W. Kuhn and A. W. Tucker: Nonlinear programming; Proceedings of the Second Berkeley Symposium on Mathematical Statistics and Probability (J. Neyman, ed.), Univ. Calif. Press, Berkeley, 1951, 481–492.
6. W. Karush: Minima of functions of several variables with inequalities as side conditions; master's thesis, Dept. of Mathematics, Univ. of Chicago, 1939.
7. D. Gale, H. W. Kuhn, and A. W. Tucker: Linear programming and the theory of games; in *Activity Analysis of Production and Allocation* (T. C. Koopmans, ed.), Wiley, 1951, 317–328.
8. R. T. Rockafellar: *Conjugate Duality and Optimization*; Regional Conference Series No. 16, SIAM Publications, 1974.
9. R. T. Rockafellar: *Network Flows and Monotropic Optimization*; Wiley, 1984.
10. R. T. Rockafellar and R. J-B Wets: A Lagrangian finite generation technique for solving linear-quadratic problems in stochastic programming; Math. Programming Studies 28 (1986), 63–93.
11. R. T. Rockafellar and R. J-B Wets: *Linear-quadratic problems with stochastic penalties: the finite generation algorithm*; in: *Numerical Techniques for Stochastic Optimization Problems*, Y. Ermoliev and R. J-B Wets (eds.), Springer-Verlag Lecture Notes in Control and Information Sciences No. 81, 1987, 545-560.
12. R. T. Rockafellar: Linear-quadratic programming and optimal control; SIAM J. Control Opt. 25 (1987), 781–814.
13. R. T. Rockafellar and R. J-B Wets: Generalized linear-quadratic problems of deterministic and stochastic optimal control in discrete time; SIAM J. Control Opt. 320 (1990), 810–822.
14. R. T. Rockafellar: Computational schemes for solving large-scale problems in extended linear-quadratic programming; Math. Prog., Ser. B 48 (1990), 447–474.
15. C. Zhu and R. T. Rockafellar: Primal-dual projected gradient algorithms for extended linear-quadratic programming; SIAM J. Optimization, 1993.
16. C. Zhu: On the primal-dual steepest descent algorithm for extended linear-quadratic programming; preprint, 1992
17. C. Zhu: Solving large-scale minimax problems with the primal-dual steepest descent algorithm; preprint, 1992
18. H. G. Chen and R. T. Rockafellar: Forward-backward splitting methods in Lagrangian optimization; SIAM J. Optimization, 1993.

DYNAMIC PROGRAMMING BASIC CONCEPTS AND APPLICATIONS

K. Neumann

Institut für Wirtschaftstheorie und Operations Research
University of Karlsruhe
D-76128 Karlsruhe, Germany

Abstract. Dynamic programming deals with sequential decision processes, which are models of dynamic systems under the control of a decision maker. At each point in time at which a decision can be made, the decision maker chooses an action from a set of available alternatives, which generally depends on the current state of the system. The objective is to find a sequence of actions (a so-called policy) that minimizes the total cost over the decision making horizon.

In what follows, deterministic and stochastic dynamic programming problems which are discrete in time will be considered. At first, Bellman's equation and principle of optimality will be presented upon which the solution method of dynamic programming is based. After that, a large number of applications of dynamic programming will be discussed.

1. Deterministic dynamic programming

1.1 The standard problem of dynamic programming

Dynamic programming deals with *sequential decision processes*, which are models of dynamic systems under the control of a decision maker. At each stage or in each period of some planning horizon, the decision maker chooses an *action* from a set of available alternatives, which generally depends on the current *state* of the system. This action induces a *cost* (or a reward) and a transition to a new state of the system depending on both the action selected and the previous state. The objective is to determine a sequence of actions (a so-called *policy*) that optimizes the performance of the system (for example, minimizes the total cost) over the planning horizon. The key aspect of such problems is that decisions cannot be viewed in isolation because one must balance the desire for low present cost with the possibility of high future cost.

Consider a planning horizon consisting of n periods or stages. The state of the underlying system at the beginning of period j (or respectively at the end of period $j-1$ is described by a *state variable* x_j ($1 \leq j \leq n$). At the beginning of period 1, the system is in the given initial state $x_1 = x_a$. In period j, an *action* or *decision* u_j from an *action space* $U_j(x_j)$ depending on state x_j is chosen (we also speak of the *decision variable* or *control variable* u_j). The selection of action u_j induces a transition to a new state $x_{j+1} = f_j(x_j, u_j)$ depending on the previous state x_j and action u_j. Moreover, a *cost* $g_j(x_j, u_j)$ is incurred. The possible states x_{j+1} at the end of period j are supposed to belong to a state space X_{j+1}. At the beginning of period 1 we have $X_1 = \{x_a\}$.

Minimizing the total cost over the n periods corresponds to the optimization problem

$$
\begin{aligned}
\text{minimize} \quad & \sum_{j=1}^{n} g_j(x_j, u_j) \\
\text{s.t.} \quad & x_1 = x_a \\
& x_{j+1} = f_j(x_j, u_j) \\
& x_{j+1} \in X_{j+1} \qquad (j = 1, \ldots, n) \\
& u_j \in U_j(x_j)
\end{aligned}
\tag{1.1}
$$

which is called the *standard problem of dynamic programming*. The state and action spaces may be discrete sets, intervals on $\mathbb{R}$, or (if states and actions are of higher dimension) subsets of $\mathbb{R}^p$ or respectively $\mathbb{R}^q$. A sequence of actions $(u_1, \ldots, u_n)$ is termed a *policy*. A policy that satisfies the constraints of (1.1) and minimizes the objective function is called an *optimal policy*.

In contrast to other types of optimization problems such as linear or nonlinear programming, where a large number of different solution procedures have been developed, there is only one principal method of solving dynamic programming problems. We will merely sketch this method in what follows very briefly. In dynamic programming, it is more important to show how to model quite different kinds of optimization problems in terms of dynamic programming. This will help us to get a feeling for where to apply dynamic programming in practice.

1.2 Bellman's equation and principle of optimality

Assume that all minima appearing in what follows do exist. This is the case, for example, if all state and action spaces are (nonempty) finite sets.

Given the functions f_j and g_j as well as the state and action spaces X_{j+1} and U_j for $j = 1, \ldots, n$, problem (1.1) and its solutions depend only on the initial state x_1. We denote this optimization problem by $P_1(x_1)$. The corresponding problem that comprises only the periods $j, j+1, \ldots, n$ $(1 < j \leq n)$

and depends on initial state x_j is designated by $P_j(x_j)$. Let $(u_j^\star, u_{j+1}^\star, \dots, u_n^\star)$ be an optimal policy and $v_j^\star(x_j)$ be the minimum cost for problem $P_j(x_j)$. Then $(u_{j+1}^\star, \dots, u_n^\star)$ is an optimal policy for problem $P_{j+1}(x_{j+1})$ with initial state $x_{j+1}^\star := f_j(x_j, u_j^\star)$ and cost $v_{j+1}^\star(x_{j+1}^\star)$. If there were a "better" policy $(u_{j+1}^+, \dots, u_n^+)$ for problem $P_{j+1}(x_{j+1}^\star)$ with smaller cost $v_{j+1}^+(x_{j+1}^\star)$, then $(u_j^\star, u_{j+1}^+, \dots, u_n^+)$ would be a "better" policy for $P_j(x_j)$ with cost

$$g_j(x_j, u_j^\star) + v_{j+1}^+(x_{j+1}^\star) < g_j(x_j, u_j^\star) + v_{j+1}^\star(x_{j+1}^\star) = v_j^\star(x_j)$$

in contradiction to the optimality of $v_j^\star(x_j)$. Moreover, it holds that

$$v_j^\star(x_j) = g_j(x_j, u_j^\star) + v_{j+1}^\star(x_{j+1}^\star) = \min_{u_j \in U_j(x_j)} \{ g_j(x_j, u_j) + v_{j+1}^\star(f_j(x_j, u_j)) \}. \tag{1.2}$$

The fact that a part of an optimal policy (with respect to a fixed initial state) represents an optimal policy for the respective partial problem is known as Bellman's principle of optimality. We formulate this principle for the problems $P_1(x_1)$ and $P_j(x_j)$:

Bellman's principle of optimality. Let $(u_1^\star, \dots, u_j^\star, \dots, u_n^\star)$ be an optimal policy for problem $P_1(x_1)$ and $x_j^\star$ be the state at the beginning of period j. Then $(u_j^\star, \dots, u_n^\star)$ is an optimal policy for problem $P_j(x_j^\star)$. In other words: The decisions in periods $j, \dots, n$ of the n-period problem $P_1(x_1)$ are independent of the decisions in periods $1, \dots, j-1$ given the state x_j at the beginning of period j.

The function $v_j^\star$, which is defined on state space X_j, is termed the *value function* $(1 \le j \le n)$. For $j = n+1$ we put

$$v_{n+1}^\star(x_{n+1}) := 0 \qquad \text{for } x_{n+1} \in X_{n+1}. \tag{1.3}$$

For $X_j \subset \mathbb{R}$ it is expedient to define $v_j^\star$ on all of $\mathbb{R}$ and to set $v_j^\star(x_j) := \infty$ for $x_j \in \mathbb{R} \setminus X_j$ $(1 \le j \le n+1)$. Relation (1.2), which is valid for $j = 1, \dots, n$, is called *Bellman's equation*:

$$v_j^\star(x_j) = \min_{u_j \in U_j(x_j)} \{ g_j(x_j, u_j) + v_{j+1}^\star(f_j(x_j, u_j)) \} \quad (x_j \in X_j,\ 1 \le j \le n). \tag{1.4}$$

Bellman's equation connects two successive value functions $v_j^\star$ and $v_{j+1}^\star$ and permits us to compute function $v_j^\star$ when function $v_{j+1}^\star$ is known.

We consider some modifications of standard problem (1.1). If the objective function is to maximize instead of minimizing it, we simply replace "min" by "max" in Bellman's equation (1.4). When the objective function has the form

$$\sum_{j=1}^{n} g_j(x_j, u_j) + g_{n+1}(x_{n+1}),$$

where $g_{n+1}(x_{n+1})$ represents a terminal cost, formula (1.3) has to be replaced by

$$v_{n+1}^\star(x_{n+1}) := g_{n+1}(x_{n+1}) \qquad \text{for } x_{n+1} \in X_{n+1}.$$

If the objective function has the form

$$\prod_{j=1}^{n} g_j(x_j, u_j), \tag{1.5}$$

where all functions g_j $(j = 1, \ldots, n)$ are supposed to be positive, then in Bellman's equation (1.4) addition is replaced by multiplication and, for $j = n+1$, $v_{n+1}^\star(x_{n+1}) := 0$ is replaced by $v_{n+1}^\star(x_{n+1}) := 1$. An objective function of type (1.5) arises, for example, if the functions g_j represent reliabilities of components of a series system whose reliability is to be minimized.

1.3 Solving the standard problem

An optimal policy for standard problem (1.1) can be found by exploiting Bellman's equation (1.4). Let

$$w_j(x_j, u_j) := g_j(x_j, u_j) + v_{j+1}^\star(f_j(x_j, u_j))$$

be the expression within the braces in (1.4), and let $z_j^\star(x_j)$ be a minimizer of function $w_j(x_j, \cdot)$ on $U_j(x_j)$, that is,

$$w_j(x_j, z_j^\star(x_j)) = \min_{u_j \in U_j(x_j)} w_j(x_j, u_j) = v_j^\star(x_j) \qquad \text{for } x_j \in X_j.$$

The quantities $z_j^\star(x_j)$ and $v_j^\star(x_j)$ for $x_j \in X_j$ can be computed by evaluating Bellman's equation (1.4) *backwards* for $j = n, n-1, ..., 1$ beginning with $v_{n+1}^\star(x_{n+1}) := 0$.

The sequence of functions $(z_1^\star, \ldots, z_n^\star)$ is termed an *optimal feedback controller* or *optimal feedback policy* because an optimal decision $z_j^\star(x_j)$ in period j depends directly on the state x_j at the beginning of period j. By Bellman's principle of optimality, $z_j^\star(x_j)$ represents an optimal decision in period 1 of partial problem $P_j(x_j)$. Hence, the states $x_j^\star$ and actions $u_j^\star$ computed *forwards* as

$$\begin{aligned}
x_1^\star &= x_a \\
u_1^\star &= z_1^\star(x_1^\star), \quad x_2^\star = f_1(x_1^\star, u_1^\star) \\
&\vdots \\
u_n^\star &= z_n^\star(x_n^\star), \quad x_{n+1}^\star = f_n(x_n^\star, u_n^\star)
\end{aligned}$$

are optimal, in particular, $(u_1^*, \ldots, u_n^*)$ is an optimal policy.

Strictly speaking, the method of dynamic programming just described is a destilled form of direct enumeration obtained by exploiting Bellman's principle of optimality. Direct enumeration reveals the advantages and disadvantages of dynamic programming. *Advantages* are that the functions f_j and g_j $(j = 1, \ldots, n)$ in standard problem (1.1) need not have specific properties such as linearity or convexity. Also, if fractional values for the variables x_j and u_j do not make sense, which occurs very often in practice, the restriction to integer values does not lead to additional difficulties as it is the case in linear and nonlinear programming. In contrast, the restriction to integer variables reduces the cardinality of the state and action spaces and makes a discretization of both spaces unnecessary, and thus reduces the computing time if direct enumeration is used. *Disadvantages* result from the generally enormous computational effort: The time complexity of dynamic programming is polynomial or pseudopolynomial only in special cases. In general, the computing time and storage requirements grow exponentially in the dimension of the state and action spaces. This "curse of dimensionality" limits the application of dynamic programming to many real-life problems.

To overcome the curse of dimensionality, several proposals have been made. Ozden (1987) [18] considers the case where, for each $j = 1, \ldots, n$, the "dynamic constraint" $x_{j+1} = f_j(x_j, u_j)$ in (1.1) can uniquely be solved for u_j so that the variables u_j can be eliminated from problem (1.1). This case occurs very often in practice. Under the assumptions that the objective function satisfies certain convexity conditions and the state spaces (modified by eliminating the variables u_j) represent bounded polyhedra, Ozden (1987) [18] proposes a polynomial iteration procedure that provides an optimal solution within a prespecified precision limit. In general, if each state space is a subset of $\mathbb{R}^p$ and each dimension is discretized into m discrete values, m^p lattice points of the respective state space have to be considered in each period j. At each iteration of Ozden's method, only $p + 1$ independent lattice points in the state space (that is, a simplex in $\mathbb{R}^p$) are taken into account. Ozden's method has turned out to be a powerful solution procedure for many large nonlinear sequential decision problems. It has been applied to solving nonlinear problems in production planning and in operational planning of water resource systems.

1.4 Applications

The formulation of an optimization problem in terms of dynamic programming requires identifying the states and actions as well as feasible state and action spaces and finding the transition functions f_j and cost functions g_j $(j = 1, \dots, n)$. Since nearly every facet of life entails a sequence of decisions, there are numerous applications of dynamic programming in practice. Some of these applications are sketched in what follows.

1.4.1 Inventory control

Controlling the replenishment of inventories is one of the main problems in production management. We consider the following inventory model for a single commodity. Given a horizon of n time periods. Let $K_j \geq 0$ be the setup cost charged at the beginning of period j if a positive amount of the good is produced or ordered in that period, let $c_j > 0$ be the production cost (or purchase cost) per unit of the respective commodity in period j, and let $h_j > 0$ be the inventory holding cost per unit charged at the end of period j. Moreover, let $r_j > 0$ be the demand in period j which must be met (that is, no shortages are permitted), let u_j be the quantity produced (or ordered) at the beginning of period j, and let x_j be the inventory at the beginning of period j before producing or placing an order (or respectively at the end of period $j-1$). Suppose that replenishment occurs instantaneously (no delivery lag). The inventory problem then consists of determining how much should be produced (or ordered) at the beginning of each time period so as to minimize the total cost incurred over the n periods.

Initially, there is supposed to be no stock on hand. Then we have

$$\begin{aligned} x_{j+1} &= x_j + u_j - r_j \quad (j = 1, \dots, n) \qquad (1.6) \\ x_1 &= 0. \end{aligned}$$

Thus, the transition function f_j has the form $f_j(x_j, u_j) = x_j + u_j - r_j$. Equation (1.6) is sometimes called the *equation of inventory movements.* The cost incurred in period j is

$$\left. \begin{aligned} K_j + c_j u_j + h_j x_{j+1} \quad & \text{if } u_j > 0 \\ h_j x_{j+1} \quad & \text{if } u_j = 0 \end{aligned} \right. \qquad (j = 1, \dots, n).$$

Introducing the function

$$\delta(u) := \begin{cases} 1 & \text{if } u > 0 \\ 0 & \text{if } u = 0, \end{cases}$$

observing (1.6), and omitting the additive constant $-\sum_{j=1}^{n} h_j r_j$ in the objective function, the inventory problem corresponds to the following dynamic programming problem

$$
\begin{aligned}
\text{minimize} \quad & \sum_{j=1}^{n} \Big(h_j x_j + (c_j + h_j) u_j + K_j \delta(u_j) \Big) \\
\text{s.t.} \quad & x_1 = 0 \\
& x_{j+1} = x_j + u_j - r_j \\
& x_{j+1} \geq 0, \ u_j \geq 0 \qquad (j = 1, \dots, n).
\end{aligned} \tag{1.7}
$$

Exploiting the specific structure of problem (1.7), Wagner and Whitin have proposed an efficient algorithm for solving that problem, which can be implemented to run in $O(n \log n)$ time (cf. Federgruen and Tzur (1991) [6]). If there is a production capacity κ_j in period j that cannot be exceeded, we have the additional constraint $u_j \leq \kappa_j$ $(j = 1, \dots, n)$ in problem (1.7). The Wagner-Whitin algorithm, however, cannot be applied to that more general problem.

1.4.2 Shortest paths in networks

We are looking for the shortest paths in a network with weights $c_{ij} \in \mathbb{R}$ of the arcs (i, j) from each node to a fixed node s. Assume that the network does not contain cycles of negative length. Let r be any node of the network different from s where s is reachable from r, and let d_r be the length of a shortest path from r to s. If k is the immediate successor of r on such a shortest path (see Fig. 1.1), then the part of the path from k to s must also be a shortest path (this corresponds to Bellman's principle of optimality). Thus, we have

$$d_r = c_{rk} + d_k. \tag{1.8}$$

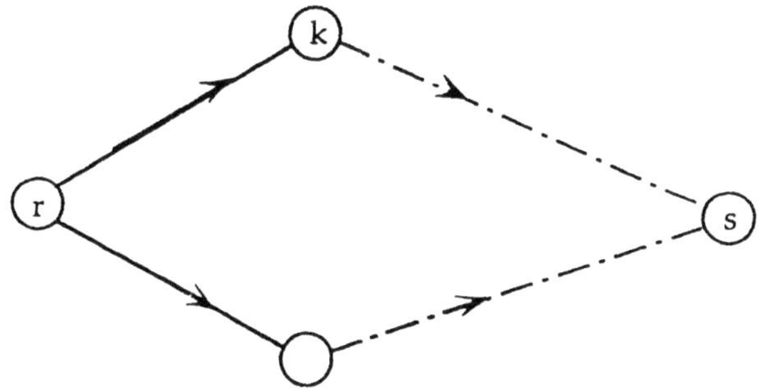

Figure 1.1

Clearly, in (1.8) k has to be a node for which $c_{rk}+d_k$ is as small as possible. Hence, the shortest-path lengths must satisfy Bellman's equation

$$d_r = \min_{k\in S(r)} \quad (c_{rk} + d_k) \tag{1.9}$$

where $S(r)$ is the set of the immediate successors of node r. In terms of dynamic programming, the nodes of the network correspond to states, a decision consists of choosing an immediate successor of the current node, and the minimum total cost corresponds to the shortest-path length. A shortest-path algorithm based upon Bellman's equation (1.9) can be implemented to run in $O(mn)$ time where m is the number of arcs and n the number of nodes in the network (cf. Gallo and Pallottino (1986, 1988) [7,8], Neumann and Morlock (1993) [17], Section 2.4).

1.4.3 Resource allocation and knapsack problems

One of the main topics of operations research is the *allocation of scarce resources.* Linear allocation problems of vast dimensions are solved by linear programming, whereas nonlinear models (of small dimensions) can be dealt with by dynamic programming. For simplicity, suppose there is only one resource of capacity $A > 0$ available. The resource can be allocated to the production of n different commodities numbered $1, 2, \ldots, n$. Commodities can only be produced in integer quantities. Producing u_j units of commodity j is supposed to consume $h_j(u_j) > 0$ units of the resource and to yield profit $g_j(u_j)$. Maximizing the total profit provides the following optimization problem

$$\begin{aligned} \text{maximize} \quad & \sum_{j=1}^{n} g_j(u_j) \\ \text{s.t.} \quad & \sum_{j=1}^{n} h_j(u_j) \le A \\ & u_j \in \mathbb{N}_0 \quad (j = 1, \ldots, n) \end{aligned} \tag{1.10}$$

where $\mathbb{N}_0$ is the set of the nonnegative integers. To formulate (1.10) as a dynamic programming problem, we make the problem dynamic artificially. To do so we consider the decisions of allocating the available capacity of resource to the commodities $1, 2, \ldots, n$ to be decisions in n successive stages or periods. We introduce nonnegative quantities x_j successively according to

$$\begin{aligned} x_1 &= A \\ x_{j+1} &= x_j - h_j(u_j) \quad (j = 1, \ldots, n). \end{aligned}$$

x_j represents the remaining capacity of resource for commodities $j, \ldots, n$. Then problem (1.10) can be rewritten as

$$\begin{aligned} \text{maximize} \quad & \sum_{j=1}^{n} g_j(u_j) \\ \text{s.t.} \quad & x_1 = A \\ & x_{j+1} = x_j - h_j(u_j) \\ & 0 \le x_{j+1} \le A \qquad (j = 1, \ldots, n) \\ & u_j \in \mathbb{N}_0 \end{aligned} \tag{1.11}$$

A special case of the allocation problem (1.10) in which the profit and consumption functions g_j and h_j are both linear is known as the *integer knapsack problem.* Interpret A as the maximum weight of a knapsack. Each unit of commodity j has weight a_j and yields profit c_j. We are interested in the most profitable way to pack the knapsack. This leads to the problem

$$\begin{aligned} \text{maximize} \quad & \sum_{j=1}^{n} c_j u_j \\ \text{s.t.} \quad & \sum_{j=1}^{n} a_j u_j \le A \\ & u_j \in \mathbb{N}_0 \quad (j = 1, \ldots, n). \end{aligned}$$

If we rewrite this problem as a dynamic programming problem of type (1.11), the action spaces take the form

$$U_j(x_j) = \{\, 0, 1, \ldots, \lfloor x_j / a_j \rfloor \,\} \quad (j = 1, \ldots, n)$$

instead of $\mathbb{N}_0$, where $\lfloor z \rfloor$ is the largest integer $\le z$.

When at most one unit of each commodity can be packed into the knapsack, we speak of the *0-1 knapsack problem.* For the 0-1 knapsack problem, the action spaces are

$$U_j(x_j) = \begin{cases} \{0, 1\} & \text{if} \quad x_j \ge a_j \\ \{0\} & \text{if} \quad x_j < a_j \end{cases} \qquad (j = 1, \ldots, n).$$

If the quantities $A, a_1, \ldots, a_n$ are all integers, the dynamic programming technique for the integer and $0-1$ knapsack problems can be implemented in $O(nA)$ time, that is, we have a pseudopolynomial algorithm (cf. Nemhauser and Wolsey (1988) [15], Sections II.5.5 and II.6.1, Neumann and Morlock (1993) [17], Section 5.1.5).

1.4.4 Machine scheduling

Suppose there are n jobs numbered $1, 2, \dots, n$ to be processed by a single machine, which can execute at most one job at a time. Job j requires a processing time $t_j \geq 0$. The jobs are assumed to be executed without interruption and without idle times between them with the first job beginning at time zero. Moreover, each job is supposed to be available for processing at time zero. Then any given sequence of jobs induces a well-defined completion time C_j for each job $j = 1, \dots, n$. Let g_j be a nondecreasing real-valued function where $g_j(t)$ represents the cost arising when job j is completed at time t. Suppose that the objective is to find a sequence of jobs that minimizes the total cost $\sum_{j=1}^n g_j(C_j)$. Moreover, assume that precedence constraints for the jobs are given, where "job j precedes job k" means that job j has to be completed before job k can be begun. Such a precedence relation can be represented by an acyclic directed graph G, where the jobs correspond to the nodes and job j precedes job k precisely if node k is reachable from node j in G where $k \neq j$.

Let $J \subseteq \{1, 2, \dots, n\}$ be any job subset and $t_J := \sum_{j \in J} t_j$. Let

$$v^\star(J) := \min_{\pi(J)} \sum_{j \in J} g_j(C_j)$$

be the minimum total cost for job subset J, where the minimum is taken over all possible sequences $\pi(J)$ for set J. In particular, $v^\star(\{1, \dots, n\})$ is the minimum objective function value of our scheduling problem. For $J = \emptyset$, we put $v^\star(\emptyset) := 0$.

A job subset J is called *feasible* if $j \in J$ implies that all predecessors of j belong to J. In what follows, only feasible job subsets need to be considered, which are termed *feasible sets* for short. Let $S(J)$ be the set of all jobs from J without successor (that is, the set of the "sinks" of the subgraph of G corresponding to J). Obviously, J is a feasible set of cardinality $|J| = k$ exactly if $J \backslash \{j\}$ with $j \in S(J)$ is a feasible set of cardinality $k - 1$ $(1 \leq k \leq n)$.

Interpret the scheduling problem to be solved as a dynamic programming problem as follows. The possible states at stage k correspond to the feasible sets of cardinality k. A transition from a state at stage $k - 1$ to a stage k corresponds to a transition from feasible set $J \backslash \{j\}$ with $j \in S(J)$ to feasible set J where $|J| = k$. Given "state" $J' := J \backslash \{j\}$ with job sequence $\pi(J')$, an action corresponds to choosing job $j \in S(J)$ and job sequence $\pi(J'), j$ (that is, job j succeeds immediately to sequence $\pi(J')$) for the "new state" $J := J' \cup \{j\}$. Bellman's equation for stage k has the form

$$v^\star(J) = \min_{j \in S(J)} \{ g_j(t_J) + v^\star(J \backslash \{j\}) \} \quad \text{for all feasible } J \text{ with } |J| = k \qquad (1.12)$$

where $g_j(t_J)$ represents the cost arising when job j succeeds immediately to job sequence $\pi(J')$. Bellman's equation (1.12) has to be evaluated successively for $k = 1, 2, \ldots, n$ and at each stage k, for all feasible sets J of cardinality k. At stage n, we obtain the minimum total cost of the scheduling problem, $v^\star(\{1, \ldots, n\})$, and, if for each feasible set J we store the optimal job sequence $\pi^\star(\{1, \ldots, n\})$ that provides $v^\star(J)$, an optimal job sequence $\pi^\star(\{1, \ldots, n\})$.

If $K = O(2^n)$ is the number of all feasible sets, the time complexity of the dynamic programming procedure is $O(nK) = O(n2^n)$. In practice, K is generally very much smaller than 2^n, so that we can hope to solve much larger problems than in the absence of precedence constraints for the jobs. For details of an efficient implementation of a dynamic programming algorithm we refer to Baker and Schrage (1978) [1] and Lawler (1979) [14].

1.4.5 *Optimal water use for electricity generation by mixed power systems with thermal and hydro plants*

In this section we closely follow Efthymoglou (1987) [5]. Suppose there is a system of several thermal plants for generating electricity ordered in merit according to increasing unit fuel cost (i.e. the cost per kWh). Let $C(Q)$ be the minimum system's hourly fuel cost of supplying a load of Q (kW). The function $C(\cdot)$ is increasing and convex. Let $h(Q)$ be the number of hours in some time period (for example, a month) that the actual load is $\geq Q$, also called the load-duration curve. The total thermal energy (in kWh) that can be supplied to meet demand up to load level Q is

$$E(Q) = \int_0^Q h(q)dq.$$

Then the minimum fuel cost of producing $E(Q)$ is

$$K(E(Q)) = \int_0^Q C'(q)h(q)dq$$

where C' is the derivative of function C. The latter equation is based on the fact that the change in a system's fuel cost resulting from a marginal increase in the installed thermal capacity is determined by the change of the system's hourly fuel cost and the number of hours that this marginal capacity is operated.

In practice, function h is generally given by a piecewise linear function, and functions C and K are mostly approximated by polynomials based upon econometric estimation.

Let E_d be the demand for energy and E_m be the maximum energy that can be supplied by the thermal plants. Assume there are hydro plants available that can cover the excess energy demand over the thermal capacity of the system, $e := \max(0, E_d - E_m)$. Moreover, the hydro plants can also be used to substitute for thermal plants with large unit fuel cost. If u is the quantity of hydro energy used to substitute thermal energy, then the thermal plants supply the rest $E_m - u$ resulting in marginal fuel cost savings of $K'(E_m - u)$. Thus, the total fuel cost savings is

$$g(u) = \int_{E_m - u}^{E_m} K'(E)dE.$$

If function K is approximated by a quadratic function, then g is also quadratic, say

$$g(u) = au - bu^2 \tag{1.13}$$

where $a > 0$ and $b > 0$.

For simplicity, assume that the hydro system consists of one large plant with water storage capacity and known water inflows. Moreover, hydro energy generation is assumed to be proportional to water use (that is, enhancement gains from higher water head offset discounting). Suppose the planning horizon (say, one year) is divided into n periods (say, 12 months). Let e_j be the given excess energy demand over the thermal capacity of the system in period j, which will be covered by hydro plant operation. Let w_j be the hydro energy of the water inflows during period j obtained from electricity utility's records. Moreover, let x_j be the hydro energy reserves stored in the reservoirs at the beginning of period j, where the initial and final hydro energy stored in the reservoirs, x_I and x_F, respectively, are known. Also, there is a given maximum storing capacity of the reservoirs, x_{max}, measured from the minimum operational level. Furthermore, let u_j be the hydro energy used to substitute thermal energy during period j. Then the problem of finding the substitute quantities of hydro energy in the individual periods that maximize the total fuel cost savings over the planning horizon is

$$\begin{aligned}
\text{maximize} \quad & \sum_{j=1}^{n} (a_j u_j - b_j u_j^2) \\
\text{s.t.} \quad & x_1 = x_I \ , \ x_{n+1} = x_F \\
& x_{j+1} = x_j + w_j - e_j - u_j, \quad j = 1, \ldots, n \\
& 0 \le x_j \le x_{max}, \quad j = 2, \ldots, n \\
& 0 \le u_j \le \max(0, x_j + w_j - e_j), \quad j = 1, \ldots, n
\end{aligned} \tag{1.14}$$

where $a_j > 0$ and $b_j > 0$ are the coefficients in the total-fuel-cost-savings function g_j for period j (compare (1.13)).

The above model and its dynamic programming solution have been used to determine optimal operating rules and reservoir levels for water use in the Greek power system (cf. Efthymoglou (1987) [5]), where oil-fired and lignite-fired plants constitute the thermal power plants. In addition, gas-turbine plants are held in cold reserve, which will be used if available steam and hydro plants fail to meet loads because of unexpected events. The results of econometric estimation and optimal solution to the dynamic programming problem (1.14) have been used to obtain short-run marginal costs and water values. Also, efficient pricing of both electricity and water supply for irrigation and other uses can be derived.

Within the area of optimization of electric power systems, dynamic programming can also be used for solving (subproblems of) the unit commitment problem without and with long-term energy constraints, cf. Braun (1992) [3], Handschin and Slomski (1989) [10], Sanders and Linke (1992) [19], and Slomski (1990) [21].

1.4.6 Irrigation system management with depleting groundwater

The problem of measuring the economic benefits of irrigation system development over a depleting aquifer along with related methodology for detailed long-range farm planning can be formulated as a dynamic programming problem, cf. Stoecker et al. (1985) [22]. In this model, management issues such as distribution system configuration, drilling policy, area developed for irrigation, and crop production are considered.

At first, parametric linear programming is used to maximize periodic profits subject to specified values of state variables related to annual water use and irrigation system capacity. The results of parametric linear programming are then used in a dynamic programming model to determine the optimal allocation of water and irrigation resources over time. The state variables include the remaining saturated thickness (measured in cm of water saturated sand), the number of operating irrigation wells, and the area developed for irrigation. The common options open to the farmer (decision variables) include the conversion to dryland crop production, the restaging of existing wells (if indicated by the state of the aquifer), the drilling of additional wells, and the replacement of all or part of the existing distribution system (depending on the age of the system) or the expansion of the distribution system. Within the capacity of an irrigation system, the farmer can choose to use varying quantities of groundwater.

This model has been used for farms in the Texas High Plains. Results indicate that the economic benefits of modern water and energy efficient irrigation systems may come from the expansion of current irrigation intensity rather than from an extended period of irrigation when water is initially scarce relative to land.

1.4.7 Determining optimal runway exit locations

Another problem dynamic programming has been applied to is the issue of improving the operational use of runways, cf. Sherali et al. (1992) [20].

The efficiency of runway usage is dictated primarily by the runway occupancy time, which is the time that an aircraft spends on the runway or its vicinity until a new arrival or departure can be processed on this runway. The problem of determining the geometry and location of high speed exits on a runway to minimize the weighted runway occupancy time of a population of aircraft (where the weight of a type of aircraft is the relative frequency of runway usage for that aircraft) under various landing scenarios and frequencies of usage can be formulated as a dynamic programming problem. Both the problem of designing a new runway and modifying an existing one can be addressed. Due to the minimum separation distance recommended by the U.S. Federal Aviation Administration, the continuous location problem of siting runway turnoffs can be reduced to a finite set of possible locations.

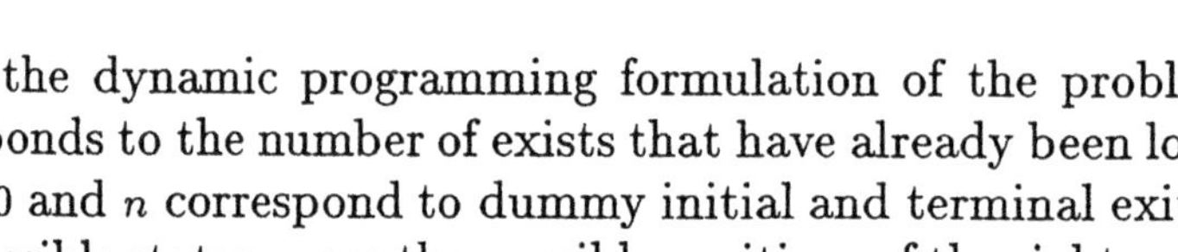

Figure 1.2

In the dynamic programming formulation of the problem, the stage j corresponds to the number of exists that have already been located ($0 \leq j \leq n$, where 0 and n correspond to dummy initial and terminal exits, see Fig. 1.2). The possible states x_j are the possible positions of the rightmost exit currently located (where $x_0 = 0$), and the decision u_j corresponds to the position of the next exit to be constructed to the right of x_j (where $u_j = L$ means that no more exit can be constructed). Obviously,

$$x_{j+1} = \begin{cases} u_j & \text{if} \quad u_j \neq L \\ x_j & \text{if} \quad u_j = L. \end{cases}$$

1.4.8 *Planning of physician requirements in developing countries*

A concluding application of deterministic dynamic programming is the planning for physician requirements in developing countries, cf. Ikem and Reisman (1990) [13]. Physician manpower requirements in most developing countries can be met by developing local training facilities, sending indigenes to train abroad, and/or recruiting expatriate physicians. Foreign training and having expatriates often involve expenditures of scarce foreign exchange in competition with other development projects. Thus, national planners and decision makers, as well as institutional administrators, must periodically deal with questions such as

i) how many students to admit to local institutions in any given year,
ii) how many scholarships/fellowships to award for training abroad in any given year,
iii) whether or not to start a new medical school in any given year,
iv) whether to import expatriates and how many.

The goal is to minimize the total cost over some planning horizon (say, 15 years) while meeting specific (per capita of population) norms for physicians. The cost includes the start-up and fixed cost of running medical schools, the variable cost of training indigenes abroad and the annual cost of having a pool of trained physicians to serve the population. Moreover, there is a penalty cost for overproducing and hence underemploying physicians.

Several runs of the dynamic programming model were made using different values for the input parameters as well as data from Nigeria (as a developing country) and the U.S.A. and United Kingdom (as countries where indigenes can be trained abroad).

2. Stochastic dynamic programming

In decision problems, we are often faced with making decisions based upon phenomena that have uncertainty associated with them. For example, a retailer may stock goods without knowing the customers' demand for them. Equipment is purchased without exact knowledge of its maintenance cost or its lifetime. Often such uncertainty is caused by inherent variation that can be described by a probabilistic model.

2.1 Bellman's equation

In what follows we only sketch the basic elements of stochastic dynamic programming without mathematical rigour. In particular, we do not discuss the measurability of all sets and functions and we assume that all minima exist. For a more detailed treatment we refer to Bertsekas (1987) [2] and Heyman and Sobel (1984) [11]. Let x_j be the (realized) state at the beginning of period j and u_j the decision made in period j $(1 \leq j \leq n)$. The state at the end of period j is a random variable whose (conditional) probability distribution is given by the probability density function $\varphi_j(\cdot|x_j, u_j)$ depending on x_j and u_j. In particular, we assume that all probability distributions are continuous (for discrete distributions, all integrals occurring subsequently have to be replaced by corresponding sums). The initial state $x_1 = x_a$ is supposed to be a given deterministic quantity. In period j, the transition from (realized) state x_j to (realized) state x_{j+1} if action u_j is taken results in a cost $g_j(x_j, u_j, x_{j+1})$. The state and action spaces are again denoted by X_j and $U_j(x_j)$, respectively.

Since the states represent random variables, only feedback controllers or policies $(z_1, \dots, z_n)$ make sense, where z_j is a function defined on X_j and $z_j(x_j) \in U_j(x_j)$ is the action chosen in period j at (realized) state x_j. We are looking for an optimal policy that minimizes the expected total cost over the planning horizon consisting of n periods. Let $v_j^\star(x_j)$ be the minimum expected cost of periods $j, j+1, \dots, n$ given the (realized) state x_j at the beginning of period j. Then Bellman's equation is

$$v_j^\star(x_j) = \min_{u_j \in U_j(x_j)} \int_{X_{j+1}} [g_j(x_j, u_j, x_{j+1}) + v_{j+1}^\star(x_{j+1})] \varphi_j(x_{j+1}|x_j, u_j) dx_{j+1} \tag{2.1}$$

for $x_j \in X_j$, $1 \leq j \leq n$, where $v_{n+1}^\star(x_{n+1}) := 0$ for $x_{n+1} \in X_{n+1}$. Let

$$\bar{g}_j(x_j, u_j) = \int_{X_{j+1}} g_j(x_j, u_j, x_{j+1}) \varphi_j(x_{j+1}|x_j, u_j)\, dx_{j+1}$$

be the expected cost in period j given the (realized) state x_j at the beginning of period j and action u_j in period j. Then Bellman's equation (2.1) can be rewritten as

$$v_j^\star(x_j) = \min_{u_j \in U_j(x_j)} [\ \bar{g}_j(x_j, u_j) + \int_{X_{j+1}} v_{j+1}^\star(x_{j+1}) \varphi_j(x_{j+1}|x_j, u_j)\, dx_{j+1}\ ]. \tag{2.2}$$

If function $v_{j+1}^\star$ is known, evaluation of (2.2) provides the functions $v_j^\star$ and $z_j^\star$, where $z_j^\star(x_j)$ is a minimizer of function $w_j(x_j, \cdot)$ on $U_j(x_j)$ and

$$w_j(x_j, u_j) := \bar{g}_j(x_j, u_j) + \int_{X_{j+1}} v_{j+1}^\star(x_{j+1}) \varphi_j(x_{j+1}|x_j, u_j)\, dx_{j+1}.$$

In this manner, the functions $v_j^\star$ and $z_j^\star$ can be computed backwards for $j = n, n-1, \ldots, 1$, which generally requires a large computational effort. $(z_1^\star, \ldots, z_n^\star)$ is then an optimal feedback policy.

If the probability distributions of states are discrete instead of continuous, the value of the probability density function $\varphi_j(x_{j+1}|x_j, u_j)$ has to be replaced by the probability that the state at the end of period j takes the value x_{j+1} given x_j and u_j $(j = 1, \ldots, n)$ and the integrals in the above formulas have to be replaced by corresponding sums.

2.2 Markov decision processes

In this section we assume that the state and action spaces are finite and independent of the time period, say

$$X := \{1, \ldots, m\}$$
$$U(i) := \{u_{i1}, \ldots, u_{is_i}\} \quad \text{for} \quad i = 1, \ldots, m.$$

Let $x_j = i$ be the (realized) state at the beginning of period j. Suppose the selection of a decision $u_{i\sigma}$ causes a transition to state $x_{j+1} = k$ at the beginning of period $j+1$ with probability $p_{ik}(u_{i\sigma})$ where

$$\sum_{k=1}^{m} p_{ik}(u_{i\sigma}) = 1 \quad \text{for} \quad i = 1, \ldots, m.$$

The transition from state i to state k caused by action $u_{i\sigma}$ results in a cost $g(i, u_{i\sigma}, k)$. Then

$$\bar{g}(i, u_{i\sigma}) := \sum_{k=1}^{m} p_{ik}(u_{i\sigma}) g(i, u_{i\sigma}, k)$$

is the expected cost per period if we start at state i and select action $u_{i\sigma}$.

A sequence of random variables $\xi_1, \xi_2, \ldots$ that take values in X and whose "stochastic behaviour" is specified by the (homogeneous) transition probabilities

$$P(\xi_{j+1} = k | \xi_j = i) = p_{ik} \quad (i, k \in X; j = 1, 2, \ldots)$$

independent of the "past history" $\xi_1, \ldots, \xi_{j-1}$ is called a (homogeneous) *Markov chain.* If the transition probabilities depend on the decision chosen in the respective period, we speak of a *Markov decision process.*

Since all quantities $X, U(i), p_{ik}(u_{i\sigma})$, and $g(i, u_{i\sigma}, k)$ that specify a Markov decision process are independent of period j (we then speak of a *stationary problem*), it is often expedient in practice to discount the associated cost. If r

is the interest rate per period, then $\beta := 1/(1+r)$ is called the *discount factor* ($0 < \beta \leq 1$). Let $v_j^\star(i)$ be the present value of the minimum expected cost of periods $j, j+1, \ldots, n$ if we start in state i at the beginning of period j. Then *Bellman's equation* (2.2) takes the form

$$v_j^\star(i) = \min_{\sigma=1,\ldots,s_i} \{ \bar{g}(i, u_{i\sigma}) + \beta \sum_{k=1}^{m} p_{ik}(u_{i\sigma}) v_{j+1}^\star(k)\} \quad (i \in X,\ 1 \leq j \leq n) \tag{2.3}$$

where $v_{n+1}^\star(i) := 0$. Let $\sigma_j^\star(i)$ be a minimizer in equation (2.3), that is, an optimal decision at state i in period j. The quantities $v_j^\star(i)$ and $\sigma_j^\star(i)$ for $i = 1, \ldots, m$ can be computed by evaluating Bellman's equation (2.3) backwards for $j = n, n-1, \ldots, 1$ beginning with $v_{n+1}^\star(i) := 0$.

Sometimes the planning horizon is long and, perhaps, of somewhat uncertain length. In that case, one often considers an *infinite planning horizon* as a surrogate. To guarantee that the total cost of the infinite-horizon problem is finite we assume that $\beta < 1$.

A first method for solving the infinite-horizon problem consists of using only finitely many periods and enlarging the number of periods successively. Let $v_j^+(i) := v_{n-j+1}^\star(i)$ be the present value of the minimum expected cost of a j-period Markov decision process starting in state i at the beginning of period 1. Then Bellman's equation (2.3) becomes

$$v_j^+(i) = \min_{\sigma=1,\ldots,s_i} \left\{ \bar{g}(i, u_{i\sigma}) + \beta \sum_{k=1}^{m} p_{ik}(u_{i\sigma}) v_{j-1}^+(k)\right\} \quad (i \in X,\ j \geq 1) \tag{2.4}$$

where $v_0^+(i) := 0$. Let $\sigma_j^+(i)$ be a minimizer in equation (2.4). If we evaluate equation (2.4) successively for $j = 1, 2, \ldots$ and compute the quantities $v_j^+(i)$ and $\sigma_j^+(i)$ for $i = 1, \ldots, m$, we speak of the *value-iteration technique*. It can be shown that for $j \to \infty$, the values $v_j^+(i)$ converge to $v^+(i)$, where $v^+(i)$ is the minimum discounted expected cost of the infinite-period Markov decision process that starts in state $i \in X$. Moreover, there is a *stationary optimal feedback policy* $\sigma^+ = (\sigma^+(1), \ldots, \sigma^+(m))$ with the following meaning: If we choose the decision $\sigma^+(i)$ once the system is in state i in any period, then the minimum discounted expected cost of the infinite-horizon problem is $v^+ = (v^+(1), \ldots, v^+(m))$. v^+ and σ^+ satisfy the equation

$$v^+(i) = \bar{g}(i, u_{i\sigma^+(i)}) + \beta \sum_{k=1}^{m} p_{ik}(u_{i\sigma^+(i)}) v^+(k) \quad \text{for } i = 1, \ldots, m. \tag{2.5}$$

In practice, Bellman's equation (2.4) is evaluated for $j = 1, \ldots, n$ where the stopping index n is such that

$$\left| \frac{v_n^+(i) - v_{n-1}^+(i)}{v_n^+(i)} \right| < \varepsilon$$

for all $i = 1, \ldots, m$ and $\varepsilon > 0$ is a prescribed tolerance. Then v_n^+ and σ_n^+ are used as approximations for the optimal quantities v^+ and σ^+, respectively.

A second method for solving the infinite-horizon problem exploits the fact that there exists an optimal policy σ^+ which is stationary and satisfies equation (2.5). This so-called *policy-improvement technique* starts with an initial policy, which is improved successively until an optimal policy is reached.

Each iteration of the policy-improvement technique consists of two steps. The first step, called *policy evaluation,* starts with a policy $\sigma = (\sigma(1), \ldots, \sigma(m))$ and computes the values $v(1), \ldots, v(m)$ by solving the system of linear equations

$$v(i) = \bar{g}(i, u_{i\sigma(i)}) + \beta \sum_{k=1}^{m} p_{ik}(u_{i\sigma(i)}) v(k) \qquad \text{for } i = 1, \ldots, m. \tag{2.6}$$

System (2.6) corresponds to (2.5) and $v(i)$ is the discounted expected cost of the infinite-period Markov decision process if we start in state i and use policy σ. The second step, called *policy improvement,* computes a new policy $\sigma' = (\sigma'(1), \ldots, \sigma'(m))$ such that

$$\bar{g}(i, u_{i\sigma'(i)}) + \beta \sum_{k=1}^{m} p_{ik}(u_{i\sigma'(i)}) v(k) = \min_{\sigma=1,\ldots,s_i} \left\{ \bar{g}(i, u_{i\sigma}) + \beta \sum_{k=1}^{m} p_{ik}(u_{i\sigma}) v(k) \right\}$$

for $i = 1, \ldots, m$. It can be shown that

(i) $v'(i) \leq v(i)$ for $i = 1, \ldots, m$ (v' is the solution of (2.6) with σ' instead of σ)
(ii) If $\sigma' = \sigma$, then σ is an optimal policy.

Case (ii) always occurs after finitely many iterations because the state and action spaces are finite. Hence, the policy-improvement technique is as follows: Pick any initial policy σ, for example, such that

$$\bar{g}(i, u_{i\sigma(i)}) = \min_{\sigma=1,\ldots,s_i} \bar{g}(i, u_{i\sigma}) \qquad \text{for} \quad i = 1, \ldots, m.$$

Perform the steps policy evaluation and policy improvement alternately until two successive policies σ and σ' coincide for the first time. Then σ is an optimal policy and (2.6) provides the minimum discounted expected cost v.

In practice, the policy-improvement technique is mostly preferred to the value-iteration technique. The policy-improvement routine always terminates after a finite number of steps in contrast to the value-iteration procedure. Moreover, policy iteration provides "better" approximate policies than value iteration in most cases if terminated prematurely. For results on the convergence of both methods and modified algorithms that combine features of both policy iteration and value iteration we refer to Heyman and Sobel (1990) [12], Section 8.6.

2.3 Applications

2.3.1 Inventory control

In this section we briefly sketch a stochastic stationary inventory model for a single commodity. For more details we refer to Bertsekas (1987) [2], Sections 1.1, 1.2, and 2.2, and Neumann and Morlock (1993) [17], Section 5.2.5.

We again consider a horizon of n time periods. Let $K \geq 0$ be the setup cost if a positive amount of the good is produced or ordered, let $c > 0$ be the production or purchase cost per unit, and let $h > 0$ be the inventory holding cost per unit and period. Excess demand in each period is backlogged and is filled when additional inventory becomes available. There is a shortage cost of $p > c$ per unit and period. The demands in periods $1, \dots, n$ are supposed to be independent identically distributed nonnegative random variables with probability density function φ (for a discrete good, the integral in the following formula (2.7) has to be replaced by a corresponding sum). We also use a discount factor β per period ($0 < \beta \leq 1$).

The initial stock level x_1 at the beginning of period 1 is a given deterministic quantity. The stock level X_{j+1} at the end of period j is a random variable ($1 \leq j \leq n$). Let $u_j \geq 0$ be the quantity ordered or produced at the beginning of period j without delivery lag, let x_{j+1} be the realized inventory level at the end of period j, and let r_j be the realized demand in period j. Then

$$x_{j+1} = x_j + u_j - r_j \quad (j = 1, \dots, n)$$

(compare (1.6)). Let $L(x_j + u_j)$ be the expected inventory holding plus shortage cost in period j. Then

$$K\delta(u_j) + cu_j + L(x_j + u_j)$$

is the expected total cost in period j where again

$$\delta(u) := \begin{cases} 1 & \text{if } u > 0 \\ 0 & \text{if } u = 0. \end{cases}$$

Let $C_j^\star(x_j)$ be the present value of the minimum expected total cost of periods $j, j+1, \dots, n$ given the stock level x_j at the beginning of period j. Then Bellman's equation reads as follows

$$C_j^\star(x_j) = \min_{u_j \geq 0} \left\{ K\delta(u_j) + cu_j + L(x_j + u_j) + \beta \int_0^\infty C_{j+1}^\star(x_j + u_j - r_j)\varphi(r_j)dr_j \right\} \quad (2.7)$$

where $1 \leq j \leq n$ and $C_{n+1}^\star(x_{n+1}) = 0$. It can be shown that an optimal inventory policy has the following form:

$$\text{At the beginning of period } j = 1, \dots, n \quad \begin{cases} \text{order } S_j - x_j & \text{if } x_j < s_j \\ \text{do not order} & \text{if } x_j \geq s_j. \end{cases}$$

This is a so-called *(s,S) policy.* Even if the exact values of s_j and S_j are unknown, it is important to know that one need only consider policies of the (s, S) type.

A nonstationary inventory model, where all costs and the probability distribution of demand may depend on time period j, is discussed in Denardo (1982) [4], Chapter 7, and Neumann (1977) [16], Section 11.5.

2.3.2 Linear systems and quadratic cost

In practice, the "functional constraint" often has the form of a linear system

$$x_{j+1} = A_j x_j + B_j u_j + w_j \quad (j = 1, \ldots, n)$$

where x_j and u_j are of dimension p and q, respectively, A_j are $p \times p$ matrices and B_j are $p \times q$ matrices, and the "disturbances" w_j are independent random vectors with given probability distributions that do not depend on x_j and u_j and have zero expectation. The state and action spaces are $\mathbb{R}^p$ and $\mathbb{R}^q$, respectively (unconstrained states and controllers). The cost for period j given realized state x_j and chosen action u_j is to be

$$x_j^T Q_j x_j + u_j^T R_j u_j$$

where Q_j are symmetric positive semidefinite $p \times p$ matrices and R_j are symmetric positive definite $q \times q$ matrices. Again, we want to find an optimal feedback controller $(z_1^\star, \ldots, z_n^\star)$ that minimizes the expected total cost over the n periods or stages.

This is a so-called *regulation problem,* where we want to keep the state of the system close to the origin and large deviations from the origin induce a high penalty whereas small deviations induce a relatively small penalty. Such problems are common in automatic control of a motion or a process.

It can be shown (see Bertsekas (1987) [2], Section 2.1) that there is an optimal controller of the form

$$z_j^\star(x_j) = G_j x_j \quad (j = 1, \ldots, n) \tag{2.8}$$

where the "gain matrices" G_j are given by

$$G_j := -(R_j + B_j^T H_{j+1} B_j)^{-1} B_j^T H_{j+1} A_j$$

and the symmetric positive definite matrices H_j can be computed successively as follows:

$$H_{n+1} = O$$
$$H_j = Q_j + A_j^T \left[H_{j+1} - H_{j+1} B_j (R_j + B_j^T H_{j+1} B_j)^{-1} B_j^T H_{j+1}\right] A_j \quad (j = n, n-1, \ldots, 1).$$

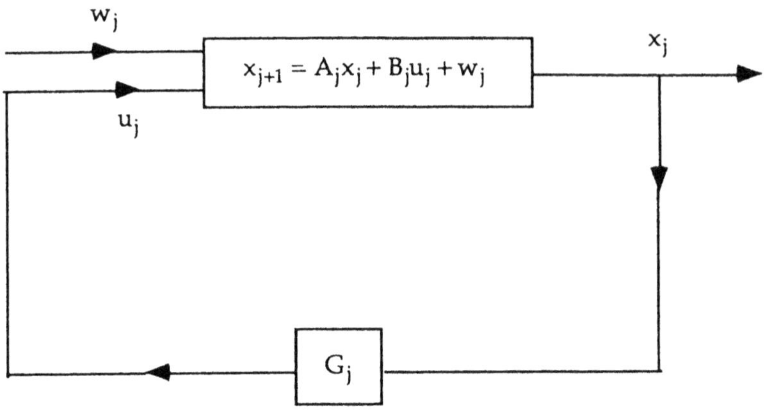

Figure 2.1

The linear controller (2.8) can be computed and implemented in engineering applications with relative ease. The current state x_j is being fed back as input by the feedback gain matrix G_j (see Fig. 2.1). The linearity of the controller is still maintained for problems where the system state x_j is not completely observable (imperfect state information), see Bertsekas (1987) [2], Sections 3.1 and 3.2.

2.3.3 Long-term scheduling of hydro-dominated power systems

Owing to uncertainty of water inflows, demands, and market prices, it is more realistic to regard the long-term scheduling of hydro-thermal power systems as a problem of stochastic dynamic programming. In Gjelsvik et al. (1992) [9] models and solution methods for long-term scheduling problems in hydro-dominated power systems are reviewed, which are frequently used in Norway. For simplicity, we briefly sketch such a model (taken from Gjelsvik et al. (1992) [9]) with a single reservoir and associated power station, where the water inflow is supposed to be stochastic.

Consider a planning horizon of n periods (for example, 3 to 5 years where a period corresponds to a week). Let x_j be the volume of water stored in the reservoir at the beginning of period j, and let u_j be the release from the reservoir, w_j^c be the controlled water inflow to the reservoir and w_j^u $(j = 1, \ldots, n)$ be the uncontrolled water inflow each during period j. The uncontrolled inflow goes directly to the power station and cannot be stored. w_j^c and w_j^u $(j = 1, \ldots, n)$ are assumed to be independent random variables. It is customary to use 30 to 50 years of observed data for the inflow from which a discrete probability distribution for each of the $2n$ inflow variables is extracted, where each variable takes only a few (say, 5 to 10) values.

Assume there is a fixed power demand D_j for period j and a power market where power can be sold or bought. Possible market transactions may refer, for example, to thermal generation, power pool, and rationing. Let K be the number of possible market transactions, and let y_j^k be the amount of electric energy traded in the kth transaction during period j and c^k be the price per unit of electric energy in the kth transaction, where $c^k > 0$ if power is bought and $c^k < 0$ if power is sold ($k = 1, \ldots, K$). Moreover,

$$\alpha^k := \begin{cases} 1 & \text{if the } k\text{-th transaction is buying} \\ -1 & \text{if the } k\text{-th transaction is selling.} \end{cases}$$

The objective function to be minimized is supposed to be

$$E\left[\sum_{j=1}^{n}\sum_{k=1}^{K} c^k y_j^k - g(x_{n+1})\right]$$

where E stands for the expected value and $g(x_{n+1})$ is the estimated value of stored water at the end of the planning horizon. The constraints of the dynamic programming problem are as follows

$$\begin{aligned}
&x_1 = x_I \\
&x_{j+1} = x_j + w_j^c - u_j \\
&0 \le x_j \le \bar{x} \\
&0 \le u_j \le \bar{u}_j && j = 1, \ldots, n \qquad (2.9) \\
&0 \le y_j^k \le \bar{y}_j^k \quad (k = 1, \ldots, K) \\
&\min(u_j + w_j^u, Q_{max}) + \sum_{k=1}^{K} \alpha^k y_j^k = D_j
\end{aligned}$$

In (2.9), x_I is the initial amount of water stored in the reservoir, upper limits are indicated by bars, Q_{max} is the maximum discharge, and, for simplicity, amount of water is measured in electric energy units (that is, the energy conversion factor is assumed to be equal to one). The second equation in (2.9) represents the reservoir balance and the last equation the power balance. In addition to the decision variable u_j (water release) and state variable x_j (amount of water stored), the amounts of electric energy traded, y_j^k ($k = 1, \ldots, K$), represent variables.

To evaluate Bellman's equation for one period, a minimization problem with constraints (2.9) has to be solved. It is expedient to include the reservoir and power balance equations in the function to be minimized by means of Lagrange multipliers. For more details and the more general case where several reservoirs and associated power stations are present we refer to Gjelsvik et al. (1992) [9].

2.3.4 Optimal maintenance and replacement strategies

Determining proper maintenance and replacement strategies is very important in practice. Preventive maintenance can decrease the likelihood of unplanned failures of equipment. This is particularly important in situations in which failures could result in loss of life (say, when planes, automobiles, or life-sustaining medical equipment are involved).

As an example, we consider a production process where some machine deteriorates rapidly in both quality and output under heavy usage. Assume that the machine is inspected periodically (say, at the end of each day) and, after inspection, the machine is in one of the four possible states "as good as new", "operable – minor deterioration", "operable – major deterioration", or "inoperable". Depending on the current state, one of three possible actions "do nothing", "overhaul" (i.e., return to state "operable – minor deterioration"), or "replace" (i.e., return to state "as good as new") can be taken. Each of the two actions "overhaul" and "replace" is supposed to take one day to complete.

The selection of an action induces a transition from the current state to another state with a certain probability. For example, if the machine is inoperable, the action "replace" leads to the state "as good as new" with probability one and to the remaining states with probability zero each. If the machine is as good as new and nothing is done, there is a transition to each of the four possible states with some probability between zero and one, where the four transition probabilities sum up to unity. Moreover, when the system is in some state and an action is taken, an expected cost is incurred, which is the sum of the maintenance cost, the cost of lost profit (if one of the actions "overhaul" or "replace" is selected, there is no production during one day), and possibly the expected cost due to producing defective items (if the machine is not in the state "as good as new" and nothing is done).

The problem of finding an optimal maintenance and replacement policy corresponds to a Markov decision process. Assuming an infinite planning horizon, the problem can be solved by means of the value-iteration or policy-improvement technique as discussed in Section 2.2.

References

1. Baker, K.R., Schrage, L.E.: Finding an Optimal Sequence by Dynamic Programming: An Extension to Precedence-Constrained Tasks; Operations Research 26 (1978) 111-120
2. Bertsekas, D.P.: *Dynamic Programming - Deterministic and Stochastic Models*; Prentice-Hall, Englewood Cliffs, 1987
3. Braun, H.: Unit Commitment and Thermal Optimization – Problem Statement; SVOR/ASRO Tutorial on Optimization in Planning and Operation of Electric Power Systems, Thun 1992, Switzerland
4. Denardo, E.V.: *Dynamic Programming - Models and Applications*; Prentice-Hall, Englewood Cliffs, 1982
5. Efthymoglou, P.G.: Optimal Use and the Value of Water Resources in Electricity Generation; Management Science 33 (1987) 1622-1634
6. Federgruen, A., Tzur, M.: A Simple Forward Algorithm to Solve General Dynamic Lot Size Models with n Periods in $O(n \log n)$ or $O(n)$ Time; Management Science 37 (1991) 909-925
7. Gallo, G., Pallottino, S.: Shortest Path Methods - A Unifying Approach; Math. Programming Study 26 (1986) 38-64
8. Gallo, G., Pallottino, S.: Shortest Path Algorithms; Annals of Operations Research 13 (1988) 3-79
9. Gjelsvik, A., Rotting, T.A., Roynstrand, J.: Long-Term Scheduling of Hydro-Thermal Power Systems; in Broch, E., Lysne, D.K. (eds.): Proceedings of the Second International Conference on Hydro Power, A.A. Balkema, Rotterdam (1992) 539-546
10. Handschin, E., Slomski, H.: Unit Commitment in Thermal Power Systems with Long-Term Energy Constraints; Power Industry Computer Application Conference, Seattle (1989) 211-217
11. Heyman, D.P., Sobel, M.J.: *Stochastic Models in Operations Research*; Vol. II. McGraw-Hill, New York, 1984
12. Heyman, D.P., Sobel, M.J.(eds.): Stochastic Models; Handbooks in Operations Research and Management Science, Vol. 2. North-Holland, Amsterdam, 1990
13. Ikem, F.M., Reisman, A.M.: An Approach to Planning for Physician Requirements in Developing Countries Using Dynamic Programming; Operations Research 38 (1990) 607-618
14. Lawler, E.L.: Efficient Implementation of Dynamic Programming Algorithms for Sequencing Problems; Report BW 106/79, Mathematisch Centrum, Amsterdam, 1979
15. Nemhauser, G.L., Wolsey, L.A.: *Integer and Combinatorial Optimization*; John Wiley & Sons, New York, 1988
16. Neumann, K.: *Operations Research Verfahren*; Band II. Carl Hanser, München, 1977
17. Neumann, K., Morlock, M.: *Operations Research*; Carl Hanser, München, 1993
18. Ozden, M.: A Dynamic Planning Technique for Continuous Activities under Multiple Resource Constraints; Management Science 33 (1987) 1333-1347

19. Sanders, H.-H., Linke, K.: Experiences with Optimization Packages for Unit Commitment; SVOR/ASRO Tutorial on Optimization in Planning and Operation of Electric Power Systems, Thun 1992, Switzerland

20. Sherali, H.D., Hobeika, A.G., Trani, A.A., Kim, B.J.: An Integrated Simulation and Dynamic Programming Approach for Determining Optimal Runway Exit Locations; Management Science 38 (1992) 1049-1062

21. Slomski, H.: Optimale Einsatzplanung thermischer Kraftwerke unter Berücksichtigung langfristiger Energiebedingungen; Ph.D. Thesis, University of Dortmund, 1990

22. Stoecker, A.L., Seidmann, A., Lloyd, G.S.: A Linear Dynamic Programming Approach to Irrigation System Management with Depleting Groundwater; Management Science 31 (1985) 422-434

INTERIOR-POINT METHODOLOGY FOR LINEAR PROGRAMMING: DUALITY, SENSITIVITY ANALYSIS AND COMPUTATIONAL ASPECTS

B. Jansen [1] **C. Roos** [2] **T. Terlaky** [3]

Faculty of Technical Mathematics and Computer Science
Delft University of Technology
P.O. Box 5031, 2600 GA Delft, The Netherlands

J.-Ph. Vial [4]

Department of Management Studies
Department of SES
University of Geneva
102 Bd Carl Vogt, CH-1211 Geneva 4, Switzerland

Abstract. In this paper we use the interior point methodology to cover the main issues in linear programming: duality theory, parametric and sensitivity analysis, and algorithmic and computational aspects. The aim is to provide a global view on the subject matter.

[1] This author completed this work under the support of research grant # 611-304-028 of NWO.

[2] This author completed this work under the support of a research grant of SHELL.

[3] On leave from the Eötvös University, Budapest, and partially supported by OTKA No. 2116.

[4] This author completed this work under the support of research grant # 12-34002.92 of the Fonds National Suisse de la Recherche Scientifique.

1 Introduction

In 1984 Karmarkar [18] made a brilliant contribution to the field of linear programming. He proposed a new polynomial algorithm. His method not only enjoyed a better complexity bound than the earlier method of Khachiyan [17] but it also showed great promises of computational efficiency. Karmarkar's paper originated an intense stream of research. It was soon discovered by Gill et al. [10] that the new method was closely related to the logarithmic barrier method of Fiacco and McCormick [7]. These latter authors used the terminology *interior point method* to describe this approach to optimization, a name that is now used to characterize the entire field.

Up to recently, the research has been mainly devoted to the design of new variants of the initial algorithm that would achieve great theoretical and practical efficiency. Some of the algorithms have been implemented and subjected to numerical testing. It appears now [27] that the state of the art implementations compete favorably with the most advanced implementations of the Simplex algorithm, especially for large scale problems. So interior point methods are not anymore of pure academic interest: they are of the utmost interest for practitioners too.

In contrast with the vast majority of the literature on interior point methods devoted to algorithms, very few papers deal with the theory of linear programming itself. However, some recent contributions [13] show that it is possible to derive the duality theory of linear programming entirely from the interior point methodology. As shown in [15], the interior point methodology also provides a sound basis for parametric analysis. Thus it is possible now to develop a full theory of linear programming, including duality, sensitivity analysis and algorithms, on the basis of the interior point methodology.

The paper has been written for those people who have some background in the field of linear programming but who are not familiar with the recent developments in the field of interior point methods. Our aim is to provide a quick and solid insight in this field, both from the theoretical and the algorithmic point of view. The paper gathers under the same cover various major facets of the theory. We wanted the paper to be self-contained, so that the reader may find the basic results and methods. We present only one specific algorithm with its convergence analysis. We chose this algorithm because its implementation would not be very different from the state of the art implementations of the commercial codes OB1 and OSL.

The paper is organized as follows. In Section 2 we present the duality theory for linear programming based on interior point methodology. This section is a condensed version of [13]. The main ingredient is that the optimal solutions are characterized by a unique partition of the set of indices. This partition is called the optimal partition. The results of Section 2 are used in Section 3 to give a new view at the topics of parametric and sensitivity analysis. The optimal value of the linear program is a piece–wise multilinear function of the objective and right hand side coefficients: each piece is characterized by an optimal partition of the set of indices. In contrast with the Simplex method which focuses on bases, an interior point algorithm detects the optimal partition. This turns out to be an asset for parametric and sensitivity analysis. The presentation in Section 3 is based on the paper [15]. In Section 4 we present a primal-dual interior point algorithm. It is very much in the spirit of the work of [7]. It uses the concept of logarithmic barrier function. The convergence analysis is quite simple: it is based on the observation that each iteration decreases the barrier function by a sizable amount. Our presentation is a simplified version of [16]. In Section 5 we discuss some of the implementation issues that make the theoretical algorithm efficient in solving practical problems. We also present some computational experience. Section 5 relies on the large body of literature in the field, but more specifically on [27, 44, 3]. In Section 6 we draw some conclusions, indicating some of the very promising developments in the area of convex programming.

For more reading the interested readers may be referred to the extensive bibliography of Kranich [22], and also to the forthcoming book of Den Hertog [14] which gives a nice and uniform treatment of many interior point algorithms for linear, convex quadratic and smooth convex programming.

Notations Some notation is of great help in the presentation of interior point methods. It is now standard in the field. Given a vector $x \in \mathbb{R}^n$, we denote X the diagonal matrix whose diagonal elements are the components x_i of the vector x. We also use x^{-1} to designate the vector whose components are the reciprocals of the components of x. The vector e is the vector of all ones of appropriate dimension. So we have $x^{-1} = X^{-1}e$.

2 A new approach to the theory of linear programming

2.1 Introduction

Up to now, the treatment of the topic of linear programming in textbooks has been dominated by the Simplex method. This algorithmic tool not only provides a mean of computing an optimal solution of the problem, but it can also be used to derive duality properties of a more theoretical nature. The notion of *basis* is the key concept in this approach. An optimal basis is characterized by simultaneous feasibility in the primal and in the dual. It is well known that there may exist multiple optimal bases. In fact this situation, known as degeneracy, is the common rule in practical problems. The subsidiary issue is thus to characterize the set of all optimal solutions. In the Simplex approach it amounts to finding all optimal bases, a formidable task indeed.

The answer to the degeneracy issue is given by the Goldman Tucker theorem [11]. One way to paraphrase this result is to state that the set of optimal solutions is characterized by a partition of the set of indices of the nonnegative variables. Unfortunately the original proof of the theorem does not lead to an algorithm. Up to now the result was considered of theoretical interest alone.

The recent theory of interior point methods for linear programming ignores bases. The founding concept there, is the notion of *analytic center* and *central path*. The central path always ends at the center of the optimal face. This center is one of the possible solutions of the Goldman-Tucker theorem. The components of the center define the *optimal partition* as the set of indices of the non zero components. The aim of this section, and of the following one, is to show that the notion of optimal partition can take over the role of the concept of optimal basis, both in the theory and in the more practical issue of parametric analysis. The main advantage of this concept is that it uniquely characterizes the set of optimal solutions of a given problem.

Interior point methods, just like the Simplex, were first proposed as an algorithmic tool for computing an optimal solution. It is quite natural to look for an idealized smooth version of this discrete algorithmic process. In other words, we want to consider a smooth curve that would correspond to the trajectory of an interior point algorithm with infinitely small steps. Such a curve has strong properties that may be extrapolated to its end point, i.e., to an optimal solution. In this way, we may prove important properties of the optimal set. This is the program we want to carry on. There is however an obvious limitation to this approach. Interior point methods deal with interior points. So we must assume that interior feasible solutions exist and thus restrict our conclusions to the class of problems for which this assumption holds. Fortunately enough, we shall show that it is possible to embed any arbitrary linear programming problem into a larger one for which the assumption holds. We shall then derive the consequences for the original problem of the properties at optimality of the larger one.

We shall break our analysis into different steps. First we shall exhibit the idealized trajectory, known as the *central path*. To this end we shall use the logarithmic barrier approach of Fiacco and McCormick [7]. Next we shall extrapolate the properties of the curve to its limit point and obtain the fundamental duality results. Finally we shall discuss the embedding technique.

2.2 Problem definition and assumptions

Before going into the problem, we want to stress a well-known fact. Any time we write down a linear programming problem, we implicitly define a related problem known as the dual. The standard approach is to concentrate on the primal and show that the analysis leads to results pertaining also to the dual. It might be judicious to jointly consider the pair of dual problems from the outset. It is not the common practice, but we shall follow it in the hope of using simpler and more powerful arguments.

As it is usual in the literature on interior point methods, we shall develop the theory of linear programming for the following pair of asymmetrical dual linear programming (LP) problems where

$(P)\quad \min\{c^T x \,:\, Ax = b, x \geq 0\},$

and

$(D)\quad \max\{b^T y \,:\, A^T y + s = c, s \geq 0\},$

where A is an $m \times n$ matrix, c, x, $s \in \mathbb{R}^n$, and b, $y \in \mathbb{R}^m$. The feasible regions of (P) and (D) are denoted $\mathcal{P}$ and $\mathcal{D}$, respectively.

If $x \in \mathcal{P}$ and $x > 0$, we shall say that x is a *positive vector* in $\mathcal{P}$; and if

$(y, s) \in \mathcal{D}$ with $s > 0$, then we shall say that s is a *positive vector* in $\mathcal{D}$.

The two problems (P) and (D) are related on their feasible sets by the obvious relation

$$c^T x - b^T y = x^T (c - A^T y) = x^T s \geq 0. \tag{1}$$

This inequality has the immediate consequence that if $x^T s = 0$, then both the primal and the dual solutions are optimal. Therefore it is legitimate to study the primal-dual problem

$(PD)\quad \min\{x^T s \,:\, Ax = b,\, x \geq 0, A^T y + s = c,\, s \geq 0\}.$

This is just formal writing since it is clear by (1) that the primal-dual nonlinear programming problem is separable into two independent linear programming problems. We shall denote $\mathcal{H} = \mathcal{P} \times \mathcal{D} \subset \mathbb{R}^n \times \mathbb{R}^n$ the set of primal dual feasible pairs

$$\mathcal{H} := \{(x, s) \,:\, Ax = b,\, x \geq 0, A^T y + s = c,\, s \geq 0\}.$$

We shall denote the set of positive feasible pairs by

$$\mathcal{H}^0 := \{(x, s) \,:\, Ax = b,\, x > 0, A^T y + s = c,\, s > 0\}.$$

Let us introduce two assumptions, one technical, one essential.

Assumption 1 *A has full row rank m.*

Assumption 2 *$\mathcal{H}^0$ is nonempty, i.e., both $\mathcal{P}$ and $\mathcal{D}$ contain a positive vector.*

Assumption 1 enforces a one-to-one correspondence between the y and the s in $\mathcal{D}$. With little abuse of notation, we shall refer to any pair $(y, s) \in \mathcal{D}$ by $y \in \mathcal{D}$ or $s \in \mathcal{D}$. This assumption is not restrictive: if it is not fulfilled, we can simply remove constraints from the primal formulation until the constraint matrix has full rank. Assumption 1 greatly simplifies the formulas.

Assumption 2 is nontrivial but it is necessary for the development of the theory. We shall refer to it as the *interior point assumption* or the *basic assumption.*

2.3 The logarithmic barrier approach

2.3.1 The barrier problem

Let us introduce a *logarithmic barrier function* that repels the primal-dual feasible pairs from the boundaries of the positive orthant in $\mathbb{R}^n \times \mathbb{R}^n$. In this way, we follow the approach of Fiacco and McCormick [7], by simply extending their method to the primal-dual framework. So let $\mu > 0$ be the barrier parameter and let $f_\mu : \mathcal{H}^0 \to \mathbb{R}$ be defined by

$$f_\mu(x,s) = \frac{x^T s}{\mu} - \sum_{i=1}^{n} \ln x_i s_i. \tag{2}$$

We shall consider the *barrier problem*

$(PD_\mu) \quad \min\{f_\mu(x,s) \,:\, (x,s) \in \mathcal{H}^0\}$

and study the properties of its solutions.

We can give an alternative and convenient expression of the barrier function. Let $g_\mu \,:\, \mathbb{R}_+ \to \mathbb{R}$ be the function in the variable t defined by

$$g_\mu(t) = \frac{t}{\mu} - \ln t.$$

Letting $t_j = x_j s_j > 0$, we get

$$f_\mu(x,s) = \sum_{i=1}^{n} g_\mu(t_i).$$

The function g_μ has some nice features that will be of great help in proving that f_μ achieves its minimum value on $\mathcal{H}^0$. To refer easily to them, we state them formally as a simple lemma.

Lemma 1 *Let $\mu > 0$. g_μ is strictly convex on its domain of definition $(0,\infty)$; $g_\mu(t) \to \infty$ as $t \to 0$ or $t \to \infty$; and $g_\mu(t) \geq 1 - \ln \mu$, with equality if and only if $t = \mu$.*

To conclude this section, we point out that the barrier function f_μ is strictly convex on its domain of definition. This follows from the facts that the scalar product $x^T s$ reduces to $c^T x - b^T y$ on $\mathcal{H}$ and that the logarithm is a strictly concave function.

2.3.2 Minimizers of the barrier function

By Lemma 1, the barrier function is bounded from below by $n(1 - \ln \mu)$. Hence it has a finite infimum. But Lemma 1 has a further direct, though perhaps surprising, consequence. Namely, if one can find a point $(x, s) \in \mathcal{H}^0$ such that $x_i s_i = \mu$ for all $i = 1, \ldots n$, then each function $g_\mu(t_i)$ in the sum achieves its absolute minimum and the point is a minimizer of f_μ. Let us formalize this property in a system of equations

$$\begin{aligned} Ax &= b \\ A^T y + s &= c \\ Xs &= \mu e. \end{aligned} \tag{3}$$

We shall thereafter designate these equations as the *centering conditions.* If (3) has a solution, this solution is a minimizer of f_μ. Moreover, since f_μ is strictly convex on $\mathcal{H}^0$, the minimizer, and hence the solution of (3) is unique. For reasons that will be clear later we name the solution of (3) a *μ-center.*

We need now to prove the converse statement: that f_μ has a minimizer and that this minimizer solves (3). Then we shall have fully characterized the solution of the barrier problem and we will be able to use the analytic definition of the minimizer in the further developments. We start this analysis with a lemma which states that each component x_i and s_i is simultaneously bounded above and bounded away from zero.

Lemma 2 *Let Assumption 2 hold and let σ be any real number. The level set $\{(x, s) \in \mathcal{H}^0 : f_\mu(x, s) \le \sigma\}$ is bounded. Moreover, it is uniformly bounded away from zero.*

Proof: For short, we shall name $\{(x, s) : f_\mu(x, s) \le \sigma\} \cap \mathcal{H}^0$ a *σ-level set.* This set may be empty. Let (x, s) be in the σ-level set. By Lemma 1 one has for each index j

$$\begin{aligned} g_\mu(x_j s_j) &\le \sigma - \sum_{i \ne j} g_\mu(x_i s_i) \\ &\le \sigma - (n-1)(1 - \ln \mu). \end{aligned}$$

Thus each component $g_\mu(x_j s_j)$ is bounded from above. Lemma 1 implies then that there exists a number $M > 0$ such that

$$\frac{1}{M} \le x_j s_j \le M, \quad j = 1, \ldots n,$$

for all pairs (x, s) in the σ-level set. If we can prove that the components x_j and s_j are bounded above, then the bounds on $t_j = x_j s_j$ will imply that the components are also bounded away from zero.

We first observe that the duality gap is bounded, since

$$x^T s = \sum_{i=1}^{n} x_i s_i \leq nM. \tag{4}$$

By Assumption 2 there is a point $(\bar{x}, \bar{s}) \in \mathcal{H}^0$. Let (x, s) be any point in the σ-level set. By construction $x - \bar{x}$ belongs to the null space of A and $s - \bar{s}$ to the range space of A^T. The two vectors are thus orthogonal and we have

$$(x - \bar{x})^T (s - \bar{s}) = 0.$$

Expanding the product, we get

$$\bar{s}^T x + \bar{x}^T s = x^T s + \bar{x}^T \bar{s} \leq nM + \bar{x}^T \bar{s}.$$

Since $\bar{x} > 0$ and $\bar{s} > 0$, then each component x_i and s_i must be also bounded. Since $t_i = x_i s_i > \frac{1}{M}$, we get the desired result. □

Lemma 2 asserts that each level set of f_μ is included in a compact subset of the domain of definition $\mathcal{H}^0$. Hence f_μ achieves its minimum value on the domain of definition.

What is left to prove is that the minimizer necessarily satisfies the centering conditions (3). This is done by elementary calculus and linear algebra. Since the minimum of f_μ is achieved in the relatively open set $\mathcal{H}^0$, we can disregard the nonnegativity constraint and retain the equality constraints only. To express the stationarity property of the barrier function over the equality constraints, one simply states that the gradient of f_μ is orthogonal to the constraint set. Taking the derivatives with respect to x and s respectively, we get

$$\frac{s}{\mu} - x^{-1} \in \mathcal{R}(A^T) \tag{5}$$

$$\frac{x}{\mu} - s^{-1} \in \mathcal{N}(A) \tag{6}$$

where $\mathcal{N}(A)$ is the null space of A (i.e., $\{x : Ax = 0\}$) and $\mathcal{R}(A^T)$ is the range space of A^T (i.e., $\{s : s = A^T y\}$). Let us define the vector $u = (\frac{Xs}{\mu})^{\frac{1}{2}}$. Note that

$$(\mu X S^{-1})^{\frac{1}{2}} (\frac{s}{\mu} - x^{-1}) = u - u^{-1} = (\mu X^{-1} S)^{\frac{1}{2}} (\frac{x}{\mu} - s^{-1}).$$

We thus have

$$u - u^{-1} \in \mathcal{R}(X^{\frac{1}{2}}S^{-\frac{1}{2}}A^T).$$

and

$$u - u^{-1} \in \mathcal{N}(AX^{\frac{1}{2}}S^{-\frac{1}{2}})$$

In conclusion $u - u^{-1}$ belongs to two orthogonal subspaces. Hence we have $u - u^{-1} = 0$; so $x_i s_i = \mu$ holds for all $i = 1, \ldots n$. We just proved that the minimizer satisfies the centering conditions. Hence we may state the result

Theorem 1 *Let $\mu > 0$. Then the following statements are equivalent:*

(i) both $\mathcal{P}$ and $\mathcal{D}$ contain a positive vector;

(ii) there exists a unique minimizer of f_μ on $\mathcal{H}^0$;

(iii) the centering conditions have a solution.

Proof: We pointed out that Lemma 2 implies that the barrier function achieves its minimum value. So, we established that (i) implies (ii). We also checked the equivalence between (ii) and (iii). To conclude the proof, it suffices to note that, since $\mu > 0$, the last equation in the centering conditions implies (i). Hence (iii) implies (i). □

2.3.3 The central path

From now on it will be assumed throughout that the pair of dual problems (P) and (D) satisfies Assumption 2. So, as a consequence of Theorem 1, the μ–centers exist and are unique. We shall stress hereafter their dependence on μ by denoting the μ–centers as $x(\mu)$, $s(\mu)$ and $y(\mu)$ respectively. The set $\{x(\mu) \;:\; \mu > 0\}$ will be called the *central path* of (P), and likewise $\{(y(\mu), s(\mu)) \;:\; \mu > 0\}$ the *central path* of (D).

Now that the existence of the central paths of (P) and (D) has been established it may be useful to provide some illustration of this important concept.

Example 1 By way of example we calculate the central paths for the following pair of dual linear programming problems:

$$(P) \;\; \min\{x_1 + x_2 + x_3 \;:\; -x_1 + x_2 = 0,\, x_3 = 1,\, x_1,\, x_2,\, x_3 \geq 0\},$$

$$(D) \;\; \max\{y_2 \;:\; -1 \leq y_1 \leq 1,\, y_2 \leq 1\}.$$

Note that (P) has a unique optimal solution: $x = (0, 0, 1)$, whereas (D) has multiple optimal solutions: $(y_1, 1)$, $-1 \leq y_1 \leq 1$. Given $\mu > 0$, the centering conditions are given by

$$\begin{aligned}
-x_1 + x_2 &= 0, \quad x_1, x_2 \geq 0, \\
x_3 &= 1, \\
-y_1 + s_1 &= 1, \quad s_1 \geq 0, \\
y_1 + s_2 &= 1, \quad s_2 \geq 0, \\
y_2 + s_3 &= 1, \quad s_3 \geq 0, \\
x_1 s_1 &= \mu, \\
x_2 s_2 &= \mu, \\
x_3 s_3 &= \mu.
\end{aligned}$$

Some straightforward calculations yield the following:

$$\begin{aligned}
x(\mu) &= (\mu, \mu, 1), \\
s(\mu) &= (1, 1, \mu), \\
y(\mu) &= (0, 1 - \mu).
\end{aligned}$$

So, in this example the central paths are straight half lines.

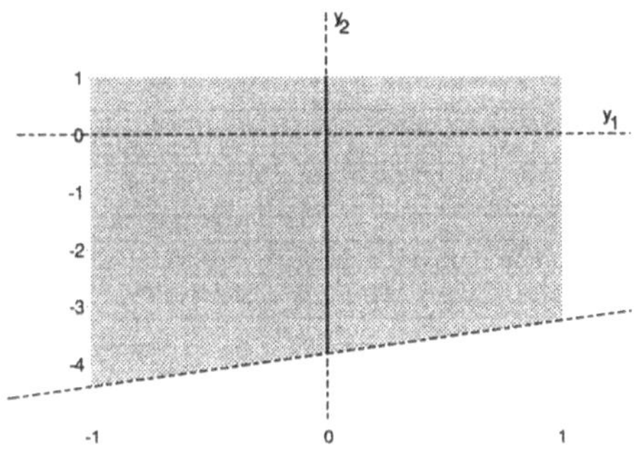

Figure 1: The dual central path in Example 1.

The central path of (D) is shown in Figure 1. Note that the limit point of the central path of (D) is the center of the optimal set of (D). □

We proceed by changing the right hand side vector in the Example 1. This makes the calculation of the central path a little more complicated.

Example 2 Consider the following pair of dual linear programming problems:

$$(P)\quad \min\{x_1 + x_2 + x_3 \;:\; -x_1 + x_2 = 1,\, x_3 = 1,\, x_1,\, x_2,\, x_3 \geq 0\},$$
$$(D)\quad \max\{y_1 + y_2 \;:\; -1 \leq y_1 \leq 1,\, y_2 \leq 1\}.$$

The centering conditions are

$$\begin{aligned}
-x_1 + x_2 &= 1, \quad x_1,\, x_2 \geq 0,\\
x_3 &= 1,\\
-y_1 + s_1 &= 1, \quad s_1 \geq 0,\\
y_1 + s_2 &= 1, \quad s_2 \geq 0,\\
y_2 + s_3 &= 1, \quad s_3 \geq 0,\\
x_1 s_1 &= \mu,\\
x_2 s_2 &= \mu,\\
x_3 s_3 &= \mu.
\end{aligned}$$

This system can be solved to yield the coordinates of the primal and dual central paths.

$$\begin{aligned}
x(\mu) &= \left(\tfrac{1}{2}(-1 + \mu + \sqrt{1+\mu^2}),\, \tfrac{1}{2}(1 + \mu + \sqrt{1+\mu^2}),\, 1\right),\\
s(\mu) &= \left(1 - \mu + \sqrt{1+\mu^2},\, 1 + \mu - \sqrt{1+\mu^2},\, \mu\right),\\
y(\mu) &= \left(-\mu + \sqrt{1+\mu^2},\, 1 - \mu\right).
\end{aligned}$$

Taking the limit for $\mu \downarrow 0$ we obtain the optimal solutions:

$$\begin{aligned}
x(0) &= (0,\, 1,\, 1),\\
s(0) &= (2,\, 0,\, 0),\\
y(0) &= (1,\, 1).
\end{aligned}$$

Note that the primal feasible region $\mathcal{P}$ is a straight half line in $\mathbb{R}^3$ in this example, and that the primal central path coincides with this line. The central path of (D) is more interesting. It is shown in Figure 2. □

We finally consider an instance for which the dual feasible region is bounded.

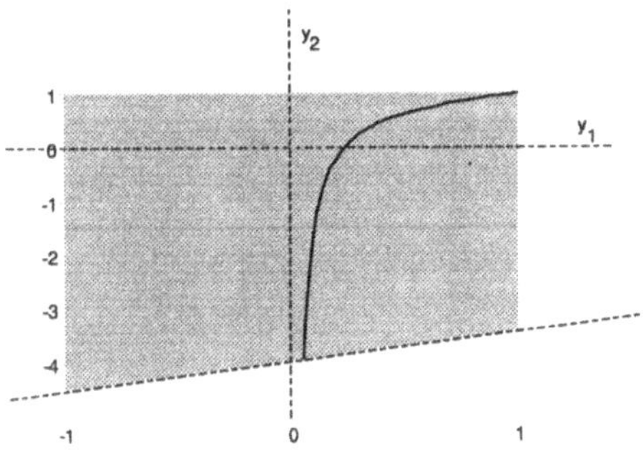

Figure 2: The dual central path in Example 2.

Example 3 Consider the following pair of dual linear programming problems:

$$
\begin{aligned}
(P)\quad & \min\{x_1+x_2+x_3+x_4 \;:\quad -x_1+x_2=1,\; -x_3+x_4=2 \\
& \qquad\qquad\qquad\qquad\qquad\quad x_1,\, x_2,\, x_3,\, x_4 \geq 0\}, \\
(D)\quad & \max\{y_1+2y_2 \;:\qquad\qquad -1\leq y_1\leq 1,\; -1\leq y_2\leq 1\}.
\end{aligned}
$$

The centering conditions are now

$$
\begin{aligned}
-x_1+x_2 &= 1, & x_1,\, x_2 &\geq 0,\\
-x_3+x_4 &= 2, & x_3,\, x_4 &\geq 0,\\
-y_1+s_1 &= 1, & s_1 &\geq 0,\\
y_1+s_2 &= 1, & s_2 &\geq 0,\\
-y_2+s_3 &= 1, & s_3 &\geq 0,\\
y_2+s_4 &= 1, & s_4 &\geq 0,\\
x_1s_1 &= \mu, &&\\
x_2s_2 &= \mu, &&\\
x_3s_3 &= \mu, &&\\
x_4s_4 &= \mu. &&
\end{aligned}
$$

This time we only give the expressions for $y(\mu)$:

$$
y(\mu) \;=\; \left(-\mu+\sqrt{1+\mu^2},\; \frac{1}{2}(-\mu+\sqrt{4+\mu^2})\right).
$$

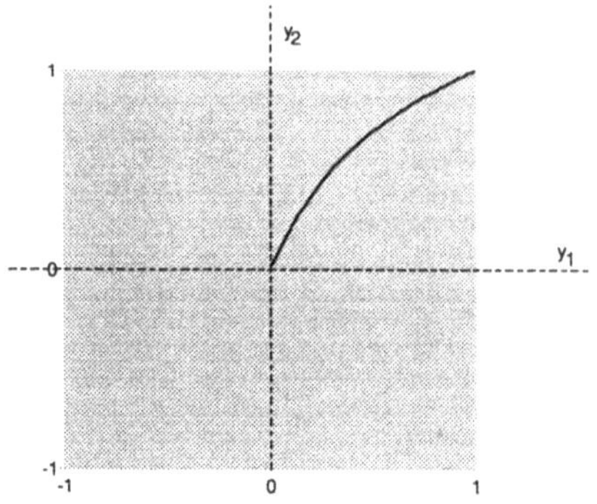

Figure 3: The dual central path in Example 3.

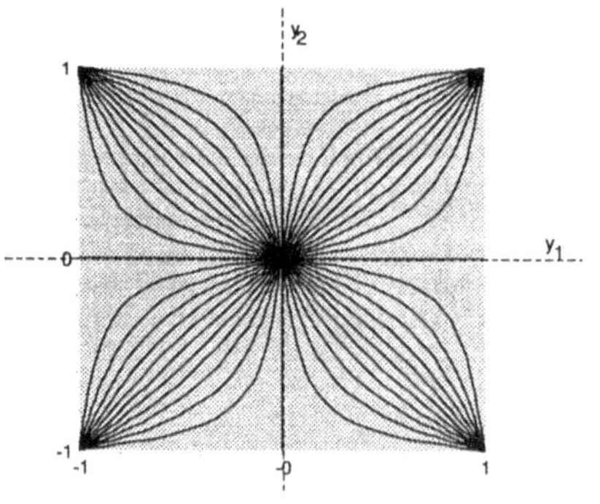

Figure 4: Dual central paths for various b in Example 3.

The central path of (D) is shown in Figure 3.

In Figure 4 we have drawn the central paths of (D) for various values of the vector b. We took $b = (\cos\phi, \sin\phi)$, $\phi = \frac{k\pi}{24}$, $k = 1, 2, \ldots, 48$. □

It must have been observed that in the examples just given the central path always converges to an optimal solution. This is due to the relation $x(\mu)^T s(\mu) = n\mu$, which implies that the duality gap goes to zero when μ approaches zero. This property, and other limiting properties will be exploited in the next section as well as in the algorithmic part of the paper. The central path has many more interesting properties which go beyond the scope of this paper. Many of these properties are related to the fact that the path is differentiable. The interested reader may be referred to the existing literature [32, 2]. At this place we only mention a property which will be needed in the next section, namely that along the central path the objective value is monotone.

2.4 Duality results

Without explicitly mentioning it, we already obtained a first duality result. We used (1) to state that if the gap $x^T s$ is zero, then the pair is optimal. This almost trivial result is known as the *weak duality theorem* of linear programming. This theorem may be strengthened to the statement that all optimal pairs have a zero duality gap. Indeed, by Theorem 1 there exists a solution to the centering conditions for any positive μ. Since

$$x(\mu)^T s(\mu) = n\mu,$$

we can obtain primal-dual feasible pairs with arbitrarily small duality gaps. It is tempting to extrapolate the result to $\mu = 0$. But we haven't proved that the central path converges as $\mu \to 0$. It is obviously true, but tedious to argue formally. However Lemma 2 gives us an answer. Let $\bar{\mu} > 0$ be some value. For all $0 < \mu \leq \bar{\mu}$ the μ-centers belong to the σ-level set, with $\sigma = n\bar{\mu}$. Since the level set is contained in a compact set there is at least one accumulation point as $\mu \to 0$. This accumulation point is a feasible pair with a zero duality gap. This result is known as the *strong duality theorem* for linear programming.

We can still extract more information from the central path. In this way we shall obtain the strongest duality theorem for linear programming, which states that among all optimal solutions (x, s) there is at least one such that for each index i exactly one of the two components of the pair (x_i, s_i) is zero and the other is positive. This property is known as *strict complementarity*.

Theorem 2 (Goldman-Tucker) *Let Assumption 2 hold. There exists at least one optimal pair (x^*, s^*) such that*

$$\begin{aligned} (x^*)^T s^* &= 0 \\ x^* + s^* &> 0. \end{aligned}$$

Proof: Let (x^*, s^*) be an accumulation point of the sequence $(x(\mu), s(\mu))$ as $\mu \to 0$. Without loss of generality we may assume that the whole sequence converges. By an obvious continuity argument, we get that the pair

(x^*, s^*) satisfies $(x^*)^T s^* = 0$ and is thus optimal. Let $(x(\mu), s(\mu))$ be the μ-center for some $\mu > 0$. We have the relation

$$0 = (x^* - x(\mu))^T(s^* - s(\mu)) = n\mu - ((x^*)^T s(\mu) + (s^*)^T x(\mu)).$$

Dividing throughout by $\mu = x_i(\mu)s_i(\mu) > 0$ we get

$$\sum_{i=1}^{n}\left(\frac{x_i^*}{x_i(\mu)} + \frac{s_i^*}{s_i(\mu)}\right) = n.$$

Since

$$\lim_{\mu\to 0}\frac{x_i^*}{x_i(\mu)} = \begin{cases} 1 & \text{if } x_i^* > 0 \\ 0 & \text{otherwise,} \end{cases}$$

and similarly with the s component, we get that exactly one of the two components of the pair (x_i^*, s_i^*) is zero and the other is positive. □

2.4.1 Optimal partition

A crucial notion associated to Theorem 2 is the *optimal partition* of (P) and (D). Given a strictly complementary pair (x, s) of optimal solutions for (P) and (D) we partition the index set $\{1, 2, \ldots, n\}$ according to

$$\{1, 2, \ldots, n\} = B \cup N, \tag{7}$$

with

$$B := \{i \ : \ x_i > 0\},$$
$$N := \{i \ : \ s_i > 0\}.$$

So B is the support of x and N is the support of s. The partition obtained in this way is unique, and does not depend on the given strictly complementary pair. To see this, let $(\bar{x}, \bar{s})$ be any other optimal pair of solutions. Then $\bar{x}^T s = 0$ implies that the support of $\bar{x}$ is contained in B, and $x^T\bar{s} = 0$ implies that the support of $\bar{s}$ is contained in N. In other words, every strictly complementary optimal pair (x, s) determines the same partition. This partition will be called the *optimal partition* for the problem (P) and for the problem (D).

Example 4 Figure 5 shows a network with given arc lengths, and we ask for a shortest path from node s to node t. When solving this problem with an interior point method one obtains the optimal partition (B, N). In Figure 5 the solid lines represent the arcs in B and the dashed lines the arcs in N. Any path from s to t using the arcs in B is a shortest path and all shortest paths use exclusively the arcs in B.

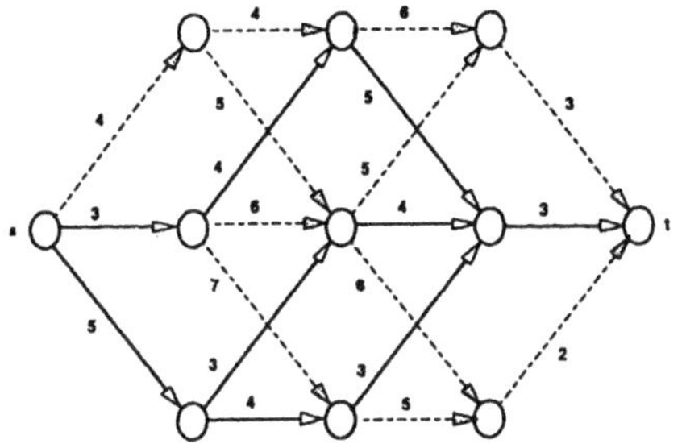

Figure 5: The optimal partition for a shortest path problem.

2.5 Analytic centers

Our statement of the Goldman-Tucker theorem asserts the existence of a strictly complementary pair. In the case of degeneracy, there are usually an infinity of them. But the pair that is exhibited in the theorem is a very special one. It is the end point of the central path and as such it is endowed with most properties of the μ-centers. Here we want to give a further characterization of the μ-centers and apply it to the limit point.

Since $(x(\mu))^T s(\mu) = n\mu$, the constraint $x^T s = n\mu$ is clearly redundant to (PD_μ). So we can reformulate the barrier problem as

$$\min\{-\sum_{i=1}^{n} \ln x_i s_i \,:\, (x, s) \in \mathcal{H}^0,\, x^T s = n\mu\}.$$

The feasible set of this problem is a polytope described as the intersection of a linear manifold with the positive orthant. The solution of (PD_μ) is the point in $\mathcal{H}$ that maximizes the product (and hence the sum of the logarithms) of the 'distances' to the boundary of the positive orthant. Such a point is known as the *analytic center* of the polytope (see Sonnevend [43]).

Our presentation is unconventional: it is more usual to introduce the notion of analytic center either for the primal or for the dual, but not jointly for both. As we have noted several times before, we can use the separability of $x^T s$ on $\mathcal{H}$ to break down the primal-dual barrier problem into two problems

$$\min\{-\sum_{i=1}^{n} \ln x_i \,:\, Ax = b,\, c^T x = n\mu + b^T y(\mu), x > 0\},$$

and

$$\min\{-\sum_{i=1}^{n} \ln s_i \ : \ A^T y + s = c, \ b^T y = -n\mu + c^T x(\mu), s > 0\}.$$

The trouble now is that the knowledge of the solution of one problem is required to solve the other one. There is thus a clear advantage to the primal dual approach. Nevertheless we can draw the useful conclusion that, once the pair of analytic centers is computed it can be interpreted as the cross product of analytic centers in the primal space and in the dual space.

Now let us extrapolate the results to the limit point as μ tends to zero. The limiting polytopes are

$(\mathcal{P}^*) := \{x \ : \ Ax = b, \ c^T x = z^*, x \geq 0\},$

and

$(\mathcal{D}^*) := \{s \ : \ A^T y + s = c, \ b^T y = z^*, s \geq 0\},$

where z^* is the optimal value of the problem. The difficulty is that the limiting polytopes have an empty interior. Indeed, those polytopes define the primal and dual optimal faces and we know that for each of them $x_N = 0$ and $s_B = 0$, where (B, N) is the optimal partition. So the logarithms are not defined for those components. We shall use an alternative definition of $(\mathcal{P}^*)$ and $(\mathcal{D}^*)$. We simply replace the constraint $c^T x = z^*$ by $x_B = 0$, and similarly with the dual we put $s_N = 0$ instead of $b^T y = z^*$. We leave to the reader to check the equivalence.

We consider now the analytic centers in reduced spaces, that is, we want to solve the problems

$$\max\{\sum_{i=1}^{n} \ln x_i \ : \ x \in \mathcal{P}^*\}$$

and

$$\max\{\sum_{i=1}^{n} \ln s_i \ : \ s \in \mathcal{D}^*\}$$

We shall now prove that the limit point of the central path is the pair of analytic centers of the limiting polytopes. The idea of this proof is due to McLinden [29, 30].

Theorem 3 *The central path converges to a unique primal-dual pair whose components are the analytic centers of* $\mathcal{P}^*$ *and* $\mathcal{D}^*$.

Proof: Let (x^*, s^*) be a limit point of the central path. (We know that there is at least one.) Let (x^0, s^0) be any pair in $\mathcal{P}^* \times \mathcal{D}^*$. Using the orthogonality relation

$$(x^0 - x(\mu))^T(s^0 - s(\mu)) = 0$$

and the optimality of the pair (x^0, s^0) we get

$$x(\mu)^T s^0 + s(\mu)^T x^0 = n\mu.$$

Dividing throughout by $\mu = x_i(\mu)s_i(\mu)$, $i = 1 \ldots n$, we obtain

$$\sum_{i=1}^{n}\left(\frac{x_i^0}{x_i(\mu)} + \frac{s_i^0}{s_i(\mu)}\right) = n.$$

Since $x_i^0 = 0$, for $i \in N$, and $s_i^0 = 0$, for $i \in B$, we also have

$$\sum_{i\in B} \frac{x_i^0}{x_i(\mu)} + \sum_{i \in N} \frac{s_i^0}{s_i(\mu)} = n.$$

Letting μ tend to zero, along an appropriate subsequence if necessary so that $(x(\mu), (x(\mu)) \to (x^*, s^*)$, we get

$$\sum_{i\in B} \frac{x_i^0}{x_i^*} + \sum_{i \in N} \frac{s_i^0}{s_i^*} = n.$$

Using the arithmetic–geometric mean inequality, we get

$$\left(\prod_{i\in B} \frac{x_i^0}{x_i^*} \prod_{i\in N} \frac{s_i^0}{s_i^*}\right)^{\frac{1}{n}} \leq \frac{1}{n}\left(\sum_{i\in B} \frac{x_i^0}{x_i^*} + \sum_{i \in N} \frac{s_i^0}{s_i^*}\right) = 1.$$

This inequality holds for arbitrary $x^0 \in \mathcal{P}^*$ and $s^0 \in \mathcal{D}^*$. Let us first take $s^0 = s^*$. We then get for all $x^0 \in \mathcal{P}^*$,

$$\prod_{i\in B} x_i^0 \leq \prod_{i \in B} x_i^*.$$

Hence x^* maximizes the product $\prod_{i\in B} x_i$ over $\mathcal{P}^*$. So x^* must be the analytic center of $\mathcal{P}^*$. Hence the central path of (P) has a unique limit point x^* and this limit is the analytic center of $\mathcal{P}^*$. In the same way, we conclude that

$$\prod_{i\in N} s_i^0 \leq \prod_{i \in N} s_i^*,$$

for all s^0 in $\mathcal{D}^*$. Thus s^* is the analytic center of $\mathcal{D}^*$ and the theorem follows. □

We mention here that the analytic centers of the optimal face will be called the *central solutions* of the primal and dual problems respectively.

2.6 Relaxation of the basic assumption

So far we obtained results for the special case when the basic, or interior point, assumption holds. To obtain similar results for arbitrary pairs of dual linear programming problems, with possibly empty interior feasible sets, we shall embed the original problem in a larger one with the required interior property. We shall apply the duality results to the larger problem and draw the implications for the original problem.

2.6.1 Transformation to a larger problem pair

Choose arbitrary positive vectors x^0, $s^0 \in \mathbb{R}^n$ and an arbitrary vector $y^0 \in \mathbb{R}^m$. Also choose "large" numbers M_p and M_d such that

$$M_p = (b - Ax^0)^T y^0 + s_{n+1}^0,$$

$$M_d = (A^T y^0 + s^0 - c)^T x^0 + x_{n+2}^0,$$

with $x_{n+2}^0 > 0$ and $s_{n+1}^0 > 0$. We shall call any such parameter pair $(M_p,\, M_d)$ *feasible* with respect to the problem pair (P) and (D). Note that in the special case that x^0 is feasible for (P) and the pair $(y^0,\, s^0)$ for (D), the above relations reduce to $M_p = s_{n+1}^0$ and $M_d = x_{n+2}^0$.

We proceed by embedding the original problem pair $(P) - (D)$ in a pair $(\tilde{P}) - (\tilde{D})$ of slightly larger linear programming problems as follows.

$$(\tilde{P}) \qquad \begin{aligned} \min\ & c^T x + M_p x_{n+1} \\ \text{s.t.}\ & Ax + (b - Ax^0)x_{n+1} = b \\ & (c - s^0 - A^T y^0)^T x - x_{n+2} = -M_d \\ & x \geq 0,\ x_{n+1} \geq 0,\ x_{n+2} \geq 0, \end{aligned}$$

$$(\tilde{D}) \qquad \begin{aligned} \max\ & b^T y - M_d y_{m+1} \\ \text{s.t.}\ & A^T y + (c - s^0 - A^T y^0) y_{m+1} + s = c \\ & (b - Ax^0)^T y + s_{n+1} = M_p \\ & -y_{m+1} + s_{n+2} = 0 \\ & s \geq 0,\ s_{n+1} \geq 0,\ s_{n+2} \geq 0. \end{aligned}$$

Let us write $\tilde{x} = (x,\, x_{n+1},\, x_{n+2})$, $\tilde{s} = (s,\, s_{n+1},\, s_{n+2})$, with $x,\, s \in \mathbb{R}^n$ and $\tilde{y} = (y,\, y_{m+1})$, $y \in \mathbb{R}^m$. Also define the $(m+1) \times (n+2)$ matrix $\tilde{A}$ and the vectors $\tilde{b} \in \mathbb{R}^{m+1}$, $\tilde{c} \in \mathbb{R}^{n+2}$, according to

$$\tilde{b} = \begin{pmatrix} b \\ -M_d \end{pmatrix}, \quad \tilde{c} = \begin{pmatrix} c \\ M_p \\ 0 \end{pmatrix}, \quad \tilde{A} = \begin{pmatrix} A & b - Ax^0 & 0 \\ (c - s^0 - A^T y^0)^T & 0 & -1 \end{pmatrix}.$$

We can then write $(\tilde{P})$ and $(\tilde{D})$ in the more compact form

$$(\tilde{P}) \quad \min\left\{\tilde{c}^T\tilde{x} : \tilde{A}\tilde{x} = \tilde{b},\, \tilde{x} \geq 0\right\},$$
$$(\tilde{D}) \quad \max\left\{\tilde{b}^T\tilde{y} : \tilde{A}^T\tilde{y} + \tilde{s} = \tilde{c},\, \tilde{s} \geq 0\right\}.$$

This augmented programming pair $(\tilde{P}) - (\tilde{D})$ has been used before in an algorithmic context [19, 21, 37, 31]. We use it here for a different, theoretical purpose.

Now consider the point $(\tilde{x}^0,\, \tilde{y}^0,\, \tilde{s}^0)$ given by

$$\begin{aligned} \tilde{x}^0 &= (x^0,\, 1,\, x^0_{n+2}), \\ \tilde{s}^0 &= (s^0,\, s^0_{n+1},\, 1), \\ \tilde{y}^0 &= (y^0,\, 1), \end{aligned}$$

One has $\tilde{A}\tilde{x}^0 = \tilde{b}$ and $\tilde{A}^T\tilde{y}^0 + \tilde{s}^0 = \tilde{c}$. Since $\tilde{x}^0$ and $\tilde{s}^0$ are positive we conclude that the problem pair $(\tilde{P}) - (\tilde{D})$ satisfies the basic assumption (Assumption 2). So all the results derived in the previous sections apply to the problem pair $(\tilde{P}) - (\tilde{D})$. What we still have to do is to establish the results for (P) and (D).

2.6.2 Constant optimal partition for large feasible pairs

We treat the pair $(\tilde{P}) - (\tilde{D})$ as being parametrized by $(M_p,\, M_d)$. Using the fundamental results for pairs of problems satisfying Assumption 2, we denote by $\pi(M_p,\, M_d) = (\tilde{B},\, \tilde{N})$ the optimal partition of the index set $\{1,\, \ldots,\, n,\, n+1,\, n+2\}$. The optimal faces of $(\tilde{P})$ and $(\tilde{D})$ are then given by

$$\begin{aligned} \tilde{\mathcal{P}}^* &= \{\tilde{x} \,:\, \tilde{A}\tilde{x} = \tilde{b},\, x_{\tilde{B}} \geq 0,\, x_{\tilde{N}} = 0\}, \\ \tilde{\mathcal{D}}^* &= \{(\tilde{y},\, \tilde{s}) \,:\, \tilde{A}^T\tilde{y} + \tilde{s} = \tilde{c},\, s_{\tilde{B}} = 0,\, s_{\tilde{N}} \geq 0\}, \end{aligned}$$

respectively. The first step in proving the equivalence between (P, D) and $(\tilde{P}, \tilde{D})$ consists in showing that for some large enough M_p and M_d the optimal partition does not change. To this end we associate to a given partition the set of parameters values M_p and M_d for which this partition is optimal. We shall prove that this set is convex: it is a relatively open rectangle, possibly degenerated or empty. This is implied by the following result.

Lemma 3 *Assume* $(M_p^i,\, M_d^i)$, $i = 1, 2$, *are two feasible pairs such that* $\pi := (\tilde{B},\, \tilde{N}) = \pi(M_p^i,\, M_d^i)$. *Let the pair* $(M_p,\, M_d)$ *be given by*

$$\begin{aligned} M_p &= (1-\lambda_p)M_p^1 + \lambda_p M_p^2, \\ M_d &= (1-\lambda_d)M_d^1 + \lambda_d M_d^2, \end{aligned}$$

where $0 \le \lambda_p,\, \lambda_d \le 1$. *Then,* $\pi(M_p,\, M_d) = \pi$.

Proof: Let $(\tilde{x}^i,\, \tilde{y}^i,\, \tilde{s}^i)$ be strictly complementary solutions for the pair $(\tilde{P}) - (\tilde{D})$ with parameter pair $(M_p^i,\, M_d^i)$, $i = 1, 2$. Then $\tilde{B}$ is the support of $\tilde{x}^i$ and $\tilde{N}$ the support of $\tilde{s}^i$. Now consider

$$\begin{aligned} \tilde{x} &:= (1-\lambda_d)\tilde{x}^1 + \lambda_d \tilde{x}^2, \\ \tilde{y} &:= (1-\lambda_p)\tilde{y}^1 + \lambda_p \tilde{y}^2, \\ \tilde{s} &:= (1-\lambda_p)\tilde{s}^1 + \lambda_p \tilde{s}^2. \end{aligned}$$

One easily verifies that $(\tilde{x},\, \tilde{y},\, \tilde{s})$ is feasible for $(\tilde{P}) - (\tilde{D})$ with parameter pair $(M_p,\, M_d)$ and, moreover, that $\tilde{B}$ is the support of $\tilde{x}$ and $\tilde{N}$ the support of $\tilde{s}$. So $(\tilde{x},\, \tilde{y},\, \tilde{s})$ is a strictly complementary optimal solution for $(\tilde{P}) - (\tilde{D})$ with parameter pair $(M_p,\, M_d)$. This implies the lemma. □

Let us consider any fixed partition $\pi = (\tilde{B},\, \tilde{N})$ of the index set $\{1,\, 2 \ldots n,\, n+1,\, n+2\}$. We define the *optimality set* of π as the set of all feasible parameter pairs $(M_p,\, M_d)$ such that $\pi(M_p,\, M_d) = \pi$. A direct consequence of Lemma 3 is that a nonempty optimality set is "rectangular" in the sense that its closure is a closed rectangle (which may be unbounded).

We draw an important conclusion from this. Since the number of partitions of the index set $\{1,\, 2,\, \ldots,\, n,\, n+1,\, n+2\}$ is finite, the number of optimality sets is finite. These sets cover all feasible pairs $(M_p,\, M_d)$. Now observe that if $(M_p,\, M_d)$ is any feasible pair then also every pair $(M_p',\, M_d')$ with $M_p' \ge M_p$ and $M_d' \ge M_d$ is feasible. Therefore, one of the optimality sets must contain all "large" feasible pairs (M_p, M_d). Now letting (M_p^*, M_d^*) be any pair in this optimality set we obtain the following result.

Theorem 4 *There exists a feasible pair* (M_p^*, M_d^*) *such that for all feasible pairs* (M_p, M_d) *with* $M_p \geq M_p^*$ *and* $M_d \geq M_d^*$ *one has* $\pi(M_p, M_d) = \pi(M_p^*, M_d^*)$. □

2.6.3 The duality theorem for linear programming

In the remainder of this section the pair (M_p^*, M_d^*) will be as described in Theorem 4. Also, $\tilde{\pi}^* := (\tilde{B}^*, \tilde{N}^*)$ will denote the optimal partition for this pair and $(\tilde{x}^*, \tilde{y}^*, \tilde{s}^*)$ a strictly complementary solution. So for all pairs (M_p, M_d) with $M_p \geq M_p^*$ and $M_d \geq M_d^*$ the optimal partition is given by $\tilde{\pi}^*$. Let $z^*(M_p, M_d)$ denote the common optimal value for the augmented problems. The next result shows that $z^*(M_p, M_d)$ is a bilinear function within the optimality set of $\tilde{\pi}^*$.

Lemma 4 *Let* $M_p \geq M_p^*$ *and* $M_d \geq M_d^*$. *Then*

$$
\begin{aligned}
z^*(M_p, M_d^*) &= z^*(M_p^*, M_d^*) + x_{n+1}^*(M_p - M_p^*), \\
z^*(M_p^*, M_d) &= z^*(M_p^*, M_d^*) - s_{n+2}^*(M_d - M_d^*).
\end{aligned}
$$

Proof: Note that the feasible regions of $(\tilde{P})$ are the same for the pair (M_p, M_d^*) and the pair (M_p^*, M_d^*). Since the optimal partition is also equal for both pairs, both pairs will yield the same strictly complementary solutions of $(\tilde{P})$. Let $\tilde{x}^*$ be such a solution. From the definition of $(\tilde{P})$ it now follows that $z^*(M_p, M_d^*)$ will depend linearly on M_p. This implies the first equality in the lemma. The proof of the second equality is similar. □

The following result implies the two classical duality results for an arbitrary pair of dual linear programming problems: the strong duality theorem and the Goldman–Tucker theorem (Theorem 2). It is the main result of this section. We use the following notation:

$$
\begin{aligned}
\tilde{x}^* &= (x^*, x_{n+1}^*, x_{n+2}^*), \quad x^* \in \mathbb{R}^n, \\
\tilde{s}^* &= (s^*, s_{n+1}^*, s_{n+2}^*), \quad s^* \in \mathbb{R}^n, \\
\tilde{y}^* &= (y^*, y_{m+1}^*), \quad y^* \in \mathbb{R}^m.
\end{aligned}
$$

Theorem 5 *Let* (M_p^*, M_d^*) *and* $(\tilde{x}^*, \tilde{y}^*, \tilde{s}^*)$ *be as defined above. Then,*

(i) if $x_{n+1}^* > 0$ *and* $s_{n+2}^* > 0$*, then both* (P) *and* (D) *are infeasible;*

(ii) if $x_{n+1}^* > 0$ *and* $s_{n+2}^* = 0$*, then* (P) *is infeasible and* (D) *unbounded;*

(iii) if $x_{n+1}^* = 0$ *and* $s_{n+2}^* > 0$*, then* (P) *is unbounded and* (D) *infeasible;*

(iv) if $x_{n+1}^* = 0$ *and* $s_{n+2}^* = 0$*, then* (x^*, s^*) *is a strictly complementary optimal pair for* (P) *and* (D).

Proof: We claim that $x_{n+1}^* > 0$ if and only if (P) is infeasible. It is clear that if (P) is infeasible, then $x_{n+1}^* > 0$. To prove the converse, assume that $\overline{x}$ is feasible for (P). Choose $\overline{x}_{n+2} > 0$ such that $M_d^1 := (A^T y^0 + s^0 - c)^T \overline{x} + \overline{x}_{n+2} \geq M_d^*$. Then $\tilde{x}^0 := (\overline{x}, 0, \overline{x}_{n+2})$ will be feasible for $(\tilde{P})$ with parameter M_d^1. Let $\tilde{x}^1$ be a strictly complementary solution of $(\tilde{P})$ with the same parameter M_d^1. Then, using Lemma 4 twice, the first time with M_d^* replaced by M_d^1, we obtain

$$\begin{aligned} z^*(M_p, M_d^1) &= z^*(M_p^*, M_d^1) + x_{n+1}^1 (M_p - M_p^*), \\ &= z^*(M_p^*, M_d^*) - s_{n+2}^* (M_d^1 - M_d^*) + x_{n+1}^1 (M_p - M_p^*). \end{aligned}$$

The objective value of $(\tilde{P})$ at $\tilde{x}^0$, which is $c^T \overline{x}$, is an upper bound for $z^*(M_p, M_d^1)$, for every value of $M_p \geq M_p^*$. However, due to Lemma 4 we have $\pi(M_p, M_d^1) = \pi(M_p, M_d^*)$, and thus $x_{n+1}^1 > 0$. Therefore we have $\lim_{M_p \to \infty} z^*(M_p, M_d^1) = \infty$. This contradiction proves the claim. In the same way, one can show that $s_{n+2}^* > 0$ if and only if (D) is infeasible. So the proof of (i) is immediate.

We now prove (ii). From the above paragraph it is clear that (P) is infeasible and (D) is feasible in this case. Moreover, $\lim_{M_p \to \infty} z^*(M_p, M_d^*) = \infty$. Since $y_{m+1}^* = -s_{n+2}^* = 0$, we have $\tilde{y}^* = (y^*, 0)$. Hence, y^* is feasible for (D) and $\tilde{b}^T \tilde{y}^* = b^T y^* = z^*(M_p, M_d^*)$. Since $\lim_{M_p \to \infty} z^*(M_p, M_d^*) = \infty$ we conclude that (D) must be unbounded. The proof of (iii) is similar.

It remains to prove (iv). Since $x_{n+1}^* = 0$ and $s_{n+2}^* = 0$, x^* is feasible for (P) and y^* is feasible for (D). Furthermore, $c^T x^* = \tilde{c}^T \tilde{x}^* = \tilde{z}(M_p, M_d) = \tilde{b}^T \tilde{y}^* = b^T y^*$. Hence, we have strong duality for the problems (P) and (D). Moreover, since the pair $(\tilde{x}^*, \tilde{s}^*)$ is strictly complementary, the pair (x^*, s^*) is also strictly complementary. □

The next corollary, which is a direct consequence of the above result, contains the classical duality theorem for linear programming and the Goldman – Tucker theorem.

Corollary 1 *If both (P) and (D) are feasible, then a strictly complementary solution pair exists. Otherwise, either both problems are infeasible, or one is infeasible and the other is unbounded.*

Theorem 5 also makes clear that the notion of *optimal partition* can be extended to the case that the basic assumption is not satisfied, namely by taking the restriction of the partition $\tilde{\pi}^* := (\tilde{B}^*, \tilde{N}^*)$ defined above to the index set $\{1, 2, \ldots, n\}$. In this paper we will not exploit this fact however.

3 Parametric analysis

Much work in the field of linear programming has been devoted to parametric analysis and sensitivity analysis. See, e.g., the book of Gal [8]. This work is based on the Simplex approach to linear programming, and the use of optimal bases. Everyone who has some experience in the field of linear programming, either in theory or practice, will know that the phenomenon of nonuniqueness of an optimal basis, causes a lot of troubles. This becomes very apparent in the case of sensitivity analysis. The output of the analysis (as usually presented in textbooks) is dependent on the optimal basis which has been obtained by the solver, which maybe either a person or a computer. So, when different solvers perform the sensitivity analysis, in the way it is described in all text books, one must expect that the output will be different. Later on we will present a striking example of this very unpleasant feature. The aim of this section is to show that the use of optimal partitions is the remedy [1, 15].

The problem (P) considered so far is completely determined by the triple (A, b, c), where A is the *constraint matrix*, b the *right hand side vector* and c the *cost coefficient vector* of (P). In this section we will investigate the effect of changes in b and c on the optimal value of (P). To express the dependence on b and c we will denote the optimal value as $z(b, c)$. This function will be called the *value function* for the matrix A. In fact we will be mainly interested in parametric perturbations of b and c, as considered in the classical topic of *parametric* and *post optimal* or *sensitivity* analysis. This means that we want to study the behavior of $z(b+\beta\bar{b}, c+\gamma\bar{c})$, where $\bar{b}$ and $\bar{c}$ are given *perturbation* vectors, as a function of the parameters β and γ. It is well known, and will be proved below, that $z(b+\beta\bar{b}, c)$ is a piecewise linear convex function of β and $z(b, c+\gamma\bar{c})$ is a piecewise linear concave function of γ. In practice one is interested in the *slopes* of these functions for given values of b and c, at $\beta = 0$ and $\gamma = 0$ respectively, and also in the range of the linear piece to which b and c belong. These quantities have important economic interpretations in practical applications. Most often we then have $\bar{b} = e_i$ for some $i, 1 \leq i \leq m$, and $\bar{c} = e_j$ for some $j, 1 \leq j \leq n$. This means that one only considers changes in the i-th entry of b and the j-th entry of c.

We prefer to do the analysis in the more general framework of arbitrary perturbation vectors. This not only makes the results more general, but also helps to simplify the presentation. Our main result will be that along the linear pieces of the value functions $z(b + \beta\bar{b}, c)$ and $z(b, c + \gamma\bar{c})$ the optimal partitions of the underlying problems remain constant. In other words, when changing for example the parameter β in $z(b + \beta\bar{b}, c)$, the breakpoints in this piecewise linear function occur exactly there where the optimal partition of the underlying problem changes. In fact, we shall also show that in a breakpoint the optimal partition differs from the partitions of the surrounding linear pieces.

3.1 Properties of the value function

Let the value function $z(b, c)$ be as defined above. So, $z(b, c)$ equals the optimal value of the linear programming problem (P) determined by the triple (A, b, c). Below the matrix A will be kept constant, but the vectors b and c will be considered as variables. Therefore, in this section, we will denote (P) as $(P(b, c))$ and (D) as $(D(b, c))$. Also, the corresponding optimal partition will be denoted as $\pi(b, c)$. We will call the pair (b, c) a *feasible pair* if the problems $(P(b, c))$ and $(D(b, c))$ both are feasible. If the problem $(P(b, c))$ is unbounded then we define $z(b, c) = -\infty$, and if its dual $(D(b, c))$ is unbounded then we define $z(b, c) = \infty$. If both $(P(b, c))$ and $(D(b, c))$ are infeasible then $z(b, c)$ is undefined.

In the next two lemmas we describe some elementary facts about the value function.

Lemma 5 *The value function $z(b, c)$ is convex in b and concave in c.*

Proof: Let b^0, b^1 and c be given such that the pairs (b^0, c) and (b^1, c) are feasible. Moreover, let x^0 be an optimal solution of $(P(b^0, c))$ and x^1

an optimal solution of $(P(b^1, c))$. As a consequence we have $z(b^0, c) = c^T x^0$ and $z(b^1, c) = c^T x^1$. Now let b be a convex combination of b^1 and b^0, say $b = \alpha b^1 + (1-\alpha)b^0$, with $0 \leq \alpha \leq 1$. Then

$$x = \alpha x^1 + (1-\alpha)x^0$$

is feasible for $(P(b, c))$, as can easily be verified. Therefore, $c^T x \geq z(b, c)$. Since

$$c^T x = \alpha c^T x^1 + (1-\alpha)c^T x^0 = \alpha z(b^1, c) + (1-\alpha)z(b^0, c),$$

we conclude that

$$z(b, c) \leq \alpha z(b^1, c) + (1-\alpha)z(b^0, c).$$

This proves the first part of the lemma. The proof of the second part goes in the same way. □

Lemma 6 *Let (b^0, c^0) and (b^1, c^1) be feasible pairs such that the partitions $\pi(b^0, c^0)$ and $\pi(b^1, c^1)$ are equal. If b is any convex combination of b^0 and b^1, and c of c^0 and c^1, then $\pi(b, c)$ is the same partition. Moreover, $z(b, c)$ is linear on the line segment from (b^0, c) to (b^1, c) and also on the line segment from (b, c^0) to (b, c^1), i.e., $z(b, c)$ is a bilinear function on the rectangle $[b^0, b^1] \times [c^0, c^1]$.*

Proof: Let $\pi = (B, N)$ denote the common partition for the feasible pairs (b^0, c^0) and (b^1, c^1). So we have $\pi = \pi(b^0, c^0) = \pi(b^1, c^1)$. Let $0 \leq \lambda_1, \lambda_2 \leq 1$ be such that

$$\begin{aligned} b &= \lambda_1 b^1 + (1-\lambda_1)b^0, \\ c &= \lambda_2 c^1 + (1-\lambda_2)c^0. \end{aligned}$$

Furthermore, let x^0, s^0, y^0 denote the central solutions of (P) and (D) for the pair (b^0, c^0), and, similarly, x^1, s^1, y^1 for the pair (b^1, c^1). Now define

$$\begin{aligned} x &:= \lambda_1 x^1 + (1-\lambda_1)x^0, \\ y &:= \lambda_2 y^1 + (1-\lambda_2)y^0, \\ s &:= \lambda_2 s^1 + (1-\lambda_2)s^0. \end{aligned}$$

Then one easily checks that x is feasible for $(P(b, c))$, and (y, s) for $(D(b, c))$. Since $\pi(b^1, c^1) = \pi(b^0, c^0) = (B, N)$, it follows that $x_k > 0$ if and only if $k \in B$, and $s_k > 0$ if and only if $k \in N$. Therefore, the pair (x, s) is strictly complementary (and hence optimal) for the pair (b, c). The partition determined by this pair being (B, N), the first part of the lemma follows.

Now let us deal with the proof of the second part. Since x is optimal for $(P(b, c))$, we have $z(b, c) = c^T x$. So

$$z(b, c) = (\lambda_2 c^1 + (1 - \lambda_2) c^0)^T (\lambda_1 x^1 + (1 - \lambda_1) x^0).$$

This makes clear that $z(b, c)$ is a bilinear function of λ_1 and λ_2. Now fixing c, i.e., fixing λ_2, this expression becomes linear in λ_1, and hence $z(b, c)$ is linear on the line segment from (b^0, c) to (b^1, c). Similarly, fixing b, i.e., fixing λ_1, the expression for $z(b, c)$ becomes linear in λ_2, so $z(b, c)$ is also linear on the line segment from (b, c^0) to (b, c^1). □

In the next sections we deal with the converse implication. We will show that if $z(b, c)$ is linear on the line segment from (b^0, c) to (b^1, c) then the partition is constant on this line segment, with a possible exception in the end points of the line segment; a similar result holds if $z(b, c)$ is linear on the line segment from (b, c^0) to (b, c^1).

3.2 Perturbations in the right hand side vector

To facilitate the discussion we introduce the notation

$$b(\alpha) := \alpha b^1 + (1 - \alpha) b^0,$$

and

$$\psi(\alpha) := z(b(\alpha), c),$$

where b^0, b^1 and c are such that the pairs (b^0, c) and (b^1, c) are feasible. For each α we will denote the corresponding optimal partition by $\pi(\alpha) = (B^\alpha, N^\alpha)$. Note that the feasible region of the dual problem $(D(b(\alpha), c))$ is independent of α. It is simply given by $\{y \; : \; A^T y \leq c\}$. So the dual problem is feasible for each α. Hence, $\psi(\alpha)$ is well defined for each $\alpha \in \mathbb{R}$. Recall that $\psi(\alpha) = \infty$ if the dual problem is unbounded.

Theorem 6 *The value function $\psi(\alpha)$ is piecewise linear and convex.*

Proof: Let α move from $-\infty$ to ∞. Then $\pi(\alpha)$ runs through a set of partitions of the index set $\{1, \ldots, n\}$. Since the number of such partitions is finite, we can get a partition of the real line into a finite number of nonoverlapping intervals by opening a new interval every time when the partition

$\pi(\alpha)$ changes. Now we know from Lemma 6 that on each subinterval the value function $\psi(\alpha)$ is linear. By Lemma 5 the value function is convex. Hence it follows that the value function is piecewise linear. $\square$

Any interval of the real line on which the value function $\psi(\alpha)$ is linear and which is not a singleton, will be called a *linearity interval* of $\psi(\alpha)$. We proceed by showing that in the open part of a linearity interval the optimal partition is constant. In other words, the linearity intervals are precisely the intervals constructed in the proof of Theorem 6.

Lemma 7 *Let for some* b^0, b^1, c *the value function* $\psi(\alpha)$ *be linear for* $0 \leq \alpha \leq 1$. *So*

$$\psi(\alpha) = \alpha\psi(1) + (1 - \alpha)\psi(0).$$

Then $\pi(\alpha)$ *is independent of* α *for all values of* α *in the open interval* $(0, 1)$.

Proof: For $0 \leq \alpha \leq 1$, let the triple $(x^\alpha,\ y^\alpha,\ s^\alpha)$ be the central solution for the pair $(b(\alpha), c)$, and (B^α, N^α) the corresponding optimal partition. Now we use that y^α is feasible for both $(D(b^0, c))$ and $(D(b^1, c))$. As a consequence we have, $(b^0)^T y^\alpha \leq (b^0)^T y^0 = \psi(0)$ and $(b^1)^T y^\alpha \leq (b^1)^T y^1 = \psi(1)$. Using this and the definition of $b(\alpha)$ we may write

$$\psi(\alpha) = \alpha\psi(1) + (1 - \alpha)\psi(0) \geq \alpha(b^1)^T y^\alpha + (1 - \alpha)(b^0)^T y^\alpha = b(\alpha)^T y^\alpha = \psi(\alpha).$$

For $0 < \alpha < 1$, this implies that

$$(b^0)^T y^\alpha = \psi(0) \text{ and } (b^1)^T y^\alpha = \psi(1).$$

Now let $0 \leq \beta \leq 1$. Then we may write

$$b(\beta)^T y^\alpha = (\beta b^1 + (1 - \beta) b^0)^T y^\alpha = \beta\psi(1) + (1 - \beta)\psi(0) = \psi(\beta).$$

This proves that y^α is optimal for all problems $(D(b(\beta), c))$, with $0 \leq \beta \leq 1$. From this we derive that $(x^\beta)^T s^\alpha = 0$. Therefore, $N^\alpha \cap B^\beta = \emptyset$. This is equivalent to $N^\alpha \subseteq N^\beta$ and $B^\beta \subseteq B^\alpha$. Now taking $0 < \beta < 1$, we can apply the same argument to β and we obtain the converse inclusions $N^\beta \subseteq N^\alpha$ and $B^\alpha \subseteq B^\beta$. We conclude that $B^\alpha = B^\beta$ and $N^\alpha = N^\beta$. This proves the lemma. $\square$

From this lemma (and its proof) we draw an important conclusion, which we state as a corollary.

Corollary 2 *The set of optimal solutions of* $(D(b(\alpha), c))$ *is constant on the interval* $0 < \alpha < 1$ *and, moreover, each of these solutions is also optimal for the extreme cases* $\alpha = 0$ *and* $\alpha = 1$.

It may be useful to illustrate this phenomenon. To this end consider Figure 6. In this figure we have drawn part of the (dual) feasible region of a two–dimensional problem, and we have drawn two objective vectors b^1 and b^0. It is clear that if the (dual) objective vector is a proper convex combination $b(\alpha) := \alpha b^1 + (1-\alpha)b^0$, with $0 < \alpha < 1$, then the vertex v (determined by the constraints numbered 1 and 2) is the only optimal solution. In these cases the set B^α consists of the indices 1 and 2 only. If $\alpha = 0$ then the facet orthogonal to b^0 is the optimal set, and this set contains the vertex v. Also, if $\alpha = 1$ then the facet orthogonal to b^1 is the optimal set, which again contains the vertex v.

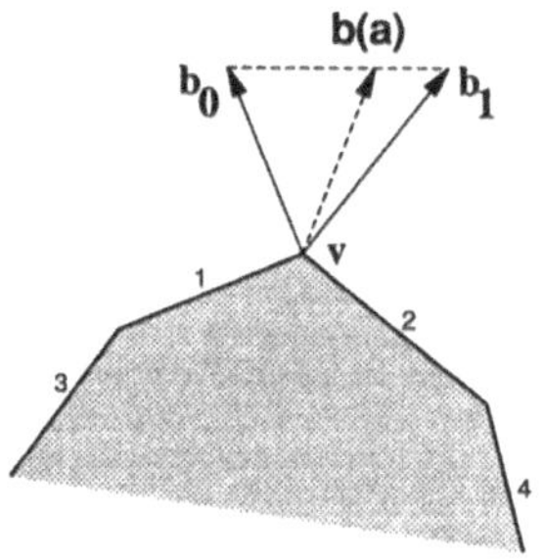

Figure 6: Illustration of Corollary 2.

In the sequel we will use the notation

$$\Delta b := b^1 - b^0.$$

As a consequence we have

$$b(\alpha) = b^0 + \alpha \Delta b.$$

Example 5 A transportation problem

Figure 7 shows a transportation problem with three supply and three market nodes. The supplies and the demands are as indicated at the nodes. Also, the cost coefficients are as shown.

Taking demand values positive and supply values negative, the vector $b = (0, -1, -5, 3, 1, 2)$ is the supply / demand vector for this problem. As usual, c denotes the cost coefficient vector. For the perturbation vector

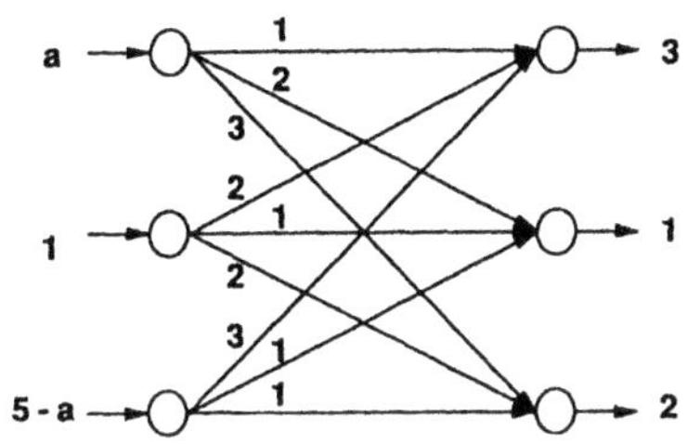

Figure 7: A transportation problem.

$\Delta b := (-1, 0, 1, 0, 0, 0)$ we have depicted the value function $z(b + \alpha\Delta b, c)$, in Figure 8. One easily sees that the pair (b, c) is feasible if and only if $0 \leq \alpha \leq 5$. For each breakpoint and for each linearity interval the cor-

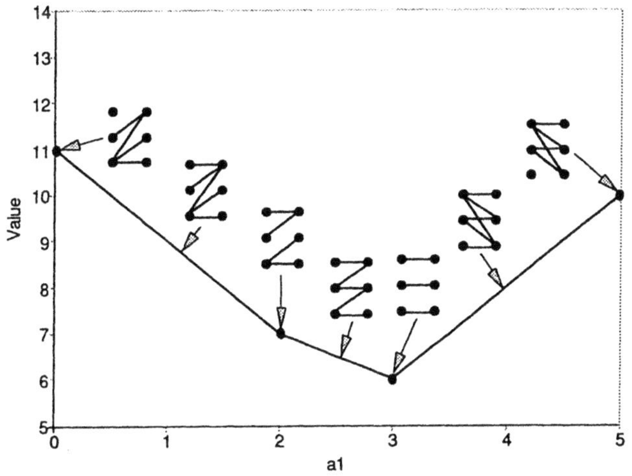

Figure 8: Value function and optimal partitions for the transport problem in Figure 7.

responding optimal partition is indicated by giving the arcs in the transportation network which (may) carry flow in an optimal solution. At this place we do not deal with the question of how to determine the optimal partitions. We only indicate that in some cases it is very easy to find the optimal partition. This is the case, e.g., if $\alpha = 3$, as easily may be verified. □

The next question we will deal with is how to find the 'surrounding' partitions of a given partition. We shall show that these partitions can be found by solving two appropriate linear programming problems. These problems are stated in terms of the given partition and the direction Δb of the perturbation in b. We need to distinguish between the case that the given partition belongs to a breakpoint and the case that the given partition belongs to a linearity interval.

To simplify the presentation we assume for the moment that $\alpha = 0$ and $\alpha = 1$ are two consecutive breakpoints of $\psi(\alpha)$. The optimal partitions occuring along the unit interval, when α runs from 0 to 1, are then $\pi(\alpha) = (B^\alpha, N^\alpha)$, and the corresponding central solutions are $x^\alpha, y^\alpha, s^\alpha$. The partitions $(B^\alpha, N^\alpha), 0 < \alpha < 1$, are all equal, so we will denote them simply by $\bar{\pi} = (\bar{B}, \bar{N})$. Similarly, the central solutions $y^\alpha, 0 < \alpha < 1$, being all equal, will be denoted by $\bar{y}$. From the proof of Lemma 7 we deduce the relation

$$\psi(\alpha) = \psi(0) + \alpha(\Delta b)^T \bar{y},\ 0 \leq \alpha \leq 1.$$

We will now show how the optimal partition in the linearity interval $0 < \alpha < 1$ can be determined from the optimal partition at the breakpoint $\alpha = 0$. To this end we define the following pair of dual linear programming problems:

$$(P_{\mapsto}^{\Delta b}) \quad \min\left\{c^T x : Ax = \Delta b,\ x_{N^0} \geq 0\right\},$$

$$(D_{\mapsto}^{\Delta b}) \quad \max\{(\Delta b)^T y \ :\ A^T y + s = c,\ s_{B^0} = 0,\ s_{N^0} \geq 0\}.$$

Note that in $(P_{\mapsto}^{\Delta b})$ the variables x_i, $i \in B^0$, are free, i.e. these variables are not required to be nonnegative. So the format of this problem is not the standard format (which requires equality constraints and nonnegative variables). As a consequence we need to define the concepts of strictly complementary solution and optimal partition for this particular case. This can be done in an obvious way. Simply replace the free part x_{B^0} of x by $x_{B^0}^+ - x_{B^0}^-$, with $x_{B^0}^+$ and $x_{B^0}^-$ nonnegative. Then we are again in the standard format. As a consequence we obtain that there exists a solution triple $(\tilde{x},\ \tilde{y},\ \tilde{s})$ such that for each $i \in N^0$ one has $\tilde{x}_i > 0$ if and only if $\tilde{s}_i = 0$. In this way we obtain a (unique) partition $(\tilde{B},\ \tilde{N})$ of the set N^0, namely by taking for $\tilde{B}$ the subset of N^0 for which the coordinates $\tilde{x}_i$ are positive and for $\tilde{N}$ the subset of N^0 for which the coordinates $\tilde{s}_i$ are positive. Then we call $(B^0 \cup \tilde{B},\ \tilde{N})$ the optimal partition of the problem. Any such solution triple with the above property will be called strictly complementary. It is clear that strict complementarity implies optimality.

It may be worthwhile to indicate that the feasible region of the dual problem $(D^{\Delta b}_{\mapsto})$ is the optimal face for the dual problem $(D(b^0, c))$. As a consequence it admits the analytic center $(y^0,\ s^0)$ of this face as a feasible solution. Recall that also $(\bar{y},\ \bar{s})$ belongs to this face.

Theorem 7 *The optimal partition for the pair of dual problems* $(P^{\Delta b}_{\mapsto})$ *and* $(D^{\Delta b}_{\mapsto})$ *is just* $(\bar{B}, \bar{N})$. *Furthermore,* $\bar{y}$ *is the central solution of* $(D^{\Delta b}_{\mapsto})$.

Proof: Let $0 < \alpha < 1$, and consider

$$x := \frac{x(\alpha) - x^0}{\alpha}.$$

Since $x^0_{N^0} = 0$ one has $x_{N^0} \geq 0$. Obviously $Ax = \Delta b$. So x is feasible for $(P^{\Delta b}_{\mapsto})$. We already observed that the dual problem $(D^{\Delta b}_{\mapsto})$ admits $(\bar{y},\ \bar{s})$ as a feasible solution. So we have found a pair of feasible solutions for $(P^{\Delta b}_{\mapsto})$ and $(D^{\Delta b}_{\mapsto})$. We conclude the proof by showing that this pair is strictly complementary and that it determines $\bar{\pi} = (\bar{B},\ \bar{N})$ as the optimal partition. Recall that the support of $x(\alpha)$ is $\bar{B}$ and the support of x^0 is B^0. So, for $i \in N^0$, we have $x_i > 0$ if and only if $i \in N^0 \setminus \bar{N}$. On the other hand, if $i \in N^0$, then we have $\bar{s}_i > 0$ if and only if $i \in \bar{N}$. This proves that the given pair of solutions is strictly complementary and that the optimal partition is just $\bar{\pi} = (\bar{B}, \bar{N})$. □

A similar result can be obtained for the pair of dual linear programming problems given by:

$$(P^{\Delta b}_{\leftarrow\!\shortmid}) \quad \min\left\{c^T x : Ax = -\Delta b,\ x_{N^1} \geq 0\right\},$$

$$(D^{\Delta b}_{\leftarrow\!\shortmid}) \quad \max\{-(\Delta b)^T y \ :\ A^T y + s = c,\ s_{B^1} = 0,\ s_{N^1} \geq 0\}.$$

Without further proof we state

Theorem 8 *The optimal partition for the pair of dual problems* $(P^{\Delta b}_{\leftarrow\!\shortmid})$ *and* $(D^{\Delta b}_{\leftarrow\!\shortmid})$ *is just* $(\bar{B}, \bar{N})$. *Furthermore,* $\bar{y}$ *is the central solution of* $(D^{\Delta b}_{\leftarrow\!\shortmid})$.

Example 6 The surrounding partitions of a breakpoint

Let the partition at the breakpoint $a_1 = 3$ be given. Then the optimal partition for the linearity interval to the left and to the right of $a_1 = 3$ are obtained by solving the problems $(P^{\Delta b}_{\mapsto})$ and $(P^{\Delta b}_{\mapsfrom})$ respectively.

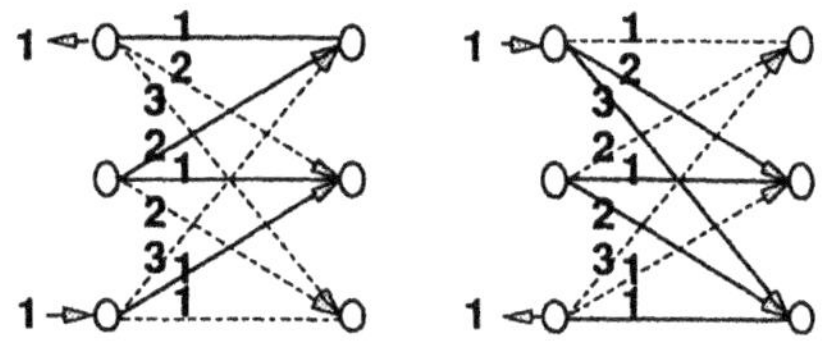

(a) (b)

Figure 9: The shortest path problem for the surrounding partitions.

These problems are now simple shortest path problems, depicted in Figure 9(a) and (b) respectively. The 'horizontal' arcs are the arcs in B^0, and the other arcs are the arcs in N^0. The arcs in B^0 may be used in both directions, so these arcs can be considered as undirected in the shortest path problem, whereas the arcs in N^0 are directed. One easily checks that the arcs which are drawn solid belong to the optimal partition for the respective shortest path problems. Adding these arcs to B^0 we obtain the optimal partitions for the two linearity intervals. See Figure 8. □

For future use we include the following

Lemma 8 $(\Delta b)^T(\bar{y} - y^0) > 0$ *and* $(\Delta b)^T(y^1 - \bar{y}) > 0$.

Proof: Recall that y^0 is the central solution if $\alpha = 0$; it is the analytic center of the optimal (dual) face for the case that $\alpha = 0$. Now Theorem 7 makes clear that when we maximize $(\Delta b)^T y$ over this face then $\bar{y}$ is the central solution. Also, as a consequence of Theorem 8, if we minimize $(\Delta b)^T y$ over the optimal (dual) face for the case that $\alpha = 0$, then the optimal partition is the optimal partition associated to the linearity interval to the left at the breakpoint $\alpha = 0$; let $\bar{\bar{y}}$ denote the central solution for the this problem. Since the value function $z(b(\alpha), c)$ has a breakpoint at

$\alpha = 0$, its left and right derivatives to α are different at $\alpha = 0$. Since these derivatives are given by $(\Delta b)^T\bar{\bar{y}}$ and $(\Delta b)^T\bar{y}$ respectively, we conclude that $(\Delta b)^T\bar{\bar{y}}$ and $(\Delta b)^T\bar{y}$ are different. This implies that $(\Delta b)^T\bar{\bar{y}} > (\Delta b)^T\bar{y}$, because $\bar{\bar{y}}$ maximizes and $\bar{y}$ minimizes $(\Delta b)^T y$ in the optimal (dual) face for $\alpha = 0$. So it follows that $(\Delta b)^T y$ is not constant in this face. Since y^0 is the analytic center of this face, we conclude that $(\Delta b)^T\bar{\bar{y}} > (\Delta b)^T y^0 > (\Delta b)^T\bar{y}$. This implies the result. □

Yet we consider the case that the optimal partition $\bar{\pi} = (\bar{B}, \bar{N})$ associated to some given linearity interval is known. We will show how the optimal partitions at the surrounding breakpoints can be found from this partition and the perturbation vector Δb. In the analysis below it will be convenient to assume that the vector b^0 belongs to the linearity interval under consideration, and that the surrounding breakpoints, if they exist, occur at $\alpha_- < 0$ and $\alpha_+ > 0$ respectively. For the present purpose we consider the following pair of dual problems.

$$(P_{\leftarrow}^{\Delta b}) \quad \min\{\alpha \;:\; Ax = b(\alpha),\, x_{\bar{B}} \geq 0,\, x_{\bar{N}} = 0\},$$

$$(D_{\leftarrow}^{\Delta b}) \quad \max\{(b^0)^T y \;:\; A^T y + s = 0,\, (\Delta b)^T y = -1,\, s_{\bar{B}} \geq 0\}.$$

Since these problems do not have the standard format we have to discuss the meaning of strictly complementary solution and optimal partition for these problems. Note that $(P_{\leftarrow}^{\Delta b})$ can be brought in the standard format by omitting the variables x_i, $i \in \bar{N}$. By omitting also the constraints in $(D_{\leftarrow}^{\Delta b})$ which are indexed by the indices in $\bar{N}$, we obtain a completely equivalent pair of dual problems, namely

$$\min\{\alpha \;:\; A_{\bar{B}} x_{\bar{B}} = b(\alpha),\, x_{\bar{B}} \geq 0\},$$

$$\max\{(b^0)^T y \;:\; A_{\bar{B}}^T y + s_{\bar{B}} = 0,\, (\Delta b)^T y = -1,\, s_{\bar{B}} \geq 0\}.$$

Let $(\tilde{B}, \tilde{N})$ be the optimal partition for this pair. Then this is a partition of the set $\bar{B}$ with the property that there exists a solution triple $(\tilde{x}, \tilde{y}, \tilde{s})$ for the original pair of problems such that for each $i \in \bar{B}$ one has $\tilde{x}_i > 0$ if and only if $\tilde{s}_i = 0$. Now it becomes natural to define $(\tilde{B}, \tilde{N} \cup \bar{N})$ as the optimal partition of the original problems. Any solution triple $(\tilde{x}, \tilde{y}, \tilde{s})$ with the above property will be called strictly complementary. It is clear that strict complementarity implies optimality.

Theorem 9 *The optimal partition for the pair of dual problems* $(P_{\leftarrow}^{\Delta b})$ *and* $(D_{\leftarrow}^{\Delta b})$ *is just* $\pi(\alpha_-)$. *Furthermore,* y^{α_-} *is the central solution of* $(D_{\leftarrow}^{\Delta b})$.

Proof: The proof follows the same line as the proof of Theorem 7. We construct in a more or less obvious way feasible solutions for both problems and prove that these solutions are strictly complementary with the correct partition. This time we observe that

$$x := x^{\alpha_-}, \; \alpha := \alpha_-$$

is feasible for the primal problem and that

$$y := \frac{y^{\alpha_-} - \bar{y}}{(\Delta b)^T(\bar{y} - y^{\alpha_-})}$$

is feasible for the second problem. The first is an obvious consequence of the fact that x^{α_-} is the central solution for the problem $(P(b(\alpha_-), c)$. The second requires a little more effort. First we deduce from Lemma 8, that $(\Delta b)^T(\bar{y} - y^{\alpha_-})$ is positive, so y is well defined. Clearly $(\Delta b)^T y = -1$. Furthermore, one has

$$\left((\Delta b)^T(\bar{y} - y^{\alpha_-})\right) A^T y = A^T(y^{\alpha_-} - \bar{y}) = \bar{s} - s^{\alpha_-}.$$

Since $\bar{s}_{\bar{B}} = 0$ and $s^{\alpha_-} \geq 0$, it follows that $\bar{s}_{\bar{B}} - s^{\alpha_-}_{\bar{B}} = -s^{\alpha_-}_{\bar{B}} \leq 0$. So the given $\tilde{y}$ is feasible for the dual problem. Finally, if $i \in \bar{B}$ then we have $x_i > 0$ if and only if $i \in B^{\alpha_-}$, and $s_i = 0$ if and only if $i \in B^{\alpha_-}$. This proves that the given pair is strictly complementary with the partition $(B^{\alpha_-}, N^{\alpha_-})$. Hence the theorem has been proved. □

A similar result can be obtained for the pair of dual linear programming problems given by:

$$(P^{\Delta b}_{\rightarrow}) \;\; \max\{\alpha \; : \; Ax = b(\alpha), \, x_{\bar{B}} \geq 0, \, x_{\bar{N}} = 0\},$$

$$(D^{\Delta b}_{\rightarrow}) \;\; \min\{(b^0)^T y \; : \; A^T y + s = 0, \, (\Delta b)^T y = 1, \, s_{\bar{B}} \geq 0\}.$$

Without further proof we state

Theorem 10 *The optimal partition for the pair of dual problems* $(P^{\Delta b}_{\rightarrow})$ *and* $(D^{\Delta b}_{\rightarrow})$ *is just* $\pi(\alpha_+)$. *Furthermore,* y^{α_+} *is the central solution of* $(D^{\Delta b}_{\rightarrow})$. □

Example 7 The surrounding partitions of a linearity interval

Let the partition (B^0, N^0) at the interval $a_1 \in (3, 5)$ be given. Then the optimal partitions for the surrounding breakpoints are obtained by solving the problems $(P^{\Delta b}_{\leftarrow})$ and $(P^{\Delta b}_{\rightarrow})$ respectively. This amounts to finding the

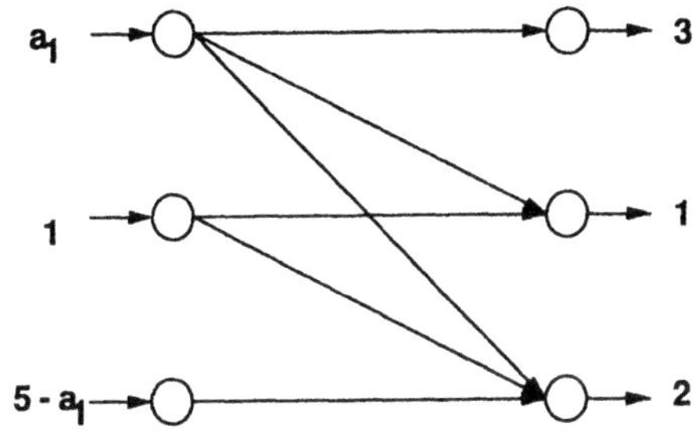

Figure 10: The shortest path problem for the surrounding partitions.

minimal and the maximal value of a_1 for which the flow problem depicted in Figure 10 is feasible. Clearly the demand in the upper right market node can only be delivered by the upper left supply node, which gives $a_1 \geq 3$. The value $a_1 = 3$ is feasible and then only the horizontal arcs will carry flow, which gives us the right partition for the breakpoint to the left of the linearity interval. On the other hand, the maximal value for a_1 is 5, and this value is feasible. Then the first two supply nodes deliver all market nodes. All arcs except the lowest horizontal arc may carry flow in this situation. Adding these arcs to the arcs in B^0 we get the partition for the breakpoint to the right of the linearity interval. □

3.3 Perturbations in the objective vector

In this section we consider the effect of variations in the vector c on the value function. It turns out that by "dualizing" the results of the previous section we obtain the appropriate results. Each result gives the optimal partition for a given pair of dual problems. The proofs are based on the same idea as for their dual counterparts. One checks that some natural candidate solutions for both problems are feasible indeed, and then shows that these solutions are strictly complementary with the correct partition. Therefore, in this section we state these results without proofs. The discussion below is facilitated by using the notation

$$\Delta c := c^1 - c^0,$$

$$c(\alpha) := c^0 + \alpha \Delta c = \alpha c^1 + (1 - \alpha)c^0,$$

and

$$\chi(\alpha) := z(b, c(\alpha)),$$

where b and c^0, c^1 are such that the pairs (b, c^0) and (b, c^1) are feasible. For each α we will denote the corresponding optimal partition by $\pi(\alpha) = (B^\alpha, N^\alpha)$.

Theorem 11 *The value function $\chi(\alpha)$ is piecewise linear and concave.*

An interval of the real line on which the value function $\chi(\alpha)$ is linear, and which is not a singleton, will be called a *linearity interval* of $\chi(\alpha)$.

Lemma 9 *Let for some b, c^0, c^1 the value function $\chi(\alpha)$ be linear for $0 \leq \alpha \leq 1$. So*

$$\chi(\alpha) = \alpha\chi(1) + (1 - \alpha)\chi(0).$$

Then $\pi(\alpha)$ is independent of α for all values of α in the open interval $(0, 1)$.

Corollary 3 *The set of optimal solutions of $(P(b, c(\alpha))$ is constant on the interval $0 < \alpha < 1$ and, moreover, each of these solutions is also optimal for the extreme cases $\alpha = 0$ and $\alpha = 1$.*

Example 8 Application to the transportation problem Using the transportation problem in Figure 7 again we give an illustration. The supply values are assumed to be 3,1 and 2 respectively. The cost coefficients are as shown in the figure, except for the arcs leaving the first supply node. We consider the situation in which the sum of these coefficients is 6, and we keep the cost coefficient $c(1, 2)$ to market 2 constant at the value 2. So the sum of the cost coefficients from supply node 1 to the market nodes 1 and 3 will be kept 4. In Figure 11 we have depicted part of the value function $z(b, c + \alpha\Delta c)$, $0 \leq \alpha \leq 5$, with $b = (-3, -1, -2, 3, 1, 2)$, $c = (1, 2, 3, 2, 1, 2, 3, 1, 1)$ and $\Delta c = (1, 0, -1, 0, 0, 0, 0, 0, 0)$. One easily sees that the pair (b, c) is feasible for every value of α. For each breakpoint and for each linearity interval the corresponding optimal partition is indicated by giving the arcs in the transportation network which (may) carry flow in an optimal solution.

□

Just as in the previous section we conclude with the problem of finding the 'surrounding' partitions of a given partition. Also in the present case these partitions can be found by solving appropriate linear programming

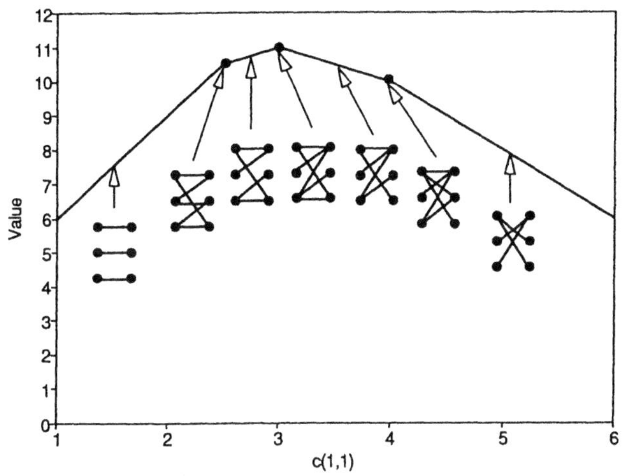

Figure 11: Value function $z(b, c+\alpha\Delta c)$, $0 \le \alpha \le 5$, and the corresponding optimal partitions.

problems. These problems are formulated in terms of the given partition and the direction Δc of the perturbation in c.

We start with the case that the given partition belongs to a breakpoint. Without loss of generality we assume again that $\alpha = 0$ and $\alpha = 1$ are two consecutive breakpoints of $\chi(\alpha)$, and that we have given either the partition $\pi(0) = (B^0, N^0)$ or the partition $\pi(1) = (B^1, N^1)$. We will show that either of these partitions, together with the perturbation vector Δc, determines the partition associated to the interval between these breakpoints, which will be denoted by $\bar{\pi} = (\bar{B}, \bar{N})$. The central solution $x^\alpha, 0 < \alpha < 1$, will be denoted by $\bar{x}$. As a consequence we have

$$\chi(\alpha) = \chi(0) + \alpha(\Delta c)^T \bar{x},\ 0 \le \alpha \le 1.$$

We consider the following pair of dual linear programming problems:

$$(P_{\mapsto}^{\Delta c}) \quad \min\left\{(\Delta c)^T x \ :\ Ax = b,\ x_{B^0} \ge 0,\ x_{N^0} = 0\right\},$$
$$(D_{\mapsto}^{\Delta c}) \quad \max\left\{b^T y \ :\ A^T y + s = \Delta c,\ s_{B^0} \ge 0\right\}.$$

Theorem 12 *The optimal partition for the pair of dual problems $(P_{\mapsto}^{\Delta c})$ and $(D_{\mapsto}^{\Delta c})$ is just $(\bar{B}, \bar{N})$. Furthermore, $\bar{x}$ is the central solution of $(P_{\mapsto}^{\Delta c})$.*

A similar result can be obtained for the optimal partition at $\alpha = 1$. Defining the pair of dual linear programming problems

$$(P_{\leftrightarrow}^{\Delta c}) \;\; \max\left\{(\Delta c)^T x \;:\; Ax = b,\, x_{B^1} \geq 0,\, x_{N^1} = 0\right\},$$
$$(D_{\leftrightarrow}^{\Delta c}) \;\; \max\left\{b^T y \;:\; A^T y + s = -\Delta c,\, s_{B^1} \geq 0\right\},$$

one has

Theorem 13 *The optimal partition for the pair of dual problems $(P_{\leftrightarrow}^{\Delta c})$ and $(D_{\leftrightarrow}^{\Delta c})$ is just $(\bar{B}, \bar{N})$. Furthermore, $\bar{x}$ is the central solution of $(P_{\leftrightarrow}^{\Delta c})$.*

Using these results one easily derives that

Corollary 4 $(\Delta c)^T(\bar{x} - x^0) < 0$ *and* $(\Delta c)^T(x^1 - \bar{x}) < 0$.

Yet we turn to the case that the optimal partition $\bar{\pi} = (\bar{B}, \bar{N})$ associated to a linearity interval is given. This is the contents of the last two results. It is convenient to assume that the vector c^0 belongs to the linearity interval under consideration, and that the surrounding breakpoints, if they exist, occur at $\alpha = -1$ and $\alpha = 1$ respectively. We consider the following pair of dual problems.

$$(P_{\rightarrow}^{\Delta c}) \;\; \max\{\alpha \;:\; A^T y + s = c(\alpha),\, s_{\bar{B}} = 0,\, s_{\bar{N}} \geq 0\},$$
$$(D_{\rightarrow}^{\Delta c}) \;\; \min\{(c^0)^T x \;:\; Ax = 0,\, (\Delta c)^T x = -1,\, x_{\bar{N}} \geq 0\}.$$

From the discussion in the previous section it will be clear how to define in a natural way the notions of strictly complementary solution and optimal partition for these problems. We now may state

Theorem 14 *The optimal partition for the pair of dual problems $(P_{\rightarrow}^{\Delta c})$ and $(D_{\rightarrow}^{\Delta c})$ is just $\pi(1)$. Furthermore, x^1 is the central solution of $(D_{\rightarrow}^{\Delta c})$.*

A similar result can be obtained for the pair of dual linear programming problems given by:

$$(P_{\leftarrow}^{\Delta c}) \;\; \min\{\alpha \;:\; A^T y + s = c(\alpha),\, s_{\bar{B}} = 0,\, s_{\bar{N}} \geq 0\},$$
$$(D_{\leftarrow}^{\Delta c}) \;\; \min\{(c^0)^T x \;:\; Ax = 0,\, (\Delta c)^T x = 1,\, x_{\bar{N}} \geq 0\}.$$

The following theorem will be no surprise.

Theorem 15 *The optimal partition for the pair of dual problems $(P_{\leftarrow}^{\Delta c})$ and $(D_{\leftarrow}^{\Delta b})$ is just $\pi(-1)$. Furthermore, x^{-1} is the central solution of $(D_{\leftarrow}^{\Delta b})$.* □

3.4 Sensitivity analysis

As stated in the introduction to this section, an important application of parametric analysis is sensitivity analysis. In practice the resulting information can be of tremendous importance. Parameter values may just be estimates; questions of the type "What if ..." are frequently encountered; and implementation of a specific solution may be difficult. Sensitivity analysis serves as a tool for obtaining information about the bottlenecks and spaces of freedom in the problem.

In standard sensitivity analysis the coefficients in the objective function and the right hand side vector are varied one at a time. So $\Delta b = e_i$ and $\Delta c = e_j$, where e_i and e_j stand for the i-th and j-th unit vector respectively. One is interested in the local behavior of the value function. This means that one would like to know the rate at which the optimal value changes as a coefficient changes (shadow prices, dual prices), and also the interval on which this rate of change is constant. Recall from the previous sections that the optimal partition exactly characterizes the linearity intervals of the value function and that slopes and intervals can be found by solving auxiliary linear programming problems.

In contrast to the optimal partition which is always unique, in degenerate linear programming problems optimal bases need not be unique, and also there may be several optimal solutions. Degeneracy is not a hypothetical situation; to the contrary, almost *any* linear programming problem arising from a practical problem is degenerate. In the next example we illustrate how this can effect the sensitivity information as given by commercial linear programming packages based on the Simplex method (the example is taken from [15]).

Example 9 We consider a problem of transporting goods from three suppliers to three markets at minimal cost. Each supplier can serve each of the markets at a transportation cost of 1 per unit good. The capacity of the suppliers is equal to 2, 6 and 5 units respectively. The markets each require at least 3 units. In Figure 12 we depict the situation. Here the circles (nodes) to the left represent the suppliers and the circles to the right the markets. From each supplier to each market there is an arrow (arc), which represents a possible transportation link. Note that this transportation problem (unlike the one in Figure 7) is unbalanced, which means that total demand does not equal total supply.

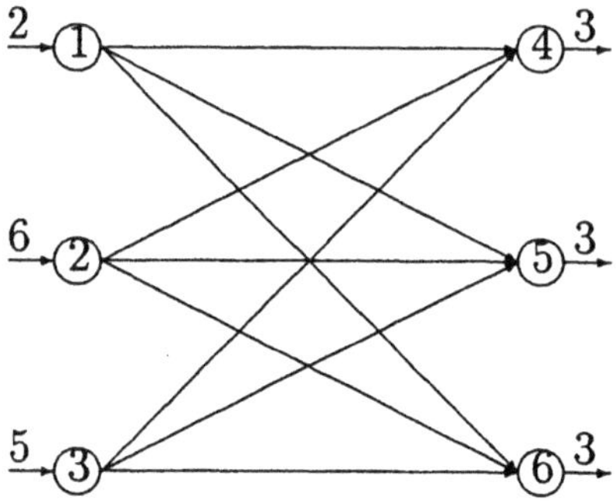

Figure 12: transportation problem.

We can formulate this problem as a linear programming problem by using the variables

x_{ij} : the amount of units transported from supplier i to market j,
s_i : excess supply at supplier i,
d_j : shortage demand at market j,

where i runs from 1 to 3 and j runs from 4 to 6. The linear programming formulation is then

$$
\begin{array}{ll}
\min & \sum_{i=1}^{3} \sum_{j=4}^{6} x_{ij} \\
\text{subject to} & x_{14} + x_{15} + x_{16} + s_1 = 2 \\
& x_{24} + x_{25} + x_{26} + s_2 = 6 \\
& x_{34} + x_{35} + x_{36} + s_3 = 5 \\
& x_{14} + x_{24} + x_{34} - d_1 = 3 \\
& x_{15} + x_{25} + x_{35} - d_2 = 3 \\
& x_{16} + x_{26} + x_{36} - d_3 = 3 \\
& x_{ij},\ s_i,\ d_j \geq 0, \quad i = 1,2,3,\ j = 4,5,6.
\end{array}
$$

We solved this problem with five commercially available Simplex–based linear programming packages and asked for sensitivity information.

The results are summarized in Tables 1 and 2. For completeness we put the correct information obtained by using our interior point implementation in the last lines. In the case of the cost coefficients, which are all equal to 1 (see Table 2), the analysis gives that these values correspond to breakpoints

LP-package	Optimal solution									Shadow prices					
	x_{14}	x_{15}	x_{16}	x_{24}	x_{25}	x_{26}	x_{34}	x_{35}	x_{36}	(1)	(2)	(3)	(4)	(5)	(6)
CPLEX	0	2	0	2	1	3	1	0	0	0	0	0	1	1	1
LINDO	2	0	0	0	0	2	1	3	1	0	0	0	1	1	1
PC-PROG	0	0	0	0	3	1	3	0	2	0	0	0	1	1	1
XMP	0	0	2	3	3	0	0	0	1	0	0	0	1	1	1
OSL	0	2	0	2	1	3	1	0	0	0	0	0	1	1	1
IPM	0.5	0.5	0.5	1.4	1.4	1.4	1.1	1.1	1.1	0	0	0	1	1	1

Table 1: Optimal solution and shadow prices in Example 3.5

of the value function. The last line gives the left and the right shadow price for each cost coefficient.

Even in this simple example, it is striking to notice that no two packages yield the same result. Of course the optimal value given is the same. However, the five simplex–based packages give four different optimal solutions and they all return different ranges for the coefficients in the objective function. Note that although *CPLEX* and *OSL* give the same optimal solution, the ranges for the objective coefficients are not all the same. Even though the shadow prices are equal in all packages, the ranges for the right hand side coefficients are different. □

4 A primal-dual interior point algorithm

The literature on interior point methods flourishes with a variety of algorithms. Not all of them are efficient. It is now considered that the best algorithms for solving real-life large scale linear programs are the so-called primal-dual methods. Extensive computational experience [25], [26], [27], [31], [34] indicates that these methods compete very favorably with the Simplex method on large scale problems. The algorithm we shall describe here is a primal–dual version of the general logarithmic barrier algorithm of Fiacco and McCormick [7]. It could be viewed as the backbone of a state of the art code, though one would have to add many tricks and modifications to make it competitive. Some of these implementation issues will be discussed in the next section. The purpose of this section is the get

LP-package	COST-ranges								
	c_{14}	c_{15}	c_{16}	c_{24}	c_{25}	c_{26}	c_{34}	c_{35}	c_{36}
CPLEX	[1,∞)	(-∞,1]	[1,∞)	[1,1]	[1,1]	[0,1]	[1,1]	[1,∞)	[1,∞)
LINDO	(−∞,1]	[1,∞)	[1,∞)	[1,∞)	[1,∞)	[1,1]	[1,1]	[0,1]	[1,1]
PC-PROG	[1,∞)	[1,∞)	[1,∞)	[1,∞)	[0,1]	[1,1]	[0,1]	[1,∞)	[1,1]
XMP			(-∞,1]	[0,1]	[0,1]	[1,1]			[1,1]
OSL	[1,∞)	[1,1]	[1,∞)	[1,1]	[1,1]	[1,1]	[1,1]	[1,∞)	[1,∞)
IPM:ranges	[1,1]	[1,1]	[1,1]	[1,1]	[1,1]	[1,1]	[1,1]	[1,1]	[1,1]
IPM:prices	[2,0]	[2,0]	[2,0]	[3,0]	[3,0]	[3,0]	[3,0]	[3,0]	[3,0]

LP-package	RHS-ranges					
	b_1	b_2	b_3	b_4	b_5	b_6
CPLEX	[0,3]	[4,7]	[1,∞)	[2,7]	[2,5]	[2,5]
LINDO	[1,3]	[2,∞)	[4,7]	[2,4]	[1,4]	[1,7]
PC-PROG	[0,∞)	[4,∞)	[3,6]	[2,5]	[0,5]	[2,5]
XMP	[0,3]	[6,7]	[1,∞)	[2,3]	[2,3]	[2,7]
OSL	[0,3]	[4,7]	$(-\infty, \infty)$	[2,7]	[2,5]	[2,5]
IPM	[0,∞)	[2,∞)	[1,∞)	[0,7]	[0,7]	[0,7]

Table 2: Ranges in Example 3.5

the reader acquainted with some of the basic ideas underlying the most advanced implementations.

Let us start with the presentation of a conceptual (but not implementable) algorithm. The idea behind this algorithm is to exploit the explicit knowledge of the value of the duality gap at points along the central path. Recall that this gap equals $n\mu$ at the μ−center $x(\mu), y(\mu), s(\mu)$. In the algorithm it is (incorrectly) presupposed that for given μ the corresponding μ−center can be easily calculated exactly. Having done this, we reduce μ with a constant factor, and we calculate the new center, and so on, until the duality gap becomes smaller than some prescribed value.

Conceptual Algorithm

Parameters

$\varepsilon > 0$ is the accuracy parameter;
$1 - \theta$ is the reduction factor for μ $(0 < \theta < 1)$;

Input

μ^0 is the initial value of the barrier parameter;
$x^0 := x(\mu^0)$ and $s^0 := s(\mu^0)$ is the initial pair of μ–centers associated with μ^0.

begin
  $x := x^0, s := s^0, \mu := \mu^0$;
  while $x^T s > \varepsilon$ **do**
    $\mu := (1 - \theta)\mu$;
    compute $x := x(\mu)$ and $s := s(\mu)$.
  end
end.

The conceptual algorithm converges in a well-defined number of steps, since each iteration reduces the duality gap by a fixed multiplicative factor $(1-\theta)$. Given the initial value $n\mu^0$ of the duality gap and its target value ε, one easily computes the required number of steps. It is common to use a simple upper bound on this number. We formalize it as a straightforward lemma.

Lemma 10 *The conceptual algorithm converges in K steps, where K is the smallest integer such that*

$$K \geq \frac{1}{\theta} \ln \frac{n\mu^0}{\varepsilon}.$$

Proof: Along the central path, one has

$$x(\mu)^T s(\mu) = n\mu.$$

After k steps, one has

$$n\mu = n(1-\theta)^k \mu^0 > \varepsilon.$$

Taking the logarithm on both sides, and using that $\ln(1-\theta) \leq -\theta$, one gets

$$-\theta k \geq k \ln(1-\theta) > \ln \frac{\varepsilon}{n\mu^0}.$$

Equivalently

$$k < \frac{1}{\theta} \ln \frac{n\mu^0}{\varepsilon}.$$

This proves the lemma by contradiction. □

Some ingredients are necessary to make the conceptual algorithm implementable. To start with, one needs a nonlinear programming algorithm to compute μ–centers. This algorithm will not compute exact solutions: so, we also need to modify the conceptual algorithm to deal with approximate centers. A nonlinear programming algorithm usually involves a search direction and a step length along that direction. If the search direction in the primal space, lies in the null space of A, primal feasibility will be maintained throughout. Similarly, if the search direction in the space of the dual slacks lies in the range space of A^T, then dual feasibility will be maintained throughout.

Prior to discussing those issues in more detail, we recall from Section 2.6.1 that without loss of generality it may be asssumed that a pair of primal and dual interior feasible (i.e., positive) solutions is available at the start of the algorithm. We also suppose that Assumption 1 holds. So there exists a one-to-one correspondence between the y and the s variables in a dual feasible solution. We recall our convention to omit then the free variable y in the definition of a dual feasible solution (y, s).

4.1 Search direction

Let us recall from Definition 2.1 that the μ–center is defined as the unique solution of the system of equations

$$\begin{aligned} Ax &= b, \quad x \geq 0 \\ A^T y + s &= c, \quad s \geq 0 \\ Xs &= \mu e. \end{aligned} \tag{8}$$

The first two equations imply primal and dual feasibility. The last equation is nonlinear: it characterizes a central point. Let (d_x, d_s) be the primal-dual search direction. As pointed out in the previous section, we may assume that the first two equations are satisfied initially. We certainly want to maintain feasibility throughout the algorithm, so we impose $Ad_x = 0$ and $A^T d_y + d_s = 0$. In contrast the last equation is almost surely not satisfied.

The method of choice is to linearize the last equation and to take d_x and d_s to solve the first order approximation. This is just Newton's method for solving the system of equations (8). Formally, the Newton search direction is defined by the system of equations

$$\begin{aligned} Ad_x &= 0, \quad x \geq 0 \\ A^T d_y + d_s &= 0, \quad s \geq 0 \\ Sd_x + Xd_s &= \mu e - Xs. \end{aligned} \tag{9}$$

The above system can be solved by direct substitution. We shall give the explicit formula in the next section. We also have that d_y appears explicitly, though by Assumption 1, the variable y is implicitly defined by the second equation. It is also possible to eliminate d_y through the second equation, but the extensive format of (9) is handier.

To measure the progress of the algorithm, we use the primal–dual logarithmic barrier function

$$f_\mu(x, s) = \frac{x^T s}{\mu} - \sum_{i=1}^{n} \ln x_i s_i. \tag{10}$$

The domain of f_μ is the set of positive points $x > 0$, $s > 0$. For the ease of the presentation we shall drop the index μ in f_μ. The first order optimality conditions of the minimization of f under the set of primal and dual constraints are precisely the system of equations (8). Consequently, the problem of computing the $\mu-$ centers boils down to the minimization of f. Let us stress that Newton's direction with respect to (8) differs from the constrained Newton direction associated with the minimization of f.

Nonetheless the direction we defined by (9) nicely relates to f. Let us compute the directional derivative of f in that direction. We get

$$Df = \nabla_x f^T d_x + \nabla_s f^T d_s = (\frac{s}{\mu} - x^{-1})^T d_x + (\frac{x}{\mu} - s^{-1})^T d_s.$$

Simple manipulations yields

$$Df = -\|u - u^{-1}\|^2, \tag{11}$$

where u is the vector defined by

$$u = (\frac{Xs}{\mu})^{\frac{1}{2}}.$$

This proves that Newton's direction associated with (8) is a descent direction for f.

4.2 Proximity measure

The quantity $\|u - u^{-1}\|$ plays a crucial role. It is clear that $Df = 0$ if and only if $u - u^{-1} = 0$, i.e., $u = e$. In other words, the directional derivative vanishes if and only if each complementary pair $x_i s_i$ is equal to μ, i.e., (x, s) is the μ–center $(x(\mu), s(\mu))$. This suggests to measure the proximity to the μ–center $(x(\mu), s(\mu))$ by

$$\delta(x, s; \mu) := \|u - u^{-1}\|.$$

Let us state a simple result that bounds the complementary pairs $x_i s_i$ by a function of δ.

Lemma 11 *Let* $\delta := \|u - u^{-1}\|$. *Then*

$$\frac{1}{\rho(\delta)^2} \leq \frac{x_i s_i}{\mu} \leq \rho(\delta)^2, \quad i = 1, \ldots n,$$

with

$$\rho(\delta) = \frac{\delta + \sqrt{\delta^2 + 4}}{2}.$$

Proof: For each $i, 1 \leq i \leq n$, we have

$$-\delta \leq u_i^{-1} - u_i \leq \delta,$$

Since u_i is positive, this is equivalent to

$$-u_i\delta \leq 1 - u_i^2 \leq u_i\delta,$$

or

$$u_i^2 - u_i\delta - 1 \leq 0 \leq u_i^2 + u_i\delta - 1.$$

One easily verifies that this is equivalent to

$$\rho(\delta)^{-1} \leq u_i \leq \rho(\delta).$$

Squaring the terms proves the lemma. □

4.3 Decrease of the barrier function

The nice result about the directional derivative of the barrier function suggests the following algorithm for computing the pair of μ–centers (or an approximation of it): the basic iteration consists of a damped step along the

Newton direction, with the condition that the new iterate remains strictly positive and the barrier function is decreased by a sufficient amount. The truly remarkable property of the logarithmic barrier method for linear programming is that it is possible to bound that decrease. More precisely, one can define a default step size that ensures that the barrier function decreases by an amount that only depends on the proximity measure. Moreover this dependence is monotone: the larger the distance, the greater is the guaranteed decrease. This property is fundamental in the convergence analysis. If the parameter μ remains unchanged, each step decreases the barrier function by a significant amount, as long as the proximity measure is above a specified level. Since, for a fixed μ, the barrier function is bounded from below, one readily gets an upper estimate for the number of steps that are required until the proximity measure falls below the specified level.

We state below the fundamental lemma that defines the default value for the step length and gives the estimate for the barrier function decrease. We omit its proof. (For a simple proof of it, we refer to [16].)

Lemma 12 *Let* $\delta := \|u - u^{-1}\| > 0$ *and let*

$$\alpha := \frac{\delta^2}{\omega(\omega + \delta^2)},$$

with $\omega := \sqrt{\|X^{-1}d_s\|^2 + \|S^{-1}d_x\|^2}$. *Then* $\omega > 0$ *and* $x + \alpha d_x > 0$ *and* $s + \alpha d_s > 0$. *Moreover*

$$\Delta f(\alpha) := f(x + \alpha d_x, s + \alpha d_s; \mu) - f_\mu(x, s) \leq -\tau + \ln(1 + \tau) < 0,$$

where $\tau := \frac{\delta}{\rho(\delta)}$.

Let us check now that the earlier claim that the bound on the increase of $-f$ (i.e., the decrease of f) is an increasing function of δ. It suffices to compute derivatives. We have

$$\frac{\mathrm{d}(\tau - \ln(1 + \tau))}{\mathrm{d}\delta} = (1 - \frac{1}{1 + \tau})\frac{\mathrm{d}\tau}{\mathrm{d}\delta} = \frac{\tau}{1 + \tau}\frac{8}{(\delta + \sqrt{\delta^2 + 4})^2\sqrt{\delta^2 + 4}} > 0.$$

The claim is proved.

4.4 Algorithm

We can now formulate the primal-dual algorithm.

Primal–Dual Barrier Algorithm

Parameters

$\varepsilon > 0$ is the accuracy parameter;
$1 - \theta$ is the reduction factor for μ; $(0 < \theta < 1.)$
$\xi > 0$ is the proximity parameter;
$\tau := \frac{2\xi}{\xi + \sqrt{4+\xi^2}}$.

Input

(x^0, s^0) is the initial pair of interior feasible solutions;
$\mu^0 := (x^0)^T s^0 / n$ is the initial value for the barrier parameter;
x^0, s^0, and μ^0 satisfy $\delta(x^0, s^0; \mu^0) \leq \xi$;

begin
  $x := x^0, s := s^0, \mu := \mu^0$;
  while $x^T s > \varepsilon$ **do**
    $\mu := (1 - \theta)\mu$;
    while $\delta(x, s; \mu) > \xi$ **do**
      find $\alpha > 0$ such that
        $f(x + \alpha d_x, s + \alpha d_s; \mu) \leq f_\mu(x, s) - \tau + \ln(1 + \tau)$;
      $x := x + \alpha d_x$;
      $s := s + \alpha d_s$;
    end
  end
end.

In the above statement of the algorithm the value of the proximity parameter ξ is user free. As we shall see it in the next section, it is perfectly admissible to take a large value for ξ, hence to update the μ parameter while the iterate is still quite far from the central path. This is much at odds with the conceptual algorithm. Yet, it is still possible to derive an upper bound for the number of iterations, as it will be shown in the next section.

4.5 Convergence analysis

We shall name "inner iterations" the steps of the algorithm that are taken between two updates of the parameter μ. In contrast, we shall name "outer iterations" the updates of μ. We start by proving a result very similar to Lemma 10. It follows from a simple bound on the duality gap. By Lemma 11 each complementary product $x_i s_i$ is bounded above by $\mu\rho(\delta)^2$, where $\delta := \|u - u^{-1}\|$. Hence

$$\frac{x^T s}{\mu} \leq n\rho(\delta)^2.$$

This inequality is weakening of the relation $x^T s = n\mu$ that holds along the central path. (It is possible to obtain a sharper bound, see [16].) At the end of the kth outer iteration, $\|u - u^{-1}\| \leq \xi$ and $\mu = (1-\theta)^k \mu^0$. Thus

$$x^T s \leq n\rho(\xi)^2 (1-\theta)^k \mu^0.$$

By the same argument as in Lemma 10, we immediately get a bound on the number of outer iterations.

Theorem 16 *The total number of outer iterations is bounded by the smallest integer K such that*

$$K \geq \frac{1}{\theta} \ln \frac{n\mu^0 \rho(\xi)^2}{\varepsilon}.$$

To complete the convergence analysis, it remains to bound the total number of inner iterations per outer iteration. Since by Lemma 12 the barrier function decreases at each iteration by at least $-\xi + \ln(1+\xi)$, we just have to bound the difference between the current value of the barrier function and its absolute minimum (for a fixed μ). It is easy to bound the barrier function from below (for a fixed μ) since the minimum is achieved at the analytic center $(x(\mu), s(\mu))$ satisfying $x_i(\mu)s_i(\mu) = \mu$. We get that this minimum value is

$$\bar{f} = \frac{x(\mu)^T s(\mu)}{\mu} - \sum_{i=1}^{n} \ln x_i(\mu) s_i(\mu) = n - n \ln \mu. \tag{12}$$

The upper bound on f is obtained through Lemma 11. This lemma bounds from below and from above each complementary product that enters the definition of f. Hence, we get an upper bound on the value of the barrier function at the begining of an outer iteration. Let us put all these elements together into a second convergence theorem.

Theorem 17 *Let* $\rho(\xi)$ *be defined as in Lemma 11 and let* $\tau = \frac{\xi}{\rho(\xi)}$. *The total number of inner iterations per outer iteration is bounded by*

$$k \leq \kappa(\theta, \xi)n,$$

where

$$\kappa(\theta, \xi) = \frac{1}{\tau - \ln(1+\tau)} \left(\frac{\rho(\xi)^2}{1-\theta} - 1 + \ln(1-\theta) + 2\ln\rho(\xi) \right).$$

Proof: Let μ be the value of the barrier parameter and let f be the value of the barrier function at the end of an outer iteration. Let $\mu^+ = (1-\theta)\mu$ and f^+ be the corresponding values at the beginning of the next outer iteration. Finally let $\bar{f}^+$ be the absolute minimum of the barrier function for the value μ^+ of the barrier parameter.

At the end of an outer iteration the proximity measure verifies $\|u - u^{-1}\| \leq \xi$. Hence by Lemma 11

$$\frac{1}{\rho(\xi)^2} \leq \frac{x_i s_i}{\mu} \leq \rho(\xi)^2.$$

We can apply these inequalities to bound f^+:

$$\begin{aligned} f^+ &= \frac{x^T s}{\mu^+} - \sum_{i=1}^{n} \ln x_i s_i \\ &\leq \frac{n\rho(\xi)^2}{1-\theta} - n \ln \frac{\mu^+}{\rho(\xi)^2(1-\theta)}. \end{aligned} \tag{13}$$

Using the value of $\bar{f}^+$ as computed above in (12), we get the upper bound

$$f^+ - \bar{f}^+ \leq n(\frac{\rho(\xi)^2}{1-\theta} - 1) + n\ln\left(\rho(\xi)^2(1-\theta)\right). \tag{14}$$

On the other hand, at each inner iteration, one has $\delta := \|u - u^{-1}\| > \xi$; thus $\frac{\delta}{\rho(\delta)} > \frac{\xi}{\rho(\xi)}$. Let $\tau := \frac{\xi}{\rho(\xi)}$. By Lemma 12, the total decrease of the barrier function after k inner iterations is at least

$$f^k - f^0 \leq k(-\tau + \ln(1+\tau)), \tag{15}$$

where f^k is the value of the barrier function at the kth inner iteration and $f^0 = f^+$. Using the two bounds (14) and (15) we get

$$k(\tau - \ln(1+\tau)) \leq n \left(\frac{\rho(\xi)^2}{1-\theta} - 1 + \ln(1-\theta) + 2\ln\rho(\xi) \right).$$

This proves the theorem. □

The strong result in Theorem 17 is that the coefficient $\kappa(\theta, \xi)$ depends solely on parameters that are freely chosen by the user. Thus the bound on the number of inner iterations per (outer) iteration grows linearly with n.

Theorems 16 and 17 can be put together to bound the total number of inner iterations until convergence.

Theorem 18 *The total number of iterations of the primal–dual barrier algorithm is bounded by*

$$K \leq \frac{n\kappa(\theta, \xi)}{\theta} \ln \frac{n\mu^0 \rho(\xi)^2}{\varepsilon}.$$

The above theorem relaxes the proximity condition that appeared in earlier results of [12] and [42] for the pure primal barrier algorithm. In these papers, the update of μ is to be done only when the iterate is very close to the central path, a condition that limits the choice of ξ to small enough values (say, smaller than 1).

4.6 Complexity analysis

We conclude this section by some remarks on the complexity of the algorithm. In the theory of complexity, one seeks to relate the computational effort to the size of the problem. The size L of a problem instance is usually measured as the number of bits that are necessary to encode the complete description of the problem. Two factors determine the total computational effort: the linear algebra operations at each iteration and the total number of iterations. The effort per iteration is dominated by the solution of the system of equations (9) to be solved. The dimension of this system is fixed. The solution involves $O(nm^2)$ elementary operations[1] (assuming $n \geq m$). Thus the complexity result hinges on the bound on the number of iterations.

In Theorem 18 the bound on the number of iterations includes three types of terms: some that are user free; some that depend on the starting and ending conditions, i.e., μ^0 and ε; and n, the number of variables. To perform the complexity analysis, one must relate the two undetermined parameters μ^0 and ε to the size L.

[1]The computational effort is also influenced by the size of the system (9). This size is not fixed, since the terms x and s change as the algorithm proceeds. A complete discussion requires complicated arguments on finite arithmetic computations. They are beyond the scope of this paper.

Let us first take care of the termination criterion. The first issue to be resolved is that the primal–dual algorithm is not an exact algorithm, since it surely ends with a small but nonzero duality gap. This is due to the fact that the iterates always stay in the interior of the feasible region, whereas the optimal solution lies at the boundary of the feasible region. The same situation prevails in the ellipsoid algorithm of Khachiyan [17], the first published algorithm that gave a polynomial bound for linear programming. There, it was proved that if $x^T s < 2^{-O(L)}$, one could construct an exact solution in polynomial time. (Actually, the amount of work is of the same order of complexity as one iteration of the barrier algorithm.) The theoretical value for ε is thus $\varepsilon = 2^{-O(L)}$. See, e.g., [39]. As to the other constant μ^0, it has been proved, see [41] and [37], that by appropriately reformulating the original problem, one can exhibit an initial central pair such that

$$x(\mu^0)^T s(\mu^0) = n\mu^0 \leq 2^{O(L)}.$$

With those results in mind, one immediately gets the complexity theorem:

Theorem 19 *For a fixed, arbitrary θ, the primal-dual barrier algorithm solves any linear programming of size L in $O(nL)$ iterations.*

5 Implementation

The algorithm of the previous section can be quite readily programmed and implemented. There is however one issue that must be solved. It concerns the computation of the starting point, since the theory assumed that an initial positive feasible point is given. This is never the case in practice. Another important issue concerns the solution technique to apply to the system of equations (9).

The requirement of a known initial positive feasible point can be met, as in the Simplex algorithm, via artificial variables with a "big M" cost coefficient in the primal and dual objectives, see [41] and [37]. In Section 2.6.1 we used this technique for a theoretical purpose. This technique was used in the early implementations but the modern codes abandonned it. Primal and dual infeasibilities are incorporated in the computation of the search direction: the resulting algorithm achieves simultaneously feasibility and optimality. Initially this approach was developed on heuristic grounds, much more than from theoretical considerations. It turned out to be very efficient. This is quite typical of actual implementations: theory provides the backbone for the design of the algorithms, but actual implementations do not follow the theory on many key points. However, quite recently this technique has been justified theoretically. See, e.g. [20, 35, 40, 45].

We shall review some of those gimmicks that are the clue to a successful implementation.

5.1 Those tricks that make it work

5.1.1 The linear algebra operations

The difficulty in solving the system of equations (9) mostly lies in the dimension of the problem. For medium size linear programming problem, the theoretical number of elementary operations to be performed to compute one solution of (9) is already large enough. Since this work must be done at least once at each iteration, the overall computing cost becomes prohibitive. The same difficulty arises in the Simplex algorithm. Fortunately enough, practical problems have a sparse data structure. In the implementations of the Simplex algorithm sparsity is exploited to cut down drastically the computational effort. A similar undeavor has been undertaken for interior point methods. The impact on computations is dramatic.

There are several ways of exploiting the sparsity of (9). One is to use a conjugate gradient scheme. This is a very efficient approach, provided one uses a good preconditioner. Unfortunately, it is not easy to find a good preconditioner at a low computational cost, that works for any type of problem. Despite some good results on specific problems, the conjugate gradient method is not the method of choice for the general–purpose state of the art codes. The other approach consists in computing the elements of the search direction directly from the explicit formulation of the solution of (9). This formulation is obtained by simple block elimination and the result is given by

$$\begin{aligned} d_y &= -(AXS^{-1}A^T)^{-1}(\mu AS^{-1}e - b), \\ d_s &= -A^T d_y = A^T(AXS^{-1}A^T)^{-1}(\mu AS^{-1}e - b), \\ d_x &= \mu s^{-1} - x - XS^{-1}d_s. \end{aligned} \tag{16}$$

The difficulty lies in solving the first equation. Once it is solved, the last two equations are obtained by simple matrix–vector operations. The product matrix $AXS^{-1}A^T$ is positive definite if A is regular. This is generally the case after preprocessing of the original data and the elimination of redundant rows. The product matrix also inherits the sparsity of A, though some problem formulations may lead to fill–in when forming the matrix. The most popular method for solving the first equation of (16) is to use a Cholesky factorization

$$AXS^{-1}A^T = LL^T.$$

The density of the Cholesky factor depends very much on the ordering of the rows of A. An extremely sparse matrix $AXS^{-1}A^T$ can produce a fully dense Cholesky factor L, for instance if the first column of $AXS^{-1}A^T$ is dense. However, it is possible to perform a simple reordering that puts this dense column in last position: the fill–in effect on the Cholesky factor is dramatically reduced. Finding the optimal ordering that minimizes the fill–in of the Cholesky factor is a very difficult problem that would cost far too much effort to solve . Rather one uses some heuristic rules, such as the *minimum degree* rule or the *minimum local fill–in* rule [6, 9, 23]. This reordering work has to be done only once, since the fill–in depends on the sparsity structure of $AXS^{-1}A^T$ and not on the numerical values of the entries.

5.1.2 Preprocessing

It is common that people who build models create linear programming problems with many redundant constraints and also dominated columns. In large problems, the equations are generated through a matrix generator by means of statements. Those statements must be general enough to cover all the relevant instances of a given family of problems. For a specific instance in the family, it is often the case that some of the statements are redundant, or that some of the variables are not necessary. The user is more concerned by producing a consistent model rather than removing all redundancy.

For this reason, commercial codes are all endowed with preprocessors that aim at eliminating redundancy, duplication and domination. The reduction in the size of the problem may be spectacular. Usually, preprocessing ensures that the constraint matrix has full row rank.

5.1.3 Step length

The theoretical algorithm provides a default value for the step length. This default value is very small. It is also suggested that a line search be performed to achieve an approximate reduction of the barrier function. One of the surprising results stemming out of practical experience is that this optimum occurs extremely close to the boundary. Taking a fixed fraction γ, close to 1, of the maximal step to the boundary, turns out to be about as efficient iteration–wise, and less time consuming. The figure that is most commonly used is .9995.

5.1.4 Barrier parameter

In the theoretical algorithm the barrier parameter is kept fixed until the proximity measure falls below a certain threshold. A more appealing strategy consists of adapting the parameter value μ at each iteration. This is based on the following analysis.

Let (d_x, d_s) solve the system of equations (9). Then

$$(x + \alpha d_x)^T(s + \alpha d_s) = x^T s + \alpha(x^T d_s + s^T d_x) = (1 - \alpha)x^T s + \alpha n\mu.$$

The term in α^2 vanishes since d_x and d_s are orthogonal. Let us consider two critical values for μ. First, if $\mu = 0$, then $(x + \alpha d_x)^T(s + \alpha d_s) = (1 - \alpha)x^T s$. This choice gives the largest duality gap reduction for a step of length α. The corresponding direction is known as the primal–dual affine scaling direction. The other choice for μ is $\mu = \frac{x^T s}{n}$. In that case, $(x + \alpha d_x)^T(s + \alpha d_s) = x^T s$: the duality gap remains constant. A step in that direction has a pure centering effect. A search direction based on a given $\mu > 0$ combines the affine scaling and the centering directions. If μ is quite smaller than $\frac{x^T s}{n}$, a situation that is likely to occur just after an update, the affine scaling direction dominates. If μ is close to $\frac{x^T s}{n}$, a situation that occurs just before an update, the centering direction dominates. A better strategy in practice is to balance these two directions in a same way at each iteration. This can be done by taking $\mu = \beta\frac{x^T s}{n}$, for some fixed $0 < \beta < 1$.

There is another interpretation of the affine scaling and centering directions. Close to the central path, the affine scaling direction is nearly tangential to the path: it is a good descent direction for the duality gap but it drifts away from the path. The centering direction keeps the duality gap unchanged but brings back the point close to the central path. A reasonable strategy is to alternate these two steps, a predictor step followed by a corrector step. In [44] the author suggested the following rule to monitor the switch. Let $\beta < 1$ be some fixed parameter. Perform a corrector step whenever

$$\min_i\{x_i s_i\} < \beta \frac{x^T s}{n}.$$

This rule gives good results with $\beta = 1 - \gamma$.

The drawback of this implementation strategy is that each corrector step requires its own factorization. A somewhat better approach consists of anticipating the corrector step. It is then combined to the predictor step. The computations require solving systems with the same matrix $AXS^{-1}A^T$, thus saving one Cholesky factorization. This strategy has been proposed first by [34], in a slightly different manner. It gives the best results in practice.

5.1.5 Infeasible start

The first enhancement aims at getting rid of the "big M" procedure. It is not very desirable from a numerical point of view to introduce the large "big M" coefficient costs; nor is it wise to insert new artificial variables with associated dense columns and rows. Practice has shown that it is possible to go without it. Let us discuss briefly a much more elegant way of dealing with infeasible start.

Recall that we introduced the search direction as the Newton direction associated with a certain system of equations. In the theoretical algorithm, we assumed that the primal and dual constraints in (8) were satisfied. Only the last equation pertaining to the complementary slackness condition is violated. Let us call "complementary slackness residuals" the vector of violation of the complementary slackness equations. The Newton direction is obtained by linearizing this equation and by taking a step that reduces the residual for the linearized equation, while preserving feasibility in the primal and the dual constraints.

We may wonder: if Newton's method is good at estimating the step that annihilates the complementary slackness residual, why not use the same method for annihilating possible residuals on the primal and the dual constraints? This common sense remark is well supported by computational experience: the infeasibilities are handled by the method, in the very same way as is the complementary equation.

The formulas for the search direction are modified as follows. Let r be the vector of residuals:

$$r = \begin{pmatrix} r_1 \\ r_2 \\ r_3 \end{pmatrix} = \begin{pmatrix} b - Ax \\ c - s - A^T y \\ \mu e - Xs \end{pmatrix}.$$

The direction is given by

$$\begin{aligned} d_y &= -(AXS^{-1}A^T)^{-1}(r_1 + AS^{-1}Xr_2 - AS^{-1}r_3), \\ d_s &= r_2 - A^T d_y, \\ d_x &= S^{-1}r_3 - XS^{-1}d_s = \mu s^{-1} - x - XS^{-1}d_s. \end{aligned} \tag{17}$$

In the case of the primal-dual algorithm, this approach was first introduced experimentally by [24]. Working with infeasible points ruins most of the arguments that sustain the analysis of the theoretical algorithm. In particular the orthogonality of d_x and d_s is lost and the quantity $\|u - u^{-1}\|$ cannot properly be interpreted as a proximity indicator. Those difficulties are more theoretical than practical. In recent contributions, Kojima et al. [20], Mizuno [35] and Zhang [45] gave variants of the practical algorithm with proven rate of convergence. It is interesting to point out that this result was obtained fairly early for the primal projective algorithm in [5].

5.1.6 Starting point

The choice of an infeasible, but positive, starting point is probably one of the most critical implementation issues. In general the algorithm can handle a very arbitrary initial starting point such as $x^0 = \kappa e$, for some arbitrary $\kappa > 0$. But on some occasions, this may turn out to be a very bad choice. A sensible alternative is the following. Let x be the least squares solution of $Ax = b$, i.e.,

$$x := A^T(AA^T)^{-1}b.$$

This solution usually has negative terms. It may also have terms that are very small in absolute value. A safe policy is to have no term less than $\sigma = \frac{\|b\|}{n}$. (Different lower bounds of the same type are conceivable.) The following rule works well:

$$x_i = \begin{cases} |x_i| & \text{if} \quad |x_i| \geq \sigma \\ \sigma & \text{if} \quad |x_i| < \sigma. \end{cases}$$

The choice for the dual is often less critical. A value $y := 0$ for the free variable is acceptable. The dual slack s is chosen in a way similar to x.

$$s_i = \begin{cases} |c_i| & \text{if} \quad |c_i| \geq \tau \\ \tau & \text{if} \quad |c_i| < \tau, \end{cases}$$

where $\tau = \frac{\|c\|}{n}$.

5.1.7 Bounds on variables and free variables

Practical problems are very seldom formulated with a constraint set of the form $\{x : Ax = b, x \geq 0\}$. There may be free variables, inequality constraints and simple bounding constraints. It is not difficult to recast the problem into the appropriate format. For instance, a free variable $\tilde{x}$ can be split into two nonnegative variables x^+ and x^-, $\tilde{x} = x^+ - x^-$. Slack variables are added to inequality constraints to make them equality. Finally the bounding constraints can be incorporated into the regular constraints. This increases the size of the problem. In particular, bounds on all the variables increase the dimension of the constraint matrix from $m \times n$ to $(m+n) \times n$. Is the impact on the computational effort per iteration as dramatic?

Let h be the bound on the x variables and let z be the associated slack

$$x + z = h, \quad x \geq 0, \, z \geq 0.$$

Let $t \geq 0$ be the dual variable associated to z. In a straight implementation of the formulas we have to form the product matrix

$$\begin{pmatrix} A & 0 \\ I & I \end{pmatrix} \begin{pmatrix} XS^{-1} & 0 \\ 0 & ZT^{-1} \end{pmatrix} \begin{pmatrix} A^T & I \\ 0 & I \end{pmatrix} = \begin{pmatrix} AXS^{-1}A^T & AXS^{-1} \\ XS^{-1}A^T & XS^{-1} + ZT^{-1} \end{pmatrix}.$$

We claim that the computational effort for solving a system of equations with the above matrix is of the same order of magnitude as solving a system involving the matrix $AXS^{-1}A^T$ only. Using the inverse partitioning formula, we get for the inverse of the right hand side

$$\begin{pmatrix} (A\Omega^{-1}A^T)^{-1} & -(A\Omega^{-1}A^T)^{-1}AXS^{-1}\Sigma^{-1} \\ -XS^{-1}\Sigma^{-1}A^T(A\Omega^{-1}A^T)^{-1} & \Sigma^{-1}XS^{-1}A^T(A\Omega^{-1}A^T)^{-1}AXS^{-1}\Sigma^{-1} + \Sigma^{-1} \end{pmatrix},$$

where

$$\begin{aligned} \Sigma &= XS^{-1} + ZT^{-1} \\ \Omega &= SX^{-1} + TZ^{-1}. \end{aligned}$$

Σ and Ω are diagonal matrices with positive entries. As far as computations are concerned, the computation of the search direction amounts to solving equations with the square matrix $A\Omega^{-1}A^T$. The diagonal matrix Ω^{-1} replaces the matrix XS^{-1} that appears in the standard formula. The claim is proved.

As to the free variables, we readily see that the splitting into two nonnegative components has no effect on the size of the $AXS^{-1}A^T$ matrix, nor on its sparsity structure. However the behavior of the algorithm may be affected: the two components tend to grow out of bound simultaneously, with a constant difference value between them. Lustig et al. [27] recommend to arbitrarily set the lowest of the two components to an arbitrary threshold value and adjust the other so as to keep the difference unchanged.

5.2 Computational experience

Many papers have reported on convincing computational experiences. The more recent one [27] is quite comprehensive: it contains detailed information on implementation issues and results on exceptionally large linear programming test problems. The main conclusion about efficiency, is that the larger the problem, the greater the advantage of interior point methods over the Simplex. This is mainly due to the fact that the total number of iterations grows extremely slowly with the size of the problem.

To get a feeling for this assertion, we reproduce below some of our own numerical results [44]. The test problems were randomly generated. They were small and dense. The algorithm we used is about the same as the one we described earlier; we did not need sparse matrix techniques. The programming language is MATLAB [36]. The problems have 20 constraints only and up to 3200 nonnegative variables. We drew 15 instances of each problem size. The convergence criterion was a maximal relative error in optimality and feasibility of 10^{-6}. The figures in Table 3 illustrate well the behavior of an interior point algorithm. A logarithmic regression analysis yielded the formula

$$\text{iter} = 5.67 + 2.24 \ln n,$$

with a correlation coefficient R=0.76. Although the problems were quite particular (small, dense, randomly generated problems), the results are very much in the line of what has been observed by other researchers.

The other result we would like to report compares two state of the art com-

Iterations	Number of variables					
	100	200	400	800	1600	3200
Maximum	19	21	22	25	30	30
Minimum	13	14	18	17	18	19
Average	16.1	16.9	19.9	20.3	22.6	23.5

Table 3: Varying the number of nonnegative variables.

mercial codes for linear programming: the Simplex code CPLEX release 2.0 and the primal–dual predictor–corrector interior point option of OSL release 2.0. These results were obtained as part of a study [3] on a decomposition approach to stochastic linear programming. The problems were randomly generated. The constraint matrix has the following structure: the first set of rows is dense and it covers all columns; there are from 10 to a 100 of them, depending on the problem size. In addition there are 100 fully dense non-overlapping blocks of equal size, from 3×5 to 39×100. We only report here the results pertaining to the direct approach. (See Table 4.) The tests were run on an IBM RS6000 320H, 25 Mhz, with 32Mb of core memory.

Problem	# rows	#columns	density	OSL(IPM) CPU in s.	CPLEX(SIMPLEX) CPU in s.
1	350	500	2%	12	17
2	850	1500	4%	70	303
3	1250	2800	3%	172	1054
4	4000	10000	1%	2230	15180

Table 4: Simplex vs. interior point.

6 Conclusions

The interior point methods for linear programming have rejuvenated the field of Linear Programming. The progresses in the performance of the algorithms steered parallel efforts on the Simplex algorithm. The two approaches can now solve problems of sizes that were considered intractable in a very recent past. It is also clear that the larger the size, the more pronounced is the advantage in favor of the interior point approach. This fact becomes to be known to researchers and practitioners. As a rule of thumb, Lustig et al. [27] indicate that interior point methods start to outperform the Simplex method when the sum of the number of rows and columns is larger than 10000.

In the face of degeneracy, a standard feature in real–life problems, the interior point algorithms end at points which approximate the analytic center of the optimal faces and not at an optimal vertex. It has been often thought that this was a disadvantage and much effort has been devoted to the recovery of an optimal basis. The theoretical analysis which is summarized in this paper shows that it might very well be considered as an asset of the method. The optimal partition associated with strictly complementary pairs, an information that interior point methods naturally provide, is the right concept to base a parametric analysis. However, if an optimal basis is really needed, it is not difficult to retrieve it from the optimal partition. In fact, theoretically this can be done in strongly polynomial time [33]. But finding the optimal partition from optimal bases requires enumeration, an NP–hard problem.

Finally, the interior point methodology provides a new and elegant approach to duality theory and parametric or sensitivity analysis. There is still much work to be done in the area of interior point methods for linear programming. In particular the related problems of finding a good starting point and of a "warm start" are open.

The impact of interior point methods goes far beyond linear programming. The seminal work of Nesterov and Nemirovski [38] shows that it is the method of choice for convex programming. This is an area of very active and promising research. To sustain this claim, we would like to mention that the results reported in Table 3 were obtained with a primal–dual interior point algorithm originally designed for general smooth convex programming [44]. This algorithm solves nonlinearly constrained problems of the same size as the linear problems reported in Table 3 with no more than three times the reported number of iterations, a striking low figure. Clearly, interior point methods should now be considered by practitioners and researchers as an important part of their tool kit in Mathematical Programming.

References

[1] I. Adler and R. D. C. Monteiro (1992), A Geometric View of Parametric Linear Programming, *Algoritmica* 8, 161–176.

[2] I. Adler and R. D. C. Monteiro (1991), Limiting Behavior of the Affine Scaling Trajectories for Linear Programming Problems. *Mathematical Programming* 50, 29–51.

[3] O. Bahn, O. du Merle, J.-L. Goffin and J.-P. Vial (1993), A Cutting Plane Method from Analytic Centers for Stochastic Programming, Technical Report 1993.5, Department of Management Studies, Faculté des SES, University of Geneva, Geneva, Switzerland.

[4] G. B. Dantzig (1963), *Linear Programming and Extensions*, Princeton University Press, Princeton, NJ.

[5] G. de Ghellinck and J.-P. Vial (1986), A Polynomial Newton Method for Linear Programming, *Algorithmica* 1, 425–453.

[6] I. S. Duff, A. Erisman and J. Reid (1986), *Direct Methods for Sparse Matrices*, Clarendon Press, Oxford, England.

[7] A. V. Fiacco and G. P. McCormick (1968), *Nonlinear Programming: Sequential Unconstrained Minimization Techniques*, John Wiley and Sons, New York, NY.

[8] T. Gal (1986), *Postoptimal Analysis, Parametric Programming and Related Topics*, Mac-Graw Hill Inc., New York/Berlin.

[9] A. George and J. Liu (1991), *Computer Solution of Large Sparse Positive Definite Systems*, Prentice Hall, Englewood Cliffs, NJ.

[10] P. E. Gill, W. Murray, M. A. Saunders, J. A. Tomlin and M. H. Wright (1986), On Projected Newton Barrier Methods for Linear Programming and an Equivalence to Karmarkar's Projective Method, *Mathematical Programming* 36, 183–209.

[11] A. J. Goldman and A. W. Tucker (1956), Theory of Linear Programming, in *Linear Inequalities and Related Systems* (H. W. Kuhn and A. W. Tucker, eds.), Annals of Mathematical Studies, No. 38, Princeton University Press, Princeton, New Jersey, 53–97.

[12] C. C. Gonzaga (1991), Large Step Path-Following Methods for Linear Programming, Part I: Barrier Function Method, *SIAM Journal on Optimization* 1, 268–279.

[13] O. Güler, C. Roos, T. Terlaky and J.-Ph. Vial (1992), Interior Point Approach to the Theory of Linear Programming, Technical Report 1992.3, Department of Management Studies, Faculté des SES, University of Geneva, Geneva, Switzerland.

[14] D. den Hertog (1993), *Interior Point Approach to Linear, Quadratic and Convex Programming – Algorithms and Complexity.* Kluwer Publishing Comp., Dordrecht, The Netherlands (forthcoming).

[15] B. Jansen, C. Roos and T. Terlaky (1992), An Interior Point Approach to Postoptimal and Parametric Analysis in Linear Programming, Technical Report 92–21, Faculty of Technical Mathematics and Informatics, Technical University Delft, Delft, The Netherlands.

[16] B. Jansen, C. Roos, T. Terlaky and J.-P. Vial (1992), Primal-Dual Algorithms for Linear Programming Based on the Logarithmic Barrier Method, Technical Report 92–104, Faculty of Technical Mathematics and Informatics, Technical University Delft, Delft, The Netherlands.

[17] L. G. Khachiyan (1979), A Polynomial Algorithm in Linear Programming, *Doklady Akademiia Nauk SSSR* 244, 1093–1096. Translated into English in *Soviet Mathematics Doklady* 20, 191–194.

[18] N. K. Karmarkar (1984), A New Polynomial-Time Algorithm for Linear Programming, *Combinatorica* 4, 373–395.

[19] M. Kojima, S. Mizuno and A. Yoshise (1989), A Primal-Dual Interior Point Algorithm for Linear Programming, in N. Megiddo, editor, *Progress in Mathematical Programming*: Interior Point and Related Methods, 29–48. Springer Verlag, New York, NY.

[20] M. Kojima, N. Megiddo and S. Mizuno (1991), A Primal-Dual Infeasible Interior Point Algorithm for Linear Programming, Research Report RJ 8500, IBM Almaden Research Center, San Jose, CA.

[21] M. Kojima, S. Mizuno and A. Yoshise (1991), A Little Theorem of the Big M in Interior Point Algorithms, Research Report on Information Sciences, No. B–239, Department of Information Sciences, Tokyo Institute of Technology, Tokyo, Japan.

[22] E. Kranich (1991), Interior Point Methods for Mathematical Programming: a Bibliography, Diskussionsbeitrag Nr. 171, Fernuniversität Hagen, Germany.

[23] J. Liu (1985), Modification of the Minimum-Degree Algorithm by Multiple Elimination, *ACM Transactions on Mathematical Software* 11, 141–153.

[24] I. J. Lustig (1990/91), Feasibility Issues in a Primal-Dual Interior-Point Method for Linear Programming, *Mathematical Programming* 49, 145–162.

[25] I. J. Lustig, R. J. Marsten and D. F. Shanno (1991), Computational Experience with a Primal-Dual Interior-Point Method for Linear Programming, *Linear Algebra and its Applications* 152, 191–222.

[26] I. J. Lustig, R. J. Marsten and D. F. Shanno (1991), Interior Method VS Simplex Method: Beyond NETLIB, *COAL Newsletter* 19, 41–44.

[27] I. J. Lustig, R. J. Marsten and D. F. Shanno (1992), Interior Methods for Linear Programming: Computational State of the Art, Technical Report SOR 92–17, Program in Statistics and Operations Research, Department of Civil Engineering and Operations Research, Princeton University, Princeton, NJ.

[28] R. J. Marsten, R. Subramanian, M. Saltzman, I. J. Lustig and D. F. Shanno (1990), Interior Point Methods for Linear Programming: Just Call Newton, Lagrange, and Fiacco and McCormick! *Interfaces* 20, 105–116.

[29] L. McLinden (1980), The Complementarity Problem for Maximal Monotone Multifunctions, in R. W. Cottle, F. Giannessi and J. L. Lions, editors, *Variational Inequalities and Complementarity Problems*, 251–270, John Wiley and Sons, New York.

[30] L. McLinden (1980), An Analogue of Moreau's Proximation Theorem With Application to the Nonlinear Complementarity Problem, *Pacific Journal of Mathematics* 88, 101–161.

[31] K. A. McShane, C. L. Monma and D. F. Shanno (1989), An Implementation of a Primal–Dual Interior Point Method for Linear Programming, *ORSA Journal on Computing* 1, 70–83.

[32] N. Megiddo (1989), Pathways to the Optimal set in Linear Programming, in N. Megiddo (editor), *Progress in Mathematical programming*: Interior Point and Related Methods, 131–158. Springer Verlag, New York.

[33] N. Megiddo (1991), On Finding Primal- and Dual-Optimal Bases, *ORSA Journal on Computing* 3, 63–65.

[34] S. Mehrotra (1992), On the Implementation of a (Primal–Dual) Interior Point Method, *SIAM Journal on Optimization* 2, 575–601.

[35] S. Mizuno (1992), Polynomiality of Kojima–Megiddo–Mizuno Infeasible Interior Point Algorithm for Linear Programming, Technical Report 1006, School of Operations Research and Industrial Engineering, Cornell University, Ithaca NY.

[36] C. Moler, J. Little and S. Bangert (1987), *PRO-MATLAB user's guide*, The MathWorks, Inc., Sherborne MA, USA.

[37] R. D. C. Monteiro and I. Adler (1989), Interior Path Following Primal–Dual Algorithms. Part I: Linear Programming, *Mathematical Programming* 44, 27–42.

[38] Y. Nesterov and A. Nemirovskii (1990), *Self-Concordant Functions and Polynomial-Time Methods in Convex Programming*, USSR Academy of Sciences, Moscow.

[39] C. H. Papadimitriou and K. Steiglitz (1982), *Combinatorial Optimization: Algorithms and Complexity*, Prentice–Hall, Englewood Cliffs, New Jersey.

[40] F. A. Potra (1992), An Infeasible Interior–Point Predictor–Corrector Algorithm for Linear Programming, Technical Report 26, Department of Mathematics, University of Iowa, Iowa City IA, USA.

[41] J. Renegar (1988), A Polynomial–Time Algorithm Based on Newton's Method for Linear Programming, *Mathematical Programming* 40, 59–94.

[42] C. Roos and J.-Ph. Vial (1990), Long Steps with the Logarithmic Penalty Barrier Function, in *Economic Decision-Making: Games, Economics and Optimization* (dedicated to Jacques H. Drèze), edited by J. Gabszevwicz, J. -F. Richard and L. Wolsey, Elsevier Science Publisher B.V., 433–441.

[43] Gy. Sonnevend (1985), An "Analytical Centre" for Polyhedrons and New Classes of Global Algorithms for Linear (Smooth Convex) Programming. In A. Prékopa, J. Szelezsán and B. Strazicky, eds. *System Modelling and Optimization: Proceedings of the 12th IFIP-Conference, Budapest, Hungary, September 1985,* Vol. 84. of *Lecture Notes in Control and Information Sciences,* Springer Verlag, Berlin, 866–876.

[44] J.-Ph. Vial (1992), Computational Experience with a Primal–Dual Interior-Point Algorithm for Smooth Convex Programming, Technical Report, Department of Management Studies, Department of SES, University of Geneva, Switzerland.

[45] Y. Zhang (1992), On the Convergence of an Infeasible Interior–Point Algorithm for Linear Programming and Other Problems, Technical Report, Department of Mathematics and Statistics, University of Maryland, Baltimore County, Baltimore, USA.

APPROACHES TO STOCHASTIC PROGRAMMING WITH APPLICATION TO ELECTRIC POWER SYSTEMS

G. B. Dantzig , G. Infanger [1]

Department of Operations Research
Stanford University
Stanford, CA 94305-4022, U.S.A.

Abstract. We demonstrate how large-scale stochastic linear programs can be efficiently solved by using a blending of classical decomposition and a relatively new technique called importance sampling. We discuss an adaptive importance sampling scheme using an additive approximation function. We show how this technique can be applied to facility expansion planning of electrical power systems. Numerical results of testproblems with numerous stochastic parameters are presented.

1. Introduction

Solutions obtained from deterministic planning models are usually unsatisfactory because they fail to hedge against unfavorable events which may occur in the future. Stochastic models address this shortcoming, but in the past have seemed to be untractable because, even for a relatively small number of parameters, subject to uncertainty the size of such problems can get very large. Stochastic problems have been studied extensively in the literature since Dantzig (1955) [7]. Different approaches attack this problem, e.g. Birge (1985) [2], Birge and Wets (1986) [4], Birge and Wallace (1988) [6], Ermoliev (1983) [14], Ermoliev and Wets (1988) [15], Frauendorfer (1988) [16], Frauendorfer (1992) [17], Frauendorfer and Kall (1988) [18], Higle and Sen (1989) [20], Kall (1979) [24], Pereira et al. (1989) [26], Rockafellar and Wets (1991) [28], Ruszczyński (1986) [29], and Wets (1984) [31].

[1] Research and reproduction of this report were partially supported by the Office of Naval Research Contract N00014-89-J-1659; the National Science Foundation Grants ECS-8906260, DMS-8913089, the Electric Power Research Institute Contract RP 8010-09, CSA-4O05335, and the Austrian Science Any opinions, findings, and conclusions or recommendations expressed in this publication are those of the authors and do not necessarily reflect the views of the above sponsors.

2. Two-stage stochastic linear programs

An important class of stochastic models are two-stage stochastic linear programs with recourse. These models are the analog extensions of deterministic dynamic systems which have a staircase structure: x denotes the first, y the second stage decision variables, A, b represent the coefficients and right hand sides of the first stage constraints and D, d represent the second stage constraints, which together with the transition matrix F, couples the two periods. In the literature D is often referred to as the technology/recourse matrix. The first stage parameters are known with certainty. The second stage parameters are random variables ω that assume certain outcomes with certain probabilities $p(\omega)$. They are known only by their probability distribution of possible outcomes at time $t = 1$, where actual outcomes will be known later at time $t = 2$. Uncertainty occurs in the transition matrix F and in the right hand side vector d. The second stage cost f and the elements of the technology/recourse matrix D are assumed to be known with certainty. We denote an outcome of the stochastic parameters with $\omega, \omega \in \Omega$, with Ω being the set of all possible outcomes. The two-stage stochastic linear program can be written as follows:

$$
\begin{array}{llcccccl}
\min z & = & cx & + & E^{\omega}(fy^{\omega}) & & & \\
\text{s.t.} & & Ax & & & = & b & \\
 & & -F^{\omega}x & + & Dy^{\omega} & = & d^{\omega} & \\
 & & x, & & y^{\omega} & \geq & 0 & \omega \in \Omega
\end{array}
$$

The problem is to find a first stage decision x which is feasible for all scenarios $\omega \in \Omega$ and has the minimum expected cost. Note the adaptive nature of the problem: while the decision x is made only with the knowledge of the distribution $p(\omega)$ of the random parameters, the second stage decision y^{ω} is made later after an outcome ω is observed. The second stage decision compensates for and adaptes to different scenarios ω.

Using discrete distributions, one can express a stochastic problem as a deterministically equivalent linear program by writing down the second stage constraints for each scenario $\omega \in \Omega$ one below the other. The objective function carries out the expected value computation by direct summation. Clearly, this formulation leads to linear programs of enormous sizes.

$$
\begin{array}{lllllllll}
\min z = & cx & + p^1 f y^1 & + p^2 f y^2 & + & \cdots & + p^K f y^K & & \\
s.t. & Ax & & & & & & = & b \\
 & -F^1 x & + Dy^1 & & & & & = & d^1 \\
 & -F^2 x & & + Dy^2 & & & & = & d^2 \\
 & \vdots & & & & \ddots & & & \vdots \\
 & -F^K x & & & & & + Dy^K & = & d^K \\
 & x, & y^1, & y^2, & & \ldots, & y^K & \geq & 0
\end{array}
$$

The method which we apply to solve large-scale stochastic linear programs uses Benders decomposition and importance sampling. The method and the underlying theory of our approach is developed in Dantzig and Glynn (1990) [8] and Infanger (1991) [23]. Dantzig and Infanger (1991) [10] report on the solution of large-scale problems. Entriken and Infanger (1990) [13] discuss how reliability constraints can be handled by additionally using Dantzig-Wolfe decomposition. In the following we give a brief review of the concept. First we introduce the formulation of a special class of multi-stage stochastic linear programs as an extension to the formulation of the two-stage stochastic linear program above. This class fits the facility expansion planning problem. Using decomposition techniques we split the problem into a series of tractable smaller problems. Using sampling techniques we compute an estimate of the expected cost and variances. Importance sampling is the key to obtaining accurate estimates, i.e. unbiased estimates with low variances, with low sample size.

3. Multi-stage stochastic linear programs

Large-scale deterministic mathematical programs, used for operations and strategic planning, often are dynamic linear programs. These problems have a staircase (multi-stage) matrix structure. In general, the size of these stochastic problems can get extremly large because the number of scenarios grows exponentially with the number of periods. We will, however, address a certain restricted class whose number of scenarios grows linearly with the number of stages: The problem (whose constraints are stated below) breaks down into two parts: a deterministic dynamic part and a stochastic part. We call the deterministic part the *master-problem*. It is a dynamic linear program with T stages. The vectors c_t and b_t, and the matrices B_{t-1} and A_t are assumed to be known with certainty.

$$
\begin{array}{llcccl}
\min & \sum_{t=1}^{T} c_t x_t & + & \sum_{t=1}^{T} E(f_t y_t^{\omega_t}) & & \\
-B_{t-1}x_{t-1} & +A_t x_t & & & = & b_t, \quad t=1,\ldots,T, \quad B_0 = 0 \\
& -F_t^{\omega_t} x_t & + & D_t y_t^{\omega_t} & = & d_t^{\omega_t}, \quad t=1,\ldots,T, \quad \omega_t \in \Omega_t \\
& x_t, & & y_t^{\omega_t} & \geq & 0
\end{array}
$$

Each stage is associated with a stochastic sub-problem. Uncertainty appears in the recourse-matrix $F_t^{\omega_t}$ and in the right hand side vector $d_t^{\omega_t}$ where ω_t, denotes an outcome of the stochastic parameters in period t, with Ω_t denoting the set of all possible outcomes in period t. The sub-problems in each stage are assumed to be stochastically independent. The sub-problem cost f_t and the technology matrix D_t are assumed to be deterministic parameters.

The facility expansion planning problem

Facility expansion planning is an example of this type of formulation. The master-problem models the expansion of the facilities over time. Decision variables are the capacity built and the capacity available at time t. The sub-problems model the operation of these capacities in an uncertain environment. Take for example the case of expansion planning of power systems: The expansion or replacement of capacities of generators and transmission lines are determined in the master problem. The capacities at each period t are made available to the system for operation. The subproblems model the power system operation, the optimal scheduling of the available capacities to meet the demand for electricity. The availabilities of generators and transmission lines and the demands are uncertain and not known at the time when the expansion decision is made.

The approach is primarily "here and now" (Dantzig and Madansky (1961) [11]) and justified by high investment cost and long lead-times for capacity expansion. However, as the operations subproblems are stochastically independent and only expected operation cost rather than the state of the system after period t affects the expansion plan (as failures of equipment get repaired, and uncertainty in the demands are interpreted as deviations from a demand path), "here and now" is equivalent to "wait and see". That means that the optimal decision in period $t+1$ depends only on the capital stock on hand at the start of period $t+1$ and is independent of any observed outcomes in period t, i.e. the same optimal capacity expansion decision would be made before as after period t operations. Thus the facility expansion plan can be laid out at the beginning for the whole planning horizon based on the expansion costs and the expected operation cost. This permits the multi-stage problem to be treated as if it were a two-stage problem. The first "stage" concerns the single decision of what facility expansion will be in all future periods without knowledge of the particular outcomes of the uncertainty parameters in future periods. The second "stage" concerns the operations problems, where the recourse decisions made depend on the realizations of the stochastic parameters. Note that for $T = 1$, the problem is exactly a two stage stochastic linear program with recourse. For $T \geq 2$ the problem is a two "stage" problem with the second stage consisting of T independent subproblems.

4. Multidimensional integration

The difficulty of solving large-scale stochastic problems arises from the need to compute multiple integrals or multiple sums. The expected value of the second stage cost in period t (we suppress the index t for this discussion), e.g. $z = E(fy^\omega) = E(C)$, is an expectation of functions $C(v^\omega), \omega \in \Omega$, where $C(v^\omega)$ is obtained by solving a linear problem. V (in general) is a h-dimensional random vector parameter, e.g. $V = (V_1, \ldots, V_h)$, with outcomes $v^\omega = (v_1, \ldots, v_h)^\omega$. For example V_i represents the percent of generators of type i down for repair or transmission lines not operating and v_i^ω the observed random percent outcome. The vector v^ω is also denoted by v, and $p(v^\omega)$ alias $p(v)$ denote the corresponding probability. Ω is the set of all possible random events and is constructed by crossing the sets of outcomes $\Omega_i, i = 1, \ldots, h$ as $\Omega = \Omega_1 \times \Omega_2 \times \cdots \times \Omega_h$. With P being the probability measure under the assumption of independence, the integral

$$E\,C(V) = \int C(v^\omega) P(d\omega)$$

takes the form of a multiple integral

$$E\,C(V) = \int \cdots \int C(v)p(v)dv_1 \ldots dv_h,$$

or, in case of discrete distributions, the form of a multiple sum

$$E\,C(V) = \sum_{v_1} \cdots \sum_{v_h} C(v)p_1(v_1) \ldots p_h(v_h).$$

In the following discussion we concentrate on discrete distributions. This is not a restriction as the approach can be easily adapted for continuous distributions. In practical applications all distributions can be approximated with sufficiant accuracy by discrete ones. Even for h as small as 20 the number of terms in the multiple sum computation gets easily out of hand and the problem is no longer practical to solve by direct summation. This is especially true because function evaluations are computationally expensive since each term in the multiple sum requires the solution of a linear program.

5. Importance sampling

Monte Carlo Methods are recommended to compute multiple integrals or multiple sums for higher h-dimensional sample spaces (Davis and Rabinowitz (1984) [12], Glynn and Iglehart (1989) [19]). Suppose $C^\omega = C(v^\omega)$ are independent random variates of $v^\omega, \omega = 1, \ldots, n$ with expectation z, where n is the sample size. An unbiased estimator of z with variance $\sigma_{\bar{z}}^2 = \sigma^2/n$, $\sigma^2 = var(C(V))$ is

$$\bar{z} = (1/n) \sum_{\omega=1}^{n} C^\omega.$$

Note that the standard error decreases with $n^{-0.5}$ and the convergence rate of $\bar{z}$ to z is independent of the dimension of the sample space h. We rewrite $z = \sum_{\omega \in \Omega} C(v^\omega) p(v^\omega)$ as

$$\sum_{\omega \in \Omega} \frac{C(v^\omega) p(v^\omega) q(v^\omega)}{q(v^\omega)}$$

by introducing a new probability mass function $q(v^\omega)$ and we obtain a new estimator of z

$$\bar{z} = \frac{1}{n} \sum_{\omega=1}^{n} \frac{C(v^\omega) p(v^\omega)}{q(v^\omega)}$$

by sampling from $q(v^\omega)$. The variance of $\bar{z}$ is given by

$$var(\bar{z}) = \frac{1}{n} \sum_{\omega \in \Omega} \left(\frac{C(v^\omega) p(v^\omega)}{q(v^\omega)} - z \right)^2 q(v^\omega).$$

Choosing

$$q^\star(v^\omega) = \frac{C(v^\omega) p(v^\omega)}{\sum_{\omega \in \Omega} C(v^\omega) p(v^\omega)}$$

would lead to $var(\bar{z}) = 0$, which means one could get a perfect estimate of the multiple sum from only one estimation. Practically however, this is useless since to compute $q(v^\omega)$ we have to know $z = \sum_{\omega \in \Omega} C^\omega p(v^\omega)$, which we eventually want to compute. The result however helps to derive a heuristic for choosing q. It should be proportional to the product $C(v^\omega)p(v^\omega)$ and should have a form that can be integrated easily. Thus a function $\Gamma(v^\omega) \approx C(v^\omega)$ is sought, which can be integrated with less cost than $C(v^\omega)$. Additive and multiplicative (in the components of the stochastic vector v) approximation functions and combinations of these are potential candidates for our approximations. In particular, we have been getting good results using $C(V) \approx \sum_{i=1}^{h} C_i(V_i)$. We compute q as

$$q(v^\omega) \approx \frac{C(v^\omega)p(v^\omega)}{\sum_{i=1}^{h}\sum_{\omega\in\Omega_i} C_i(v^\omega)}.$$

To understand the motivation for the importance sampling scheme, assume for convenience $C_i(v_i^{\omega_i}) > 0$ and let $\Gamma(v^\omega) = \sum_{i=1}^{h} C_i(v_i^\omega)$ If $\sum C(v^\omega)p(v^\omega)$ were used as an approximation of $\bar{z}$ it can be written

$$\sum_{\omega=1}^{n} \Gamma(v^\omega)p(v^\omega) = \sum_{i=1}^{h} \alpha_i \sum_{\omega=1}^{n} [\frac{C_i(v_i^\omega)}{\alpha_i}]\; p_1(v_1^{\omega_1})p_2(v_2^{\omega_2})\dots p_h(v_h^{\omega_h})$$

where $\omega = (\omega_1, \omega_2, \dots, \omega_h)$ and where we define

$$\alpha_i = \sum_{\omega_i\in\Omega_i} C_i(v_i^{\omega_i})p_i(v_i^{\omega_i}),$$

which is relatively easy to compute since it can be evaluated by summing only one of the dimensions of ω. Note that

$$\bar{p}_i(v_i^{\omega_i}) = \frac{C_i(v_i^{\omega_i})p_i(v_i^{\omega_i})}{\alpha_i} \geq 0, \;\; \omega_i \in \Omega_i$$

may be viewed as a modified probability distribution of v_i associated with term i. It is, of course, a trivial matter to directly sum each term i since each of its factors, being independent probability distributions, sum to one. Suppose, however, one does not notice this fact and decides to estimate the sum by estimating each of the h terms by Monte Carlo sampling. The i-th term would then be evaluated by randomly sampling v_i from the distribution $\bar{p}_i(v_i^{\omega_i})$ and all the rest of the components v_j of v from the distributions $p_i(v_i^{\omega_i})$.

In an analogous manner, let

$$\rho(\omega) = \frac{C(\omega)}{\Gamma(\omega)}$$

and write

$$\bar{z} = \sum C(\omega)p(\omega) = \sum_{i=1}^{h} \alpha_i \sum_{\omega=1}^{n} \rho(\omega)[\frac{C_i(v_i^\omega)}{\alpha_i}]\; p_1(v_1^{\omega_1})p_2(v_2^{\omega_2})\dots p_h(v_h^{\omega_h})$$

If our approximation $\Gamma(\omega)$ to $C(\omega)$ is any good, $\rho(\omega)$ will be roughly 1 for almost all ω. This suggests the heuristic that the sampling be carried out differently for each term i. The importance sampling scheme then is to sample v_i of the i-th term according to the distribution $\bar{p}_i(v_i^{\omega_i})$ and to sample all other components $v_j^{\omega_j}$ of the i-th term according to the distribution $p_j(v_j^{\omega_j})$.

If the additive function turns out to be a bad approximation of the cost function, as indicated by the observed variance being too high, it is easily corrected by increasing the size of the sample. This is done adaptively.

Actually we use a variant of the additive approximation function. By introducing $C(\tau)$, the cost of a base case, we make the model more sensitive to the impact of the stochastic parameters v. Our approximation function is computed as follows:

$$\Gamma(V) = C(\tau) + \sum_{i=1}^{h} \Gamma_i(V_i), \quad \Gamma_i(V_i) = C(\tau_1, \dots, \tau_{i-1}, V_i, \tau_{i+1}, \dots, \tau_h) - C(\tau)$$

We refer to this as a *marginal cost* approximation. We explore the cost function at the margins, e.g. we vary the random elements v_i to compute the cost for all outcomes v_i while we fix the other random elements at the level of the base case. τ can be any arbitrary chosen point of the set of k_i discrete values of v_i, $i = 1, \dots, h$. For example we choose τ_i as that outcome of V_i which leads to the lowest cost, ceteris paribus.

Summarizing, the importance sampling scheme has two phases: the preparation phase and the sample phase. In the preparation phase we explore the cost function $C(V)$ at the margins to compute the additive approximation function $\Gamma(V)$. For this process $n_{prep} = 1 + \sum_{i=1}^{h}(k_i - 1)$ subproblems have to be solved. Using $\Gamma(V)$ we compute the approximate importance density

$$q(v^\omega) = \frac{\Gamma(v^\omega)p(v^\omega)}{C(\tau) + \sum_{i=1}^{h} \sum_{\omega \in \Omega_i} \Gamma_i(v^\omega)p(v^\omega)}.$$

Next we sample n scenarios from the importance density and, in the sample phase, solve n linear programs to compute the estimation of $\bar{z}$ using the Monte Carlo estimator. We compute the gradient G and the right hand side g of the cut using the same sample points at hand from the expected cost calculation. See Infanger (1991) [23] for the computation of the cuts and details of the estimation process.

6. Benders decomposition

We decompose the 2-stage multi-period stochastic linear program by applying Benders (1962) [2] decomposition. See Van Slyke and Wets (1969) [30] for a reference to using Benders decomposition for stochastic linear programs.

The master problem:

$$
\begin{array}{lllllll}
z_M^L = & \min & \sum_{t=1}^T c_t x_t & + & \sum_{t=1}^T \theta_t & & \\
 & -B_{t-1}x_{t-1} & +A_t x_t & & & = & b_t, \quad t=1,\dots,T, \quad B_0 = 0 \\
 & & -G_t^l x_t & + & \alpha_t^l \theta_t & \geq & g_t^l, \quad t=1,\dots,T, \quad l=1,\dots,L \\
 & & x_t & & & \geq & 0
\end{array}
$$

where the latter constraints, called cuts, are initially absent but are inserted in later iterations. The master problem is optimized to obtain an approximate optimal feasible solution $x_t = \hat{x}_t^l$ that is used as input to the subproblems.

The sub-problems for ω_t in period t:

$$
\begin{array}{llllll}
z_t^{\omega_t}(\hat{x}_t^l) & = \min & f_t y_t^{\omega_t} & & & \\
\pi_t^{\omega_t}(\hat{x}_t^l): & & D_t y_t^{\omega_t} = & d_t^{\omega_t} + F_t^{\omega_t}\hat{x}_t^l, & \omega_t \in \Omega_t, & t=1,\dots,T \\
 & & y_t^{\omega_t} \geq & 0, & \hat{x}_t^l \text{ given}, &
\end{array}
$$

where $\pi_t^{\omega_t} = \pi_t^{\omega_t}(\hat{x}_t^l)$ are dual multipliers corresponding to the constraints and $z_t^{\omega_t} = z_t^{\omega_t}(\hat{x}_t^l)$ is the value of the objective as functions of $\hat{x}_t^l$. These are used to generate the next cut for the master.

The cuts for $t = 1, 2, \dots, T$:

$$
G_t^l = E(\pi^{\omega_t} B^{\omega_t}), \quad g_t^l = E(\pi^{\omega_t} d^{\omega_t}), \quad z_t(\hat{x}_t^l) = E(z_t^{\omega_t}), \quad \pi_t^{\omega_t} = \pi_t^{\omega_t}(\hat{x}_t^l).
$$

Lower (LB^L) and *upper* (UB^L) bounds to the problem:

$$
LB^L = z_M^L, \quad UB^L = \min\{UB^{L-1}, \sum_{t=1}^T (c_t \hat{x}_t^l + z_t(\hat{x}_t^l))\}, \quad UB^0 = \infty
$$

$\hat{x}_t^l$ is the optimal solution of the master problem in iteration l, $\pi_t^{\omega_t}(\hat{x}_t^l)$ is the optimal dual solution of subproblem ω_t, given $\hat{x}_t^l$. Note that if the subproblems are infeasible, a slightly different definition of the cut is used. $\alpha = 0$ corresponds to feasibility cuts and $\alpha = 1$ to optimality cuts. Solving the master problem in iteration l we obtain a trial solution $\hat{x}_t^l$ which we pass to the subproblems. By solving a sample of sub-problems ω_t, $\omega_t \in N_t, t = 1, \ldots, T$, according to the importance sampling scheme we compute estimates of the second stage cost z_t and estimates of the gradients G_t^l and the right hand sides g_t^l of the cuts. Cuts represent an outer linearization of the second stage cost expressed in first stage decision variables and θ_t. Note that there is one cut for each period t. The cuts are added to the master problem and the master problem is solved again. The objective function value of the master problem gives a lower bound estimate and the total expected cost of a trial solution $\hat{x}_t^l, t = 1, \ldots, T$ gives an upper bound estimate to the objective function value of the problem. If the lower and the upper bound are sufficiently close, which is tested by a Student-t test, the problem is considered to be solved. Lower and upper bounds can be seen as a sum of *i.i.d.* random terms which for sample sizes of 30 or more can be assumed normally distributed with known (derived from the estimation process) variances. A 95% confidence intervall of the optimal solution is computed. See Dantzig and Glynn (1990) [8] and Infanger (1990) [23] for details of the algorithm.

7. Implementation

This method for solving large-scale two-stage stochastic linear programs with recourse has been implemented. The code of MINOS (Murtagh and Saunders (1983) [25]) has been adapted for this purpose as a subroutine for solving both the master-problem and the sub-problems. When solving large numbers of sub-problems it is important for the performance of the algorithm to take advantage of good starting bases. Computation time can be reduced dramatically by solving *first* an expected value problem by replacing the stochastic parameters by their expectations. The expected value solution of the resulting deterministic problem is then used as a starting point for the stochastic solution. Additionally we keep cuts obtained from the expected value problem to initially guide the algorithm. It can be shown that cuts obtained from the expected value problem are valid for the stochastic problem. They are "weak" and get replaced as the algorithm proceeds. The code uses sparce matrix techniques and efficient data structures for handling large-scale problems.

Computational results of the large scale test problems are represented in Table 1. Besides the solution of the stochastic problems, the results from solving the expected value problems are also reported. We also report on the estimated expected cost if the expected value solution is used as the decision in a stochastic environment. The objective function value of the true stochastic solution has to lie between the minimum value of objective function of the deterministic problem and the expected cost of the expected value solution.

Expansion planning of multi-area power systems (Numerical results)

The expansion planning of multi-area power systems under uncertainty is an important problem in the class of facility expansion planning problems. The method allows for different formulations of the operations planning sub-problems. The power flow in the transmission lines can be represented by a network flow formulation as well as by a (linearized) power flow model. In the numerical example presented below we used a network flow formulation of the multi-area power system.

WRPM is a multi-area capacity expansion planning problem for the western USA and Canada. The model is very detailed and covers 6 regions, 3 demand blocks, 2 seasons, and several kinds of generation and transmission technologies. The objective is to determine optimum discounted least cost levels of generation and transmission facilities for each region of the system over time. The model minimizes the total discounted cost of supplying electricity (investment and operating cost) to meet the exogenously given demand subject to expansion and operating constraints. A description of the model can be found in Dantzig et. al. (1989) [9].

In the stochastic version of the model the availabilities of generators and transmission lines and demands are subject to uncertainty. There are 13 stochastic parameters per time period (8 stochastic availabilities of generators and transmission lines and 5 uncertain demands) with discrete distributions with 3 or 4 outcomes. Table 1 represents three different versions of WRPM. WRPM1 covers a planning horizon of 1 future period, WRPM2 covers 2 future periods and WRPM3, the largest problem, covers 3 future periods. There are differences in the parameters between WRPM1, WRPM2 and WRPPM3. The operating sub-problems of each period are stochastically independent. The number of universe scenarios is larger than $5 \cdot 10^6$ per period. In the deterministic equivalent formulation the problems if it were possible to state it would have more than 1.5 billion (WRPM1), 3 billion (WRPM2) and 4.5 billion (WRPM3) constraints.

For solving the problems we chose a sample size of 100. The estimate of the objective function value of the stochastic solution (289644.2 in case of WRPM1, 143109.2 in case of WRPM2 and 199017.4 in case of WRPM3) turned out to be amazingly accurate. The 95% confidence interval is computed as 0.0913% on the left side and 0.063% on the right side (WRPM1), 0.0962% on the left side and 0.1212% on the right side (WRPM2) and 0.029% on the left side and 0.067% on the right side (WRPM3). Thus the objective function value of the stochastic solution lies with 95% probability between $289379.7 \leq z \leq 289826.0$ (WRPM1), $142971.5 \leq z \leq 143282.6$ (WRPM2) and between $198959.3 \leq z \leq 199164.1$ (WRPM3). In all cases the expected cost of the expected value solution and the expected cost of the stochastic solution differ significantly. The solution time on a Toshiba T5200 laptop PC with 80387 mathematic coprocessor was 75 minutes (WRPM1), 187 minutes (WRPM2) and 687 minutes (WRPM3). During this time about 7500 (WRPM1), 15700 (WRPM2) and 26295 (WRPM3) subproblems (linear programs of the size of 302 rows and 289 columns) got solved.

	WRPM1	WRPM2	WRPM3
# iter.	139	131	197
sample size	100	100	100
solution (estimate)	289644.2	143109.2	199017.4
conf. left % (est.)	0.0913	0.0962	0.0292
conf. right % (est.)	0.063	0.1212	0.067
solution time (min)	75	187	687
problem size			
Master rows	44	86	128
Master columns	76	151	226
Master nonzeros	153	334	413
Subproblem rows	44	86	128
Subproblem columns	76	151	226
Subproblem nonzeros	153	334	413
# stoch. param.	13	26	39
# univ. scenarios	5038848	10077696	15000000

Table 1: Large test problems: computational results power planning

8. Conclusion

We have demonstrated that large-scale stochastic linear programs can be efficiently solved using Benders decomposition and importance sampling. We have formulated a class of two-stage multi-period stochastic linear problems concerning facility expansion planning. Numerical results of test problems in the area of expansion planning of electric power systems with numerous stochastic parameters indicate that very accurate solutions can be obtained using only small sample sizes.

References

1. Beale, E.M.L.: On Minimizing a Convex Function Subject to Linear Inequalities; J. Roy. Stat. Soc. 17b (1955) 173-184
2. Benders, J.F.: Partitioning Procedures for Solving Mixed-Variable Programming Problems; Numerische Mathematik 4 (1962) 238-252
3. Birge, J.R.: Decomposition and Partitioning Methods for Multi-Stage Stochastic Linear Programming; Operations Research 33 (1985) 989-1007
4. Birge, J.R. and Wets, R.J.: Designing Approximation Schemes for Stochastic Optimization Problems, in Particular For Stochastic Programs with Recourse, Math. Progr. Study 27 (1986) 54-102
5. Birge, J.R. and Wets, R.J.(eds.): Stochastic Programming I,II; Proceedings of the 5th International Conference on Stochastic Programming Ann Arbour, Michigan, August 13-18, 1989; Annals of Operations Research 30-31, 1991
6. Birge, J.R. and Wallace, S.W.: A Separable Piece wise Linear Upper Bound for Stochastic Linear Programs; SIAM J. Control and Optimization 26/ 3 (1988)
7. Dantzig, G.B.: Linear Programming under Uncertainty; Management Science 1 (1955) 197-206
8. Dantzig, G.B. and Glynn, P.W.: Parallel Processors for Planning Under Uncertainty, Ann. of OR 22 (1990) 1-21
9. Dantzig, G.B., Glynn, P.W., Avriel, M., Stone, J., Entriken, R., Nakayama, M.: Decomposition Techniques for Multi-Area Generation and Transmission Planning under Uncertainty, EPRI report 2940-1, 1989
10. Dantzig G.B. and Infanger, G.: Large-Scale Stochastic Linear Programs: Importance Sampling and Benders Decomposition; Technical Report SOL 91-4, Department of Operations Research, Stanford University 1991
11. Dantzig, G.B. and Madansky M.: On the Solution of Two-Staged Linear Programs under Uncertainty; in J. Neyman (ed.): Proc 4th Berkeley Symp. on Mathematical Statistics and Probability I, (1961) 165-176
12. Davis, P.J., and Rabinowitz, P.: *Methods of Numerical Integration*; Academic Press, London, 1984

13. Entriken, R. and Infanger, G.: Decomposition and Importance Sampling for Stochastic Linear Models; Energy, The International Journal 15/7-8 (1990) 645-659

14. Ermoliev, Y.: Stochastic Quasi-gradient Methods and Their Applications to Systems Optimization, Stochastics 9 (1983) 1-36

15. Ermoliev,Y. and R.J. Wets (eds.): *Numerical Techniques for Stochastic Optimization*; Springer Verlag 1988

16. Frauendorfer, K.: Solving SLP Recourse Problems with Arbitrary Multivariate Distributions - The Dependent Case, Mathematics of Operations Research, 13/3 (1988) 377-394

17. Frauendorfer, K. (1992): *Stochastic Two-Stage Programming*; Lecture Notes in Economics and Mathematical Systems 392, Springer-Verlag 1992

18. Frauendorfer K. and Kall, P.: Solving SLP Recourse Problems with Arbitrary Multivariate Distributions - The Independent Case; Problems of Control and Information Theory 17/4 (1988) 177-205

19. Glynn, P.W. and Iglehart, D.L.: Importance Sampling for Stochastics Simulation; Management Science 35 (1989) 1367-1392

20. Higle, J.L. and Sen, S.: Stochastic Decomposition: An Algorithm for Two Stage Linear Programs with Recourse; Math. of OR 16/3 (1991) 650-669

21. Ho, J.K. and Loute, E.: A Set of Staircase Linear Programming Test Problems, Math. Programming 20 (1981) 245-250

22. Ho, J.K. and Manne, A.S.: Nested Decomposition for Dynamic Models, Math. Progr. 6 (1974) 121-140

23. Infanger, G.: Monte Carlo (Importance) Sampling within a Benders Decomposition Algorithm for Stochastic Linear Programs; Ann. of OR 39 (1992)

24. Kall, P.: Computational Methods for Two Stage Stochastic Linear Programming Problems, Z. angew. Math. Phys. 30 (1979) 261-271

25. Murtagh, B.A. and Saunders, M.A.: MINOS User's Guide, SOL 82-20, Department of Operations Research, Stanford University, Stanford CA 94305, 1983

26. Pereira, M.V., Pinto, L.M.V.G., Oliveira, G.C. and Cunha, S.H.F.: A Technique for Solving LP-Problems with Stochastic Right-Hand Sides, CEPEL, Centro del Pesquisas de Energia Electria, Rio de Janeiro, Brazil, 1989

27. Pereira, M.V., Pinto, L.M.V.G.: Stochastic Dual Dynamic Programming; Technical Note, DEE- PUC/RJ - Catholic University of Rio de Janeiro, Caixa Postal 38063 Gávea, Rio de Janeiro RJ CEP 22452 Brazil, 1989

28. Rockafellar, R.T. and Wets, R.J.: Scenario and Policy Aggregation in Optimization under Uncertainty, Math. of OR 16 (1991) 241-266

29. Ruszczyński, A.: A Regularized Decomposition Method for Minimizing a Sum of Polyhedral Functions, Math. Progr. 35 (1986) 309-333

30. Van Slyke and Wets, R.J.: L-Shaped Linear Programs with Applications to Optimal Control and Stochastic Programming; SIAM J. of Appl. Math. 17 (1969) 638-663

31. Wets R.J.: Programming under Uncertainty: The Equivalent Convex Program; SIAM J. on Appl. Math. 14 (1984) 89-105

Chapter II

UNIT COMMITMENT

UNIT COMMITMENT AND THERMAL OPTIMIZATION - PROBLEM STATEMENT

H. Braun

BASF AG
D 6700 Ludwigshafen, Germany

Abstract. The demands made on the electrical energy supply are manifold. Consumers expect an inexpensive, adequate supply at all times with high degree of reliability and closely defined quality criteria (voltage, frequency). National and international institutions require attainment of objectives and compliance with regulations in line with the energy and environmental policies. To an increasing extent, long-term quantitative requirements are also bearing an influence on the operation and development of supply systems. In the FRG this applies to coal with precisely defined obligation to take supplies, it applies in general to natural gas with its 'take or pay rulings and as a rule, it applies to agreements with zoned tariffs and minimum periods of use. Hydrothermal systems with annual reservoirs have long been familiar with the constraints imposed by yearly reservoir management. The inclusion of other regenerative sources of energy, such as the sun and wind, makes new demands on development and operation of the systems. The way in which the supply sector is developing is characterized by a trend for both the systems themselves and the demands from outside made on them to become more and more complex. With the introduction of a more liberal energy and power supply market (e.g. 'Third Party Access') there will be an additional increase in the pressure of costs for the energy supply companies. Consequently, the question of optimal system operation is gaining in importance. Heuristic approaches based on the experience of the load dispatcher are no longer adequate. Mathematical, computer-aided approaches are increasingly being used. However, a closed solution to the overall range of problems involved in system optimization cannot be accomplished in the foreseeable future; it is necessary to break them down into individual problems : i) expansion and design planning, ii) revision planning, iii) fuel resource scheduling and reservoir management, iv) weekly and daily unit commitment, v) load distribution, vi) load management, vii) voltage / reactive power optimization. Due to the long periods of time considered in long-term planning (up to 20 years for development planning and 1-5 years for revision and resource scheduling) stochastic influences such as the failure mode of the components and frequency distribution of the load and sources of supply have a strong bearing. This must be taken into account in modelling and method selection. In the case of short-term planning the emphasis is the exact reproduction of the operating behavior of the individual components. A deterministic consideration is adequate. During the course of this presentation the incidental conditions affecting the individual problems will be described. The question of how to link the problem areas will be looked at more closely. Modelling approaches which have been adopted in operational practice will be illustrated.

1 Introduction

A basic requirement for the high productivity which has developed mainly in the industrialized western countries is the utilization of the existing primary energy sources. Apart from the availability of food and information, the availability of energy is the prerequisite for human life per se.

In the period from 1960 to 1989, the annual world primary energy demand has increased from 34500 TWh to 83000 TWh [1] (see Fig. 1).

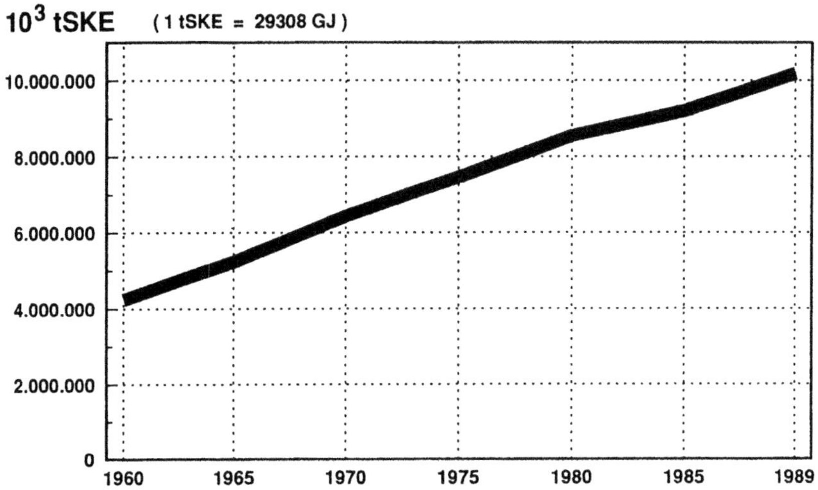

Figure 1: World energy consumption 1960 - 1989

At present, the global consumption is 16MWh per head; for industrialized countries it ranges from 20 to 80 MWh/year and for developing countries it ranges from 2 to 5 MWh/year. It must be assumed that the primary energy demand will still increase quite markedly due to the population development and the economic growth in developing and threshold countries (see Fig. 2).

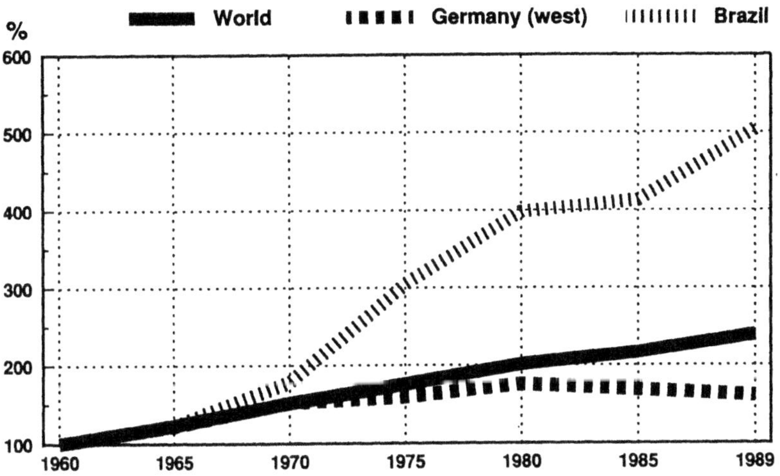

Figure 2: World energy consumption 1960 - 1989 (Index 1960 = 100 %)

Current forecasts [2] predict rates of increase of 1.5% to 2%.

An even greater increase can be found in the development of demand for electrical energy. Growth rates of 4% are predicted in this field. The global development of demand is shown in Fig. 3 [1].

At present, the consumption per head is 2200 kWh/year worldwide. However, large differences can be observed: For the United States, the value is around 19000 kWh/year, for Brazil 1700 kWh/year and for India around 300 kWh/year. As in the case of the primary energy sources, the highest growth rates must be expected in the third-world countries (see Fig. 4).

This continuing increase in energy demand is contrasted by clearly visible negative effects. This applies particularly to the combustion of fossil type energy sources. Limits to the damage which can be inflicted on the environment are indicated both by the local environmental damage and the global greenhouse effect which is attributable to the CO_2 emissions to a significant extent. Additionally almost none of the primary energy sources, on which today's energy supply is based, are renewable and their scope is therefore limited.

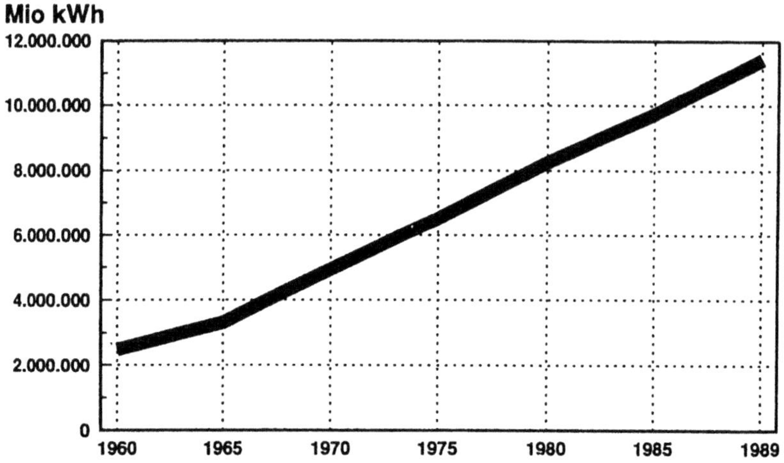

Figure 3: World electrical consumption 1960 - 1989

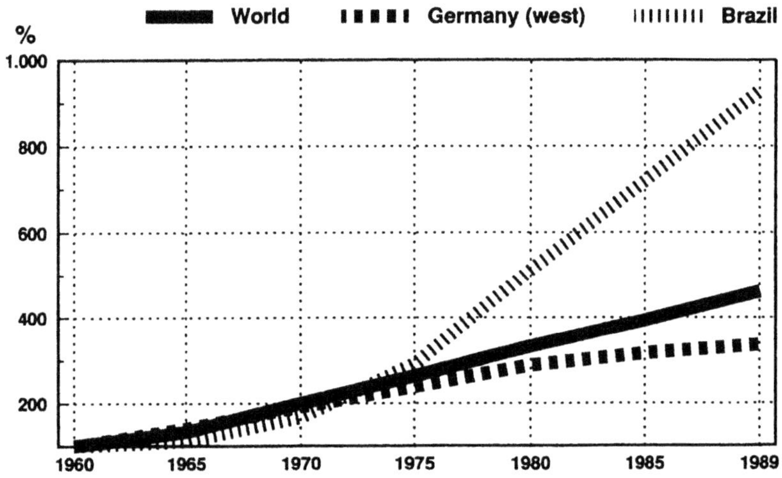

Figure 4: Electrical consumption 1960 - 1989 (Index 1960 = 100 %)

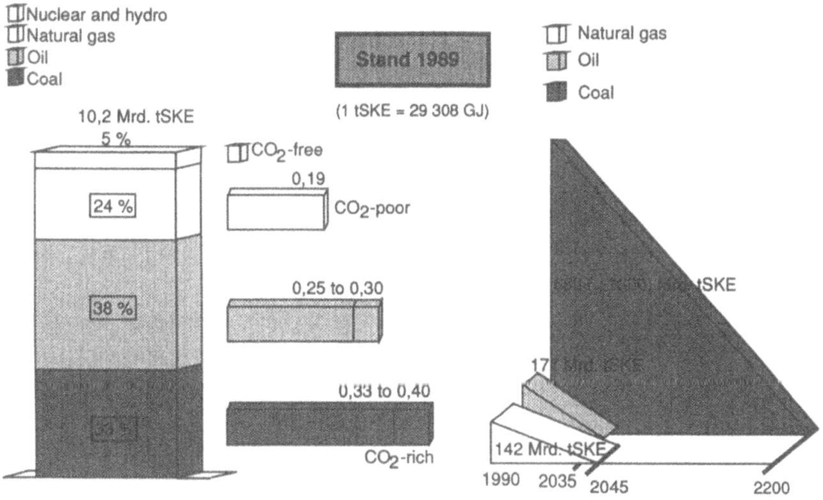

Figure 5: Left side: World wide energy demand and CO_2-emission rates kg CO_2 / kWh in 1989 — Right side: Resources of fossil fuel and usable range

Fig. 5 shows the estimates on the reserves of the 1988 consumption: 50 to 60 years for oil and natural gas. Even if the resources are considered which can be recovered with a probability of only 50 to 95% the scopes can only be extended by another 40 to 50 years [2]. The figure illustrates that around 50% of our current primary energy basis will be used up in less than one century and moreover that these are exactly the primary energy sources which will increasingly be used because of the CO_2 emissions. Thus, in the future the power supply industry will be confronted by the following, in some cases contradictory general economic and ecological conditions:

- The rise in global energy demand,
- the scarcity of essential resources, particularly of mineral oil and natural gas,
- the limits to which the environment can be burdened by anthropogenic emissions.

In view of this situation the following actions are required:

- More efficient energy conversion,
- more efficient use of energy,
- substitution of the non-renewable primary power sources by renewable ones,
- reduction in the emissions of primary energy sources with particular impact on the environment.

The power engineer has the central task of creating systems and components and to operate these in such a way that this requirement for action is met whilst at the same time taking into consideration a balanced measure of economy and ecology.

2 The planning tasks in the electric energy supply

The supply of electricity represents a complex task. The most important planning activities can be arranged in a three-dimensional planning space spanned by system, time and requirements [3] (see Fig. 6).

The time horizon to be investigated can be divided into operation planning and expansion planning. The system domain contains all relevant subsystems of energy extraction, conversion and distribution including the area of application. It also includes delivery contracts for fuels and electrical power. The requirements set for the supply of electrical power are of a technical, operational, legal and economic type. In general, the target of planning is to minimize the costs for construction and operation of the power supply system. The economic efficiency thus becomes the objective function of the planning task in normal operation. The other requirements are represented by the constraints. When there are operational disturbances the order of priority of the requirements is different. In such a case the restoration of a reliable (n-1) operation and minimization of the undelivered energy takes precedence over economic efficiency.

For normal operation other planning objectives such as minimization of fuel demand or minimization of emissions are also conceivable. Such objectives can be set, for example by governmental laws.

A self-contained solution to all tasks in the system planning space is neither possible nor useful. It is the task of the planning engineer to carry

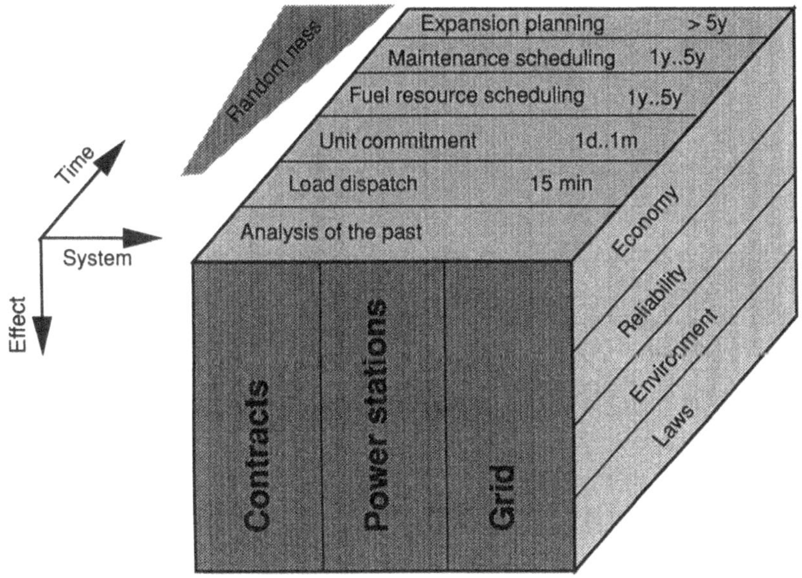

Figure 6: Planning space

out the correct decoupling operations and to determine the corresponding part-solutions.

Since the decisions concern periods of different lengths it is possible to split up the overall task in the time domain. This provides the hierarchical planning structure (see [4], [5], [6], [7], [8]) represented in Fig. 7.

The splitting-up is essentially based on separating long-term from short-term planning. The remarks which follow are restricted to the area of operational planning. However, the planning methods developed for this domain also represent an important aid to the problems of how to plan the expansion of the supply system.

Long-term operational planning covers the period from one month to more than five years. As a rule, the week or the month is selected as time reference. Short-term operational planning takes into consideration a maximum period of 10 days and is divided into steps of one hour down to 15 minutes in time.

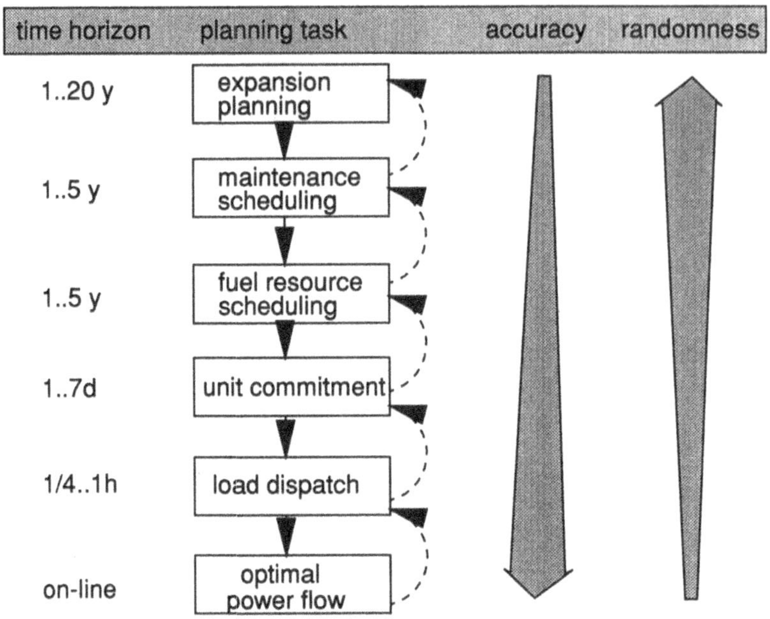

Figure 7: Hierarchical structure of the planning task

All planning activities are based on forecasts and predictions. Both the mean and the maximum forecasting errors increase with the length of the period of forecasting. Forecasting to a longer planning horizon, for example a year, is no longer useful for some quantities.

The first uncertainty to be mentioned is the probability of failing units, especially in predominantly thermal systems. The reliability of thermal units varies as a function of their type, their rated output and their age. It is distinctly less than 1. Forced outages of units have an effect on the fuel utilization of the failed unit itself but also on the units used as standby. They also require the provision of spinning reserve. Since the failed unit, as a rule, had better generation costs than the standby plant there are increases in costs which can be within the percent range depending on the structure of the power system [9].

A further stochastic influence can be found in the error of the long-term load forecast. It was shown in [10] that the additional costs for planning the daily capacity due to inaccurate load forecasting can be in the range of some tenths percentage points of standard deviation of the load forecasting error. Similar values apply to the long-term planning tasks.

In systems with hydraulic power stations, the stochastic nature of the natural water inflows plays an important role. The inflows can only be forecasted in an inaccurate way, especially in the long term. Also, they are subject to annual forecasting errors of over 10%. Since there is a severe difference between the operating costs of hydraulic and of thermal power stations the resultant influence on the operating costs is also very high [9].

Since uncertainties have a higher influence especially in the long-term planning tasks, stochastic planning methods are already used in this area. The short-term operational planning, in contrast, is based on known, determined input variables. Long-term and short-term operational planning can be subdivided even further with reference to the problems set:

- Maintenance scheduling establishes the planned outages for maintenance purposes for each power station unit.
- Fuel resource scheduling makes cost-optimized use of fuels and delivery contracts whilst meeting all long-term constraints conditions.
- Unit commitment determines the optimum start-up and shut-down times for the individual units. Here, all technical/operational constraints must be met and the inputs from fuel resource scheduling must be taken into consideration.
- Load dispatch evaluates the optimal output of the individual units. Here too, energy constraints from fuel resource scheduling must be taken into consideration.
- The area of short-term operational planning also includes the use of controllable loads. These are essentially electrical storage heaters which must be allocated to a time window for the charging-up process [11].

As a rule, the electrical grid is not taken into consideration in the long-term planning tasks. It is only in unit commitment and load dispatch where various attempts to include the grid are known:

- Limited exchange capacities between the different areas,
- loss coefficients,

- linearization of the load flow equations and quadratic modelling of the active-power losses [10],
- optimization of voltage/reactive power ([12], [13]).

The order of priority of the different planning activitites is fixed. The higher-priority tasks determine inputs for the lower-priority tasks. However, the interaction of the short-term planning with the long-term planning must also be ensured. As soon as more accurate input data are available or the actual schedule deviates greatly from the planned schedule the long-term planning tasks must be rerun.

3 Structure of the power systems

The electrical power generated worldwide was 11500 TWh in 1989, of which 18% was hydroelectric power, 17% was nuclear power and the remaining 65% was produced by conventional thermal generation. Thermal generation thus clearly dominates with respect to energy, but 80% of the countries covered by UN statistics operate hydroelectric power stations [14] (see Fig. 8).

An inquiry by CIGRE in 1989, in which 30 electricity supplyers were approached, showed that 40% of the systems are to be classified as 'purely' thermal systems (less than 15% hydroelectric power), 40% as hydro-thermal systems (between 15 and 85% hydroelectric power) and 20% as 'purely' hydraulic systems [5]. The number of generating units ranges from a minimum of 5 hydro stations and 5 thermal generators to a maximum of 450 hydro and 400 thermal stations. This paper deals exclusively with optimization tasks from the area of thermal systems. However, the objectives are similar to the hydro-thermal systems where, as a rule, the complexity of tasks is even greater.

4 Long-term planning

4.1 Maintenance scheduling

Power station units must be routinely taken out of operation for preventative maintenance at certain time intervals which depend on the unit type, age and output. These outages result in additional personnel and material

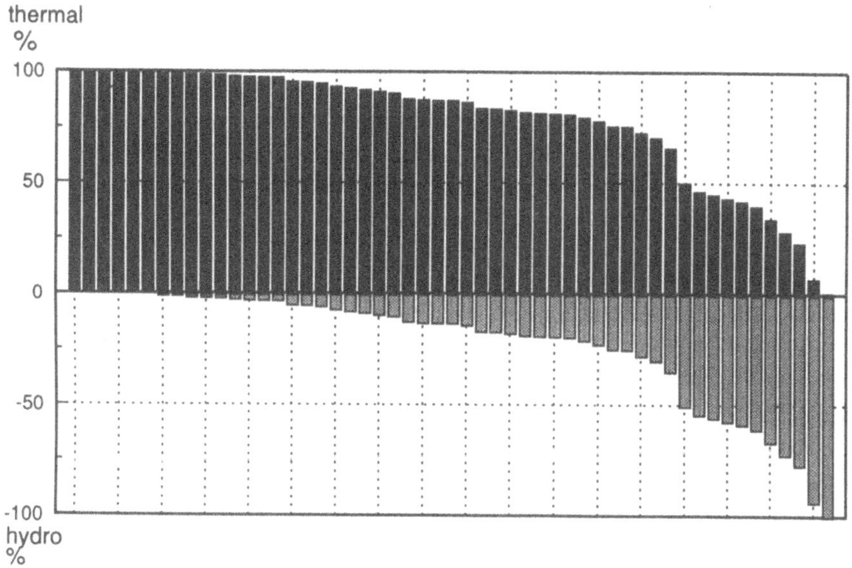

Figure 8: Electricity production 1989 (Shares of thermal and hydro electric generation)

costs for the work to be performed during maintenance on one hand and on the other hand, additional energy costs arise for replacing the shut-down plants by more expensive units. The costs mentioned first depend on the extent of the work to be performed. It is determined by manufacturer's information, operator experience and legal or insurance-related factors. The additional energy costs, in contrast, are determined by the load situation during the maintenance shutdown and the fuel costs of the stand-by unit; they are thus immediately dependent on the time of shutting-down. Maintenance scheduling is the task of determining the optimum shut-down times for all maintenance related issues of the supply system. Here, the following constraints must be met:

- Meeting the demand for power at any time,
- maintaining a predetermined minimum and maximum interval from the preceding study,
- maintenance outage of the unit,

- restrictions in the availability of resources (for example personnel),
- restrictions with respect to the simultaneous maintenance of certain power station units, for example units from one geographic region in the case of transmission bottlenecks,
- restricted fuel reserves in the case of nuclear power station units,
- keeping adequate reserves in stock. The usual requirement is a total reserve capacity of the largest unit not undergoing maintenance outage.

In the methods developed until today two approaches have been used for the objective functions:

- Minimization of the annual production costs,
- maximization of system reliability.

In principle, these two approaches can be merged into one another via the assessment of unused energy. In [15] an approach is developed in which the minimization of energy costs is selected as objective function and the level of reliability is formulated as a constraint.

The method is based on the mixed integer linear programming (MILP) method. The sequence upon which this method is based is shown in Fig. 9.

The time horizon is one year which is split into time intervals of one week. The optimum maintenance schedule is determined in two steps:

- In a first step, all theoretically conceivable maintenance states which meet all constraints are determined for each week. For each of the permissible states, the expected value of the weekly energy costs and the level of reliability are determined by means of a stochastic system-operating cost calculation [9]. If the level of reliability is too low, the state is discarded.
- The remaining states are coupled via a mixed integer linear computing model for the annual period under consideration. The problem is solved with a standard algorithm of the mixed integer linear programming method.

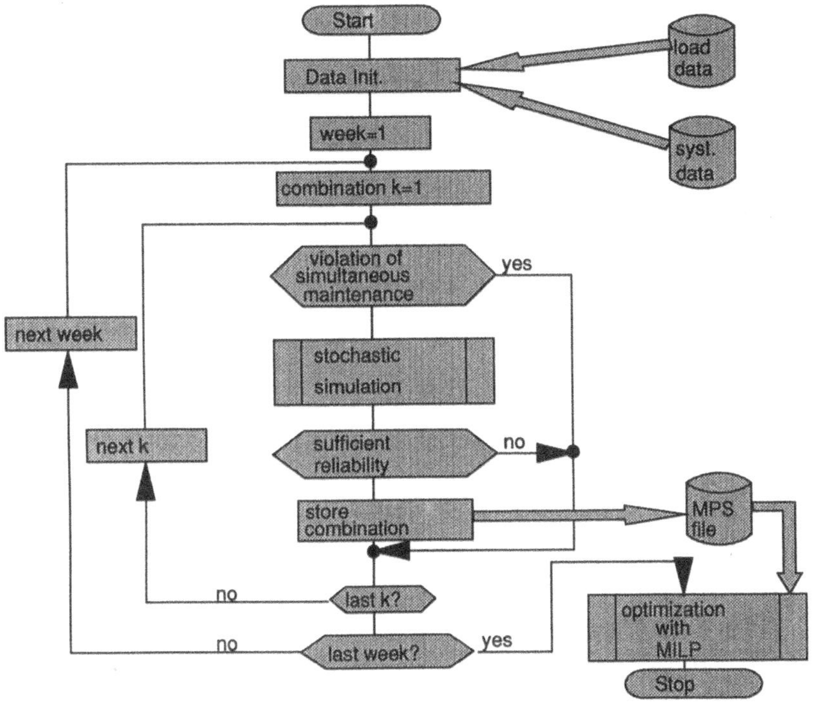

Figure 9: Flowchart for maintenance scheduling

Since a large part of the violations of constraints is already recognized when the MILP model is set up the number of maintenance states can be greatly reduced. The mathematically highly efficient enumeration of the states leads to low computing times which are within a range of 500CPU seconds (CYBER 175) in a system comprising 15 units. In addition, the algorithm allows the power system to be split up and the maintenance plans of the subsystems to be determined sequentially. This reduces the computing times to about one fifth [15]. The deviation from the mathematical optimum is less than 0.01% and is thus negligible. The method can thus be used for very large systems. In local german power systems studies have been undertaken and the savings in fuel costs amounted to DM 3 to 5 million as compared to a good heuristic method; this corresponds to about 0.2 to 0.3% of the fuel costs [15].

4.2 Fuel resource scheduling

The electricity supply systems are increasingly characterized by long-term energy conditions and price structures. This not only applies to hydro-thermal or hydro-electric systems with their annual reservoirs but also to 'purely' thermal systems. As a rule, these long-term price structures have the aim of reducing the sales risk of the fuel or electricity suppliers and of ensuring a uniform capacity utilization of its extraction, transportation or conversion facilities which is as high as possible. The following are examples of long-term quantity conditions:

- Minimum purchase liability (take-or-pay quantity) in the case of natural gas. The minimum purchase quantity must be paid even when it has not been used. It is determined on the basis of the maximum capacity per hour. As a rule, the take-or-pay quantity is obtained by multiplying the maximum capacity per hour by a full-load utilization period (e.g. 7500 hours). If the minimum quantity is not reached, the contracts frequently provide for a catching-up period of two or three years but the maximum quantity per hour is not increased.
- Minimum purchase liability in the case of electrical energy imports. This mainly applies to the area of industrial supplementary electricity supply.
- Purchase liability for domestic hard coal in Germany. This liability which is based on the 'Drittes Verstromungsgesetz' prescribes a minimum purchase of domestic coal for a 15-year, 5-year and 1-year period. 1-year and 5-year quantities may be varied within certain limits. The total quantity over 15 years must be accurately maintained. When the purchase liability has been met the utility receives permission - under certain conditions - to use less expensive imported coal.
- Restricted energy quantity in nuclear units.

There can be long-term price structures both in fuels and in electricity purchases. An example of such a price structure is shown in Fig. 10.

In this example the purchase price changes not only at a fixed time t_1 but also after certain quantities of energy have been used. Constant-price energy areas are called zones; they are normally the size of 1000 to

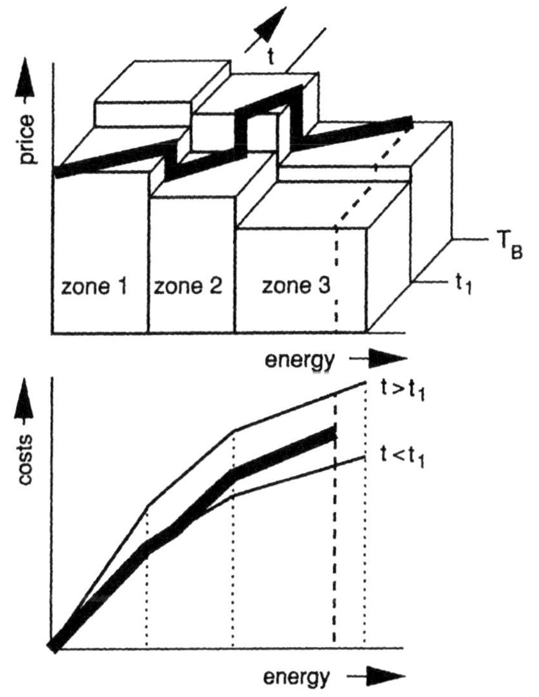

Figure 10: Structure of zoned contract

3000 full-load hours. With such price structures the electricity received in the current time interval influences the cost and the price structure in the remaining long-term planning period. It is absolutely necessary to take into consideration these effects of energy purchase in unit commitment and load dispatch; i.e. these short-term tasks must receive inputs related to energy to be used in the short-term period from a higher-priority program. It is the task of fuel resource scheduling to determine these inputs in a way that the long-term constraints are met.

A large number of papers has already been published in the field of fuel resource scheduling [4], [8], [9], [16], [17], [18], [19]; due to the complexity and system-specific structure of the task, however, no method has been really successful. Stochastic methods are in most cases used in hydro-thermal systems because of the inflow characteristic. In thermal systems deterministic methods are also used. They provide for more accurate modeling of the unit and contract characteristics and lead to time- and quantity-dependent price structures.

The time horizon of fuel resource scheduling is one year or longer. It is split into one-week or one-month intervals. The weekly pattern has the advantage that it corresponds to a natural load cycle and is consistent with the maintenance dates which are also specified in this time frame. In contrast, the monthly pattern leads to smaller optimization models but requires another intermediate step during the transition to the short-term operational planning. The fuel resource scheduling can be coupled to the short-term operational planning either via quantity inputs or via shadow prices. It was shown in [4] that coupling via shadow prices leads to better results. It was possible to lower the annual costs by up to 1% compared with energy inputs.

The constraints which are only relevant in the short-term domain are neglected in fuel resource scheduling. This includes the minimum unit outputs, minimum down time and minimum operation duration. On the other hand, the power balance and the requirement for spinning reserve must be taken into consideration.

Figs. 11 and 12 show the flowcharts of two methods for fuel resource scheduling which have been implemented. The method in Fig. 11 achieves optimization by means of successive linear programming (sLP). Costs are evaluated by means of stochastic system operation calculation. Stochastic influences such as forced unit outages and forecast errors are taken into consideration. The objective function of the optimization is the expected value of the annual costs. The weekly quantities of energy of the energy-restricted units are variables. The objective function is linearized at each iteration of the successive linear programming and the solution space becomes smaller from iteration to iteration (contraction of the solution space). After 30 to 50 iterations the process begins to oscillate around a solution. This means that the break-off criterion has been reached. Process convergence is ensured only if both the objective function and the solution space are convex. As a rule, however, this is not so in the case of zoned purchasing contracts.

If such price structures exist the process shown in Fig. 12 is better. It is based on the mixed integer linear programming method (MILP). This deterministic approach models the nonlinear cost variations with the aid of 'special ordered sets'. The load is represented as a step-shaped duration curve and the requirement for spinning reserve is taken into consideration

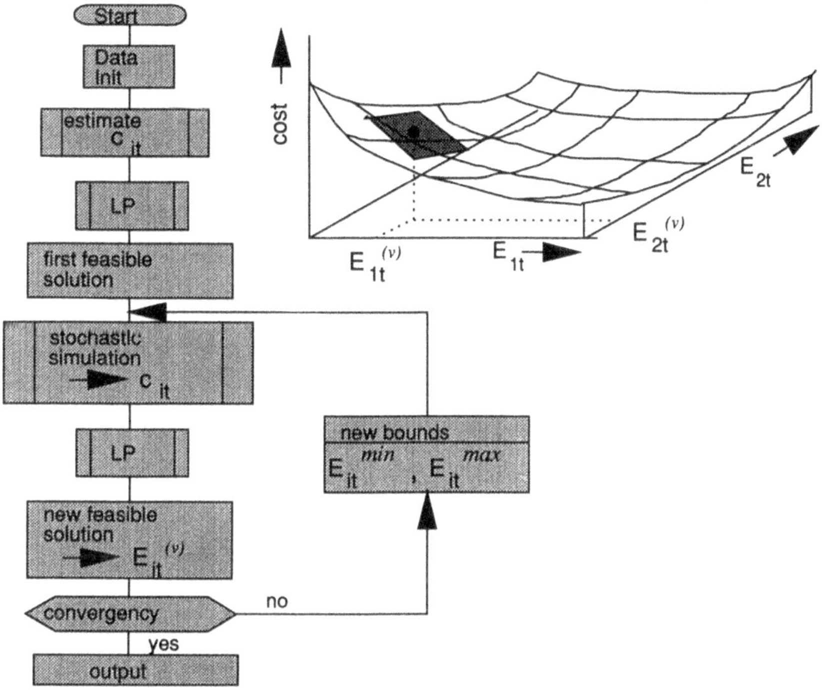

Figure 11: Fuel resource scheduling stochastic approach: successive LP

via a heuristic equation: before it is transformed into a staircase type curve the load duration curve is modified by a $\cos(\phi)$ transformation. The shadow prices which must be transferred to the short-term optimization are supplied directly by the MILP algorithm.

5 Short-term operational planning

5.1 Unit commitment

The unit commitment specifies the start-up and shut-down times of the individual generation plants. A large number of constraints must be taken into consideration. It is advisable to split them into component-dependent and system-dependent constraints.

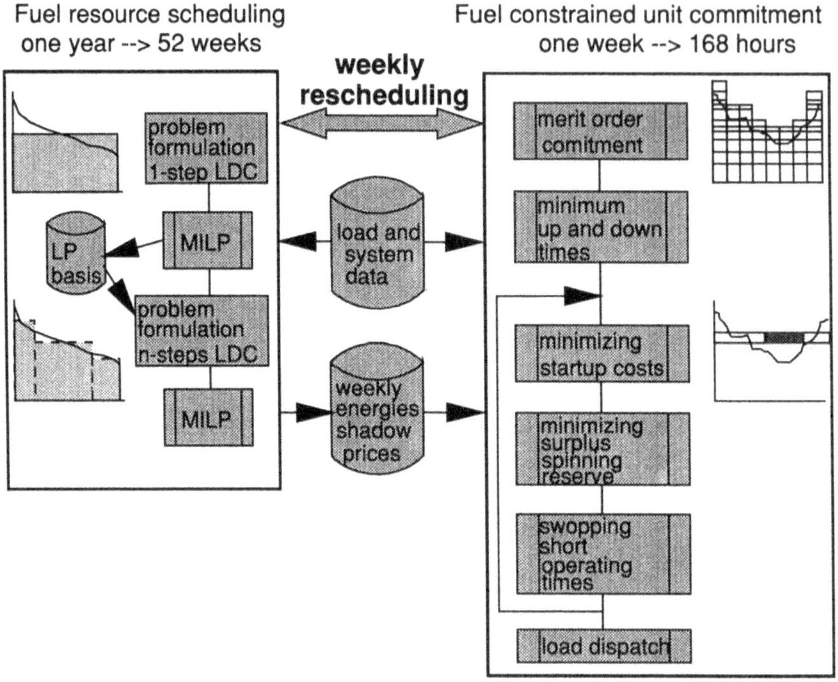

Figure 12: Fuel resource scheduling: Deterministic approach - Mixed integer/LP

5.1.1 Component-dependent constraints:

- Unit output limits: A thermal unit can either be shut down or operated in a steady-state condition in the area between its minimum output and its maximum output. The minimum output is here 50 to 70% of the rated output. The output limits relevant to the optimization can be closer together than the technical output limits of the plant. This is the case, for example, if the unit is involved in primary or secondary regulation control.

- Minimum-down and minimum-up times: The minimum-down time specifies that a unit can only be started up when its down time exceeds the minimum value. This ensures that inadmissibly high thermal stresses are prevented. These stresses can occur in the thick-walled turbine sections when the unit had an actual down-time shorter than a given minimum down-time. The minimum down-times lie between two and twelve hours. In some cases the requirement for a minimum up-time is also raised for the same reason. The minimum up-times lie between one and eight hours.

- Ramp rate limits: When optimization is performed at hourly intervals ramp rate limits are of no importance. Even in the case of optimization at 15-minute intervals usually they only need to be taken into consideration for the start-up and shut-down time intervals [10].

- Other constraints: Further requirements can occur in certain cases, e.g. scheduled mandatory service or mandatory stoppages. A further group of secondary conditions can arise from the fact that the manpower needed for start-up is not always available at the same time for several units. This means that only a certain number of units may be operated or started up at the same time in a power station.

5.1.2 System-dependent constraints

The power balance between generation and consumption which includes the transmission losses must be ensured at any time. In addition, a given spinning reserve must be ready at any time in order to meet random disturbances on the system such as forced outages or unforeseen load charges. The spinning reserve capacity is often assigned to the output of the largest unit in operation.

Other restrictions may arise from the presence of fuel constraints or zoned price structures. If such constraints and prices are of a long-term type the inputs of fuel resource scheduling must be taken into account.

If the power system includes controllable loads a large number of special constraints must be met. Section 5.2 provides a more detailed description of the relevant tasks which need to be done.

5.1.3 Objective function

The objective function is composed of the steady-state operating costs and the start-up costs. Steady-state operating costs are the energy-related costs which occur when a unit is operated at constant output. In most cases they are approximately 99% of the total operating costs. The steady-state operating costs are composed of the fuel costs (85%) and the additional, energy related costs (15%) in thermal units. The fuel cost depends on the heat rate of the plant which is a function of the delivered output. In many cases this function is not convex. The costs arising from the intermittent operation of the units are combined in the start-up costs which only amount to about 1% of the total operating costs. These costs are composed of two parts: The first one represents the fuel expensed for producing the accumulated heat and the warm-up losses. This fuel expenditure can be simulated by an e^x-function which reproduces the cooling-out behavior of the unit. The second part includes all the cost components which are independent of the down time. These are essentially the maintenance costs produced on start-up and the costs which can arise from the increased risk of forced outage in the case of the start-up process.

5.1.4 Period of observation

The length of the time horizon depends on the type of constraints. The simplest case is given when there are no long-term conditions and no minimum down time and up time periods. In this case, a period of observation of 24 hours is sufficient for work days; a period of 72 hours is appropriate for the weekend. If, however, minimum-down time and up time periods are defined the observation of a natural load cycle, e.g. one work day, is no longer sufficient since the decision for the cycle just observed can have an influence on the decisions of the subsequent cycle. A sliding optimization window must therefore be defined which should have a length of 48 hours or 72 hours, respectively.

In systems with long-term conditions even this two- or three-day rhythm is no longer sufficient. The optimization period of unit commitment must be matched to the time difference of the fuel resource scheduling. If the latter was in weekly intervals the unit commitment must also span a period of 168 hours. If the fuel resource scheduling takes place at monthly intervals a further process step must be provided which determines weekly inputs from the monthly inputs.

5.1.5 Methods

A large number of methods for unit commitment have been developed in accordance with the requirement which vary largely among different power systems (see [20] - [26]). Usually they are based on heuristic or mixed mathematical/heuristic approaches. Purely mathematical methods are not used mainly because of the computing times involved.

In the mathematical method components dynamic programming and the mixed integer linear programming can be considered due to the mixed continuous/combinational character of the task defined. For both basic approaches, further developments are known which make use of the Lagrangian Relaxation for certain steps [26].

The method described in [21] can be used for tasks in which a purely thermal system without fuel constraints is to be optimized. The concept of this method is based on the fact that the start up costs are of the order of magnitude of only 1% of the total operating costs and can therefore be taken into consideration in a simplified form. In purely thermal systems without fuel constraints only a time-coupled optimization over the entire observation period is required. This is due to the start-up costs and the minimum-up and down time periods which lead to a large number of variables to be optimized simultaneously. To simplify the unit commitment problem can therefore be modeled in such a way that the dynamic optimization calculates the optimal unit separately per time interval in an iterative process. Each time interval is considered individually and the time-coupling start-up costs and minimum-up and down time periods are handled by penalty terms in the cost functions of the individual units.

Initially the unit is committed in the individual time intervals in accordance with the following algorithm: The power system is built up step by step from the individual power station units. Each step of the optimization process corresponds to the addition of another unit. For each time interval related optimized variables are obtained. These variables are the decision variables. For each optimization the cost functions of the units are specified in the form of discrete values in a table. In step 1 of the optimization the costs of each possible output step is given by the cost table of unit 1. In step 2 the output combination leading to the lowest total cost is sought for each possible aggregate output of the two units. This process is continued until all units have been combined to form the total system (see Fig. 12). After this forward calculation has been completed the results represent the minimum aggregate cost as a function of all possible outputs of the system and the relative optimum output of the unit considered in each case for each optimization step. In order to meet the spinning reserve requirement the sum of the capacity of the units switched on must also be stored at each state point. This table is set up once for all time steps. The backtracking is carried out individually for each time interval on the basis of the loads and standby requirements which differ in each time interval. Since this first iteration did not take into account any start-up costs a unit commitment schedule with many unit-up and unit-down changes if obtained. The sum of the steady state operating costs of the units has the lowest possible value, the start-up costs, however, are very high. The unit commitment schedule is therefore quietened down step by step in the subsequent iterations. Uneconomic down and up times are eliminated by introducing down time and up time penalties. These penalties depend on the magnitude of the actual start-up costs. The process converges after 4 to 18 iterations. In an optimum commitment schedule the sum of all penalties corresponds to the sum of the calculated start-up cost.

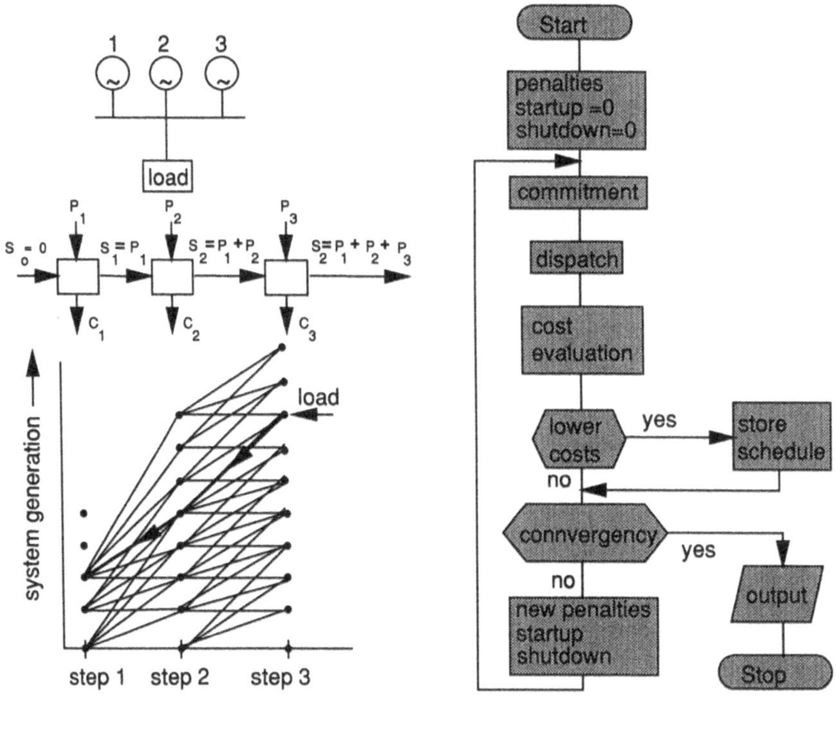

Figure 13: Unit commitment - Dynamic Programming

Fig. 13 shows the sequence of the search process.

This algorithm has a good convergence characteristic and short computing times. A comparison of methods carried out in [25] has also shown that usually with dynamic programming methods the commitment schedules with the lowest operating costs are obtained. However, the approach can no longer be used if fuel constraints are present.

Several methods for unit commitment in fuel constrained power systems have been described. The approach developed in [23] is based on the integer programming method and can only be applied to very small systems. The approach in [24] uses a network flow algorithm for load dispatch and a heuristic approach for the unit commitment. A combination of Lagrangian Relaxation and linear programming is described in [26].

In [4] a purely heuristic process is presented which has been successful for optimizing systems in which a large number of units are subject to energy restrictions. The optimization period is one week and the time reference is one hour. The algorithm works also with energy inputs or with shadow prices evaluated in the higher-level fuel resource scheduling (see Fig. 12). In the first case load dispatch is performed by means of network flow algorithm and in the second case it is carried out by means of the Lagrangian relaxation method.

5.2 Use of controllable loads

A large number of utilities try to achieve a flat load curve in order to increase the economic efficiency of the supply. This can be done if loads are in the power system where the supply of electrical power can be decoupled from the load demand during a certain time period. The prerequisites to be met by the loads are indirect storage capability and the technical possibility of centralized control. An example is the electric storage heater switched by means of audio-frequency ripple control which is widely used in Germany. The installed power of electric storage heating devices in Germany is currently around 40000 MW [11]. The load curve of power systems with electric storage heating devices is composed of two components: the basic load which cannot be influenced by the power suppliers and the controllable load which can be switched group by group by means of the above mentioned ripple control system. The switch-on time can be shifted and interrupted for a certain time without violating the requirement for adequate supply. The supplier is thus able to influence the total load characteristics in such a manner that the electric storage heating does not contribute anything to the load peak and that the demand is covered if possible from base-load units or by purchasing only in the low-cost time. There is close coupling between the commitment of the controllable load groups and the commitment of the generation plant and imported energy via the overall load curve. This coupling requires that the controllable loads are taken into consideration in commitment scheduling. This results in further constraints for the unit commitment:

- Maintaining the minimum relief time for each load group,
- maintaining the control period,
- maximum number of interruptions of the charging process,
- charging characteristic of the individual load groups.

In [11] a mathematical approach to a solution is developed which is based on the mixed integer linear programming method. The non-linear cost curve of a unit is represented by a special ordered set for each time interval. This set is coupled to a binary state variable which defines the switch-on state of the unit. The start-up costs can optionally be represented as a step function, as a linear or as an equally weighted step function. The controllable loads are simulated by means of special ordered sets of type 1. The computing times for this model can be very high. For a system with 3 units, one purchasing contract and 6 controllable load groups CPU times between 1s and 800s can be expected (CYBER 175). The last value mentioned applies when the charging process can be interrupted. An application of the method in operational practice is not known. In [27] an heuristic approach is described which is currently being used in a regional power system in Germany.

5.3 Load dispatch

After the determination of the plants by the unit commitment the load dispatch function has the task of specifying the optimum output for each unit in operation and for the purchasing contracts. Here, the output limits of the individual units and the load requirement must be considered. This is a purely continuous optimization problem. The selection of method depends on:

- the cost structure of the plants: linear / non-linear, convex / non-convex, continuously differentiable / discontinuously differentiable....
- the presence of energy quantity conditions.

5.3.1 Systems without fuel constraints

If no fuel constraints are present the time steps of the observation period are independant from each other. In the simplest case of constant incremental costs the load dispatch can be carried out by means of a hierarchy list in which the units are ordered in accordance with their incremental costs at the nominal working point. This approach is frequently used in heuristic commitment scheduling methods in order to obtain a very rapid determination of the steady-state operating costs for a particular starting combination.

If the cost functions can be modeled by convex second- or third-order polynomials, the incremental cost method (Lagrange multipliers) supplies the optimum-cost load dispatch with relatively low computing times.

If no convex analytical representation of the cost functions is possible the load dispatch can be carried out by means of dynamic programming.

5.3.2 Systems with fuel constraints

If there are fuel constraints for certain units in a power system it is no longer possible to decouple the time intervals. The load dispatch must be carried out as a closed calculation for the same period for which the fuel constraints have been evaluated: As a rule, one week periods are used or, if updating occurs within this period, the constraints are evaluated for the remainder of the week. A method which can be used is the network flow algorithm. Compared with linear programming which would also be suitable in principle the computing time is about 100 times faster [11]. In this approach, however, the operating costs must be approximated by linear or convex, piecewise linear functions.

6 Summary and outlook

The supply of electricity takes place in the complex field of mutually contradictory requirements. On one hand there is the requirement for inexpensive power supply. On the other hand requirements for reliability of supply and for the protection of the environment and of the resources are present. Due to the globalization of the markets and the introduction of competitive elements in the power supply industry (e.g. third-party access) the cost pressure on the power supplying utilities will rise. Furthermore, new environmental protection regulations and contributions must be expected in the medium term. In the long term a scarcity and increase in the fuel costs is possible.

Due to this development the significance of mathematical tools for planning the expansion and operation of power systems is still increasing. In recent years a large number of methods have been developed for the individual subtasks of:

- Maintenance scheduling,
- fuel resource scheduling,
- unit commitment and
- load dispatch.

A variety of requirements must be fulfilled:

- The solution must be optimal,
- the model must have a high accuracy,
- the computing times must be short and
- robustness, i.e. the method must converge reliably and must not produce any infeasible solutions.

Since this list of requirements is neither satisfied by purely heuristic nor by purely mathematical methods, practical applications must be based on a combination of both. Based upon estimates by users these methods result in a gain around 0.5% of the variable operating costs compared with purely heuristic approaches. It must be assumed that the heuristic method components loose their significance to the extent that the performance of the mathematical algorithms and the computer capacity increase. The mixed-integer/linear programming method, in particular, is rated as having a high development potential for deterministic tasks.

The following aspects are particularly important for the practical value of the methods and their acceptance in the electrical power industry:

- The development should take place in close collaboration with the relevant personnel of the supplier.
- Before the method is implemented extensive tests should be carried out in order to ensure that the method is robust enough.
- The man-machine interface must be designed in such a way that
  - input data can be modified easily (e.g. load data, fuel prices, input/output curves),
  - intermediate solutions are output if necessary in order to check them for plausibility and optimality,
  - the results are displayed clearly in table form and graphically,
  - optimization runs can be carried out easily for the remaining optimization period,
  - the method can also be used for subtasks in expansion planning.

For many power systems, the practical applicability of the method also depends on the extent to which the hydraulic subsystems including all special features can be taken into consideration. It may also be necessary to include the electrical grid, particularly in systems with bottlenecks in the high-voltage lines. Both aspects represent significant extensions to methods and models with an even higher degree of mathematical complexity.

References

[1] VIK-Statistik, 1991/1992

[2] N.N. : Energie und Klima ; Hrsg. Enquete-Kommission 'Vorsorge zum Schutz der Erdatmosphäre' of the German Bundestag, Economica Verlag 1990

[3] Edwin, K. : Methoden systemtechnischer Planung. Lecture manuscript, RWTH Aachen,1989

[4] Wolter, H. : Kurzfristige Kraftwerkseinsatzoptimierung in thermischen Systemen mit langfristigen Nebenbedingungen. Dissertation, RWTH Aachen, 1990

[5] Mariani, E. : Methodologies in Medium-Long Term Scheduling. CIGRE Paper No. SC 87 03 Tokyo, 1987

[6] Vemuri, S.; et al : Fuel Ressource Scheduling, Part I - Overview of an Energy Management Problem; IEEE Trans. PAS-103, No. 7 July 1984

[7] Harhammer, P. G.; Infanger, Gerd, M. : Decision support system - operation planning; Electrical Power & Energy Systems, Vol. 11 (9189), No. 3, S. 155 - 160

[8] Slomski, H. : Optimale Einsatzplanung thermischer Kraftwerke unter Berücksichtigung langfristiger Nebenbedingungen. Dissertation, Universität Dortmund, 1990

[9] Th. Schroeder: Langfristige Energieeinsatzplanung von Kraftwerkssystemen unter Berücksichtigung Stochastischer Einflüsse. Dissertation, RWTH Aachen, 1984

[10] R. Hummel : Blockeinsatzplanung und Lastaufteilung unter Berücksichtigung des Netzes. Dissertation, RWTH Aachen, 1985

[11] A. Stockem : Optimale Tageseinsatzplanung in regionalen Elektrizitätsversorgungsunternehmen unter Berücksichtigung der steuerbaren Last. Dissertation, RWTH Aachen, 1988

[12] Glavitsch, H.; Spoerry, M. : Quadratic Loss Formula for Reactive Dispatch. IEEE Trans. PAS-102 (1983), S. 3850 - 3858

[13] Lemmer, S. : Rechnergestützte Spannungs-Blindleistungssteuerung in Hochspannungsnetzen. Dissertation, RWTH Aachen, 1982

[14] United Nations Energy Statistics Yearbook 1979 .. 1987

[15] Curtius, F. : Zum Einfluß des Revisionsplans auf die Betriebskosten im Kraftwerksystem. Dissertation, RWTH Aachen, 1985

[16] Ortjohann, E. : Mathematisches Modell und Verfahren zur langfristigen Einsatzplanung thermischer Kraftwerkssysteme unter Berücksichtigung des Energiefremdbezuges aus dem Verbundnetz. Dissertation, Universität-Gesamthochschule Paderborn, 1989

[17] Möhring-Hüser, W. ; Orthjohann, E. : Verfahren zur integrierten kurz- und langfristigen Kraftwerkseinsatzplanung. Elektrizitätswirtschaft, Jg. 90 (1991), Heft 24, S. 1323 - 1329

[18] Duran, H. ; et al : Long term generation scheduling of hydro thermal systems with stochastic inflows. IFAC, Rio de Janeiro, 1985

[19] Sherkat, V. R. ; et al : Stochastic Long-Term Hydrothermal Optimization for a Multireservoir System. IEEE Trans. PAS-104, No. 8, 1985

[20] Cohen, A. I.; Sherkat, V. R. : Optimization-Based Methods for Operations Scheduling Proceedings of the IEEE, Vol. 75(1987), Nr.12, S.1574 - 1591

[21] Machate R.-D. : Wirtschaftliche Auswirkungen ungenauer Eingangsinformationen bei der kurzfristigen Einsatzplanung thermischer Kraftwerke. Dissertation, RWTH Aachen, 1979

[22] Mariani, E. : Methodologies in Short-Term Scheduling. CIGRE Paper No. SC 87 02 Tokyo, 1987

[23] Pang, C.K., et al : Pool Daily Fuel Scheduling. EPRI Report, EL-1659, Palo Alto, CA, 1981

[24] Kumar, R., et al: Fuel Ressource Scheduling, Part III - The Short-Term Problem IEEE Trans. PAS-103, No. 7 July 1984

[25] Krenz, G.: Zur Frage des Nutzens mathematischer Optimierungsverfahren bei der Tageseinsatzplanung von Kraftwerkssystemen. Dissertation, RWTH Aachen, 1983

[26] S.K. Tong; S.M. Shahidepur : Combination of Lagrangian-relaxation and linear-programming approaches for fuel-constrained unit-commitment problems. IEE PROCEEDINGS, Vol. 136, No. 3, May 1989, S.162 - 174

[27] Effler, L.; Steiner, H.; Wagner, H. : Energiebezugsoptimierung und Lastführung; etz Bd.112 (1991), Heft 9, S.442 - 447

[28] N.N. : EDV-Optimierung des Kraftwerkseinsatzes - Definitionen, Anforderungen, Verfahren. Elektrizitätswirtschaft, Jg. 89 (1990), Heft 15, S. 848 - 855

EXPERIENCES WITH OPTIMIZATION PACKAGES FOR UNIT COMMITMENT

H. Sanders, K. Linke

VEW
D-4600 Dortmund, Germany

Abstract. The power system of VEW is a thermal system with an installed capacity of about 6.500 MW at present. The optimization packages are oriented towards company-owned and jointly-owned plants and long-lasting contracts. These contracts include a lot of different conditions such as price zones and energy constraints and non-convex cost curves. The complexity of problems in optimization at VEW leads to a three-step hierarchical structure of optimization consisting of the two steps long-term and medium-term optimization in operational planning and a final step of momentary optimization in actual operation. The final step takes charge of the unit commitment and the constraints computed in the preliminary steps of optimization. With respect to the long-term optimization at VEW experiences have been acquired with different optimization strategies, which use mainly analytical methods but also heuristic techniques. The applicability of these methods has been checked critically with respect to the need of optimization in real power systems and the limits of the methods were shown. The medium-term optimization in connection with the long-term optimization has to perform among other things the difficult task to link long-term and quasi actual conditions. Extensive examinations have been carried out for the integration of both optimization steps. The on-line performed momentary optimization cooperates very closely with load-frequency-control.

1. Introduction

The optimization of power plant generation is a basic task of all modern electrical energy supply systems. Its goal is to ensure adequate supply and utilization of electricity while at the same time being:

- economical in the use of resources
- reliable and sufficient as well as
- economic.

Simultaneously, constraints of technical, contractual and govermental nature have to be taken into account.

For existing power stations only operational cost can be optimized. The investment-cost is fixed when the power stations are built. There is no possibility of influencing the depreciation, the interest, the taxes or the personnel-

and insurance-cost by optimization of generation. Also, transient operational cost and a small part of stationary operational cost, such as maintenance and lubrication cannot be optimized. The main parts of the operating cost which can be minimized for existing power stations are:

- for generation in own stations
  - fuel cost and fuel transport cost
  - unit start-up and shut-down cost
  - cost for the spinning reserve
  - cost for auxiliary materials for the cooling water system, desulphurization and catalytic NO_x-reduction
  - waste removal cost (e.g. for ash removal)
- for contracted generation
  - contracted operating cost.

The optimization procedure yields
- the optimal commitment of the individual generators and power contracts
- the loading of committed generators and contracts dependent on system load.

These two interrelated tasks are solved by the help of computer programs which operate on the basis of a power system model [1, 2]. Such a model consists of the following two components :

- the load, which must be forecast for the optimization period
- the generation system, consisting of power units and electricity supply contracts.

2. Load modelling

The question which energy source should be used to cover the expected demands for electricity in an economical way, requires various methods of planning and operational tasks with different time frames.

The main tasks of long term planning (1 to 20 years) are
- planning of power station construction and de-commissioning
- long term electricity purchase and supply contracts
- fuel balancing of conventional and nuclear plants
- profitability planning for the company itself

- maintenance planning.

The main tasks of medium term planning (1 day to 1 year) are
- operational planning of unit utilization
- maintenance planning
- power purchase and supply agreements.

A sufficiently accurate load forecast is an essential prerequisite for optimizing the output of generating stations. The demand for electricity from the network by the customers varies at different stages. For example, during a year it is dependent on the season, on the day time and on the weather. Depending on the task at hand and on its time frame the models for load forcasting use either
- electrical energy forecast during a specific time or
- electrical power forecast at a specific time.

Experience has shown that energy forecasts are more reliable than power forecasts. This is based on the fact that deviations can be detected more quickly and more accurately in energy forecasts. Energy forecasts are made for each different customer group based on regional areas and their trend factors. These factors take into account

- long term average values from the past
- population expansion
- development of the number and structure of households
- the current and expected economic situation
- saturation effects of some electrical appliances
- the increasing usage of new appliances with especially low electricity consumption.

The expected growth of large industrial consumers is forecast according to their evaluations of future economic and industrial developments.

VEW uses its own in-house developed program to transform the forecast of monthly or yearly energies into daily MW load forecasts [3]. This program uses 15 minute load values and hourly weather data for the last five years as a basis.

3. Generation modelling

The VEW power system is a thermal system with a total installed capacity of approximately 6,500 MWs at present (Figure 1). The power stations which provide this capacity are owned 35% totally and 31% partly by VEW. 34% of the capacity is owned by other utilities, and VEW has long term contracts with these utilities. The primary sources of energy are nuclear, coal and natural gas.

There are no hydro power stations in the VEW system. Therefore, problems associated with water storage or water management are not taken into account. Heating power stations are only of minor importance. Existing power stations of these types follow the demand of heat.

Coal is fired with over 40% efficiency in conventional power stations and in one combined-cycle unit of the 700 MW class. This combined-cycle unit consists of a coal fired boiler and a natural gas fired turbine for the preliminary stage and was developed with significant participation by VEW. Its calculations have to take into account constraints on the primary energy sources natural gas, German coal and imported coal.

Combined gas-steam turbines are also used in four 400 MW natural gas units. The optimization process calculates the following operational conditions:

- *Waste-Heat Operation* is the complete utilization of a 55 MW gas turbine and uses the boiler as a waste-heat boiler. It is not economical to vary the output, which is a constant 70 MWs in this form of operation.
- *Benson Operation* works with variable output between 45 and 100% of the maximum capacity in the operation-range.

A further characteristic of the VEW power system is the relatively large amount of power purchased from thermal power stations not owned by VEW. This requires special attention for modelling the electricity generation. A wide variety of contractual conditions covering the amount of power, the minimum operating time and the minimum down time must be considered when applying these contracts.

The majority of electricity supply contracts, so-called "zoned" contracts, are for either monthly or yearly zones (Figure 2). This means that during the corresponding month or year the energy price depends on the utilization of the contract. Low utilization is associated with a high energy price which is designed to cover start-up cost and poor operating efficiency at low

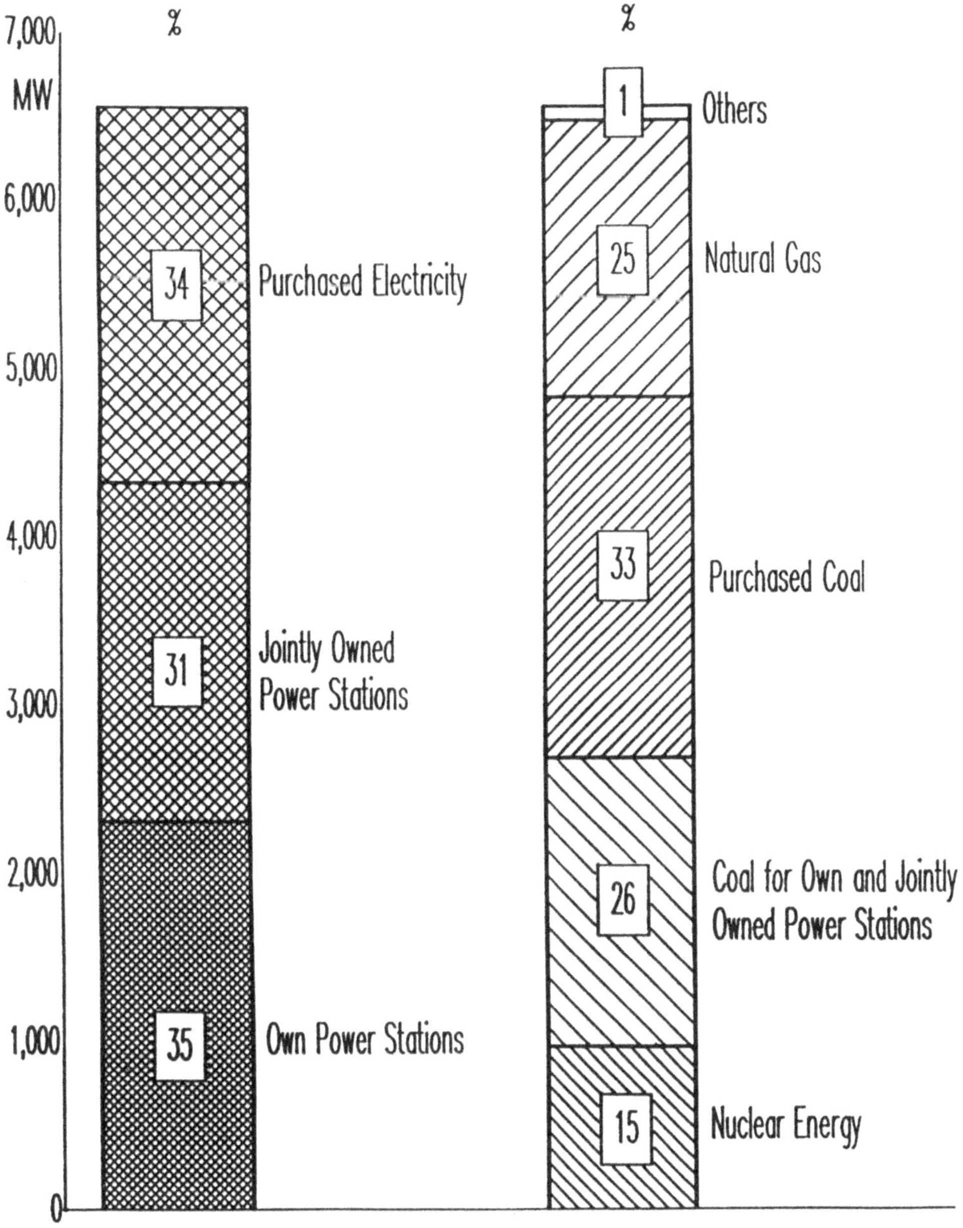

Figure 1: VEW - Generation System

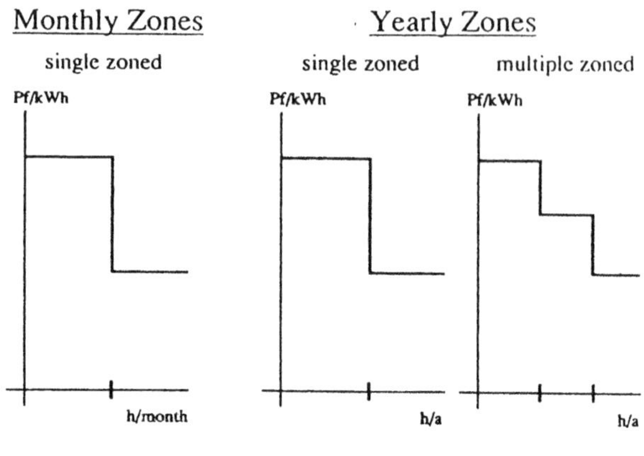

Figure 2: "Zoned" Electricity Contracts

unit output levels. Lower energy prices are charged for higher utilization compensating the cost penalties.

There are also constraints on the quantities of individual fuel types delivered to the power stations. These constraints are expressed either in absolute terms or in terms of upper and lower limits.

The German utilities have had to make contracts with the German coal industry for the 15-year period from 1981 to 1995. Different parts of the contracted quantities are legally subsidized in various ways. An annual quantity must be observed for each contract, with a yearly recognized tolerance level of +/- 15 % and a 5-yearly period tolerance level of +/- 3%. At the end of the 15-year period the whole quantity has to be exactly met. Based on the contracted German coal the German utilities have permission to import foreign coal without duties. The contracts for imported coal are valid for one year, and depend on the market-situation. In order to supply gas-fired and combined-cycle power plants, long-term gas-contracts over a 20-year period have been signed. The minimum and maximum values of annual, daily and hourly amounts are required subject to the whole period for optimization.

Apart from the constraints for primary energy, the following facts must be incorporated into the model for optimizing the generation system:

- Technical constraints for individual units and contracts
  - maximum and minimum output can change several times per year depending on various maintenance situations

- regulating range as reserve for units participation in load frequency control, which cannot be included in the optimization process
- minimum up and down times to prevent frequent start-ups and shut-downs (typical contractual minimum-operating times are 5 hours and typical contractual minimum-down-times are 8 hours)
- maintenance
- fixed output (e.g. of industrial plants)
- specific heat consumption from company or jointly owned units dependent on the power output (normally the heat consumption is calculated as a second order polynomial)
- start-up and shut-down heat losses of company or jointly owned units

- Technical and operational constraints for the whole generation system
  - load must be met every hour
  - spinning reserve must exceed the minimum requirements

- Economic and contractual constraints
  - it must be possible to change prices of primary energy for each unit during a year
  - operating cost for contracts (zone prices for electricity contracts can vary during the year depending on the price of primary energy)
  - quantity constraints
    a) quantity constraints have to be regarded for individual units as well as for groups of units and contracts.
    b) electric energy values must be observed when utilising units or contracts; minimum monthly or yearly electricity values must be observed for electricity contracts.
    c) a minimum amount of German coal must be used prior to the cheaper imported coal.

4. VEW optimization levels

The complexity of the VEW electricity supply system requires the optimization to be divided into 3 levels (Figure 3). They cover two planning levels for

- long term optimization
- medium term optimization

as well as an operational level for

- short term optimization.

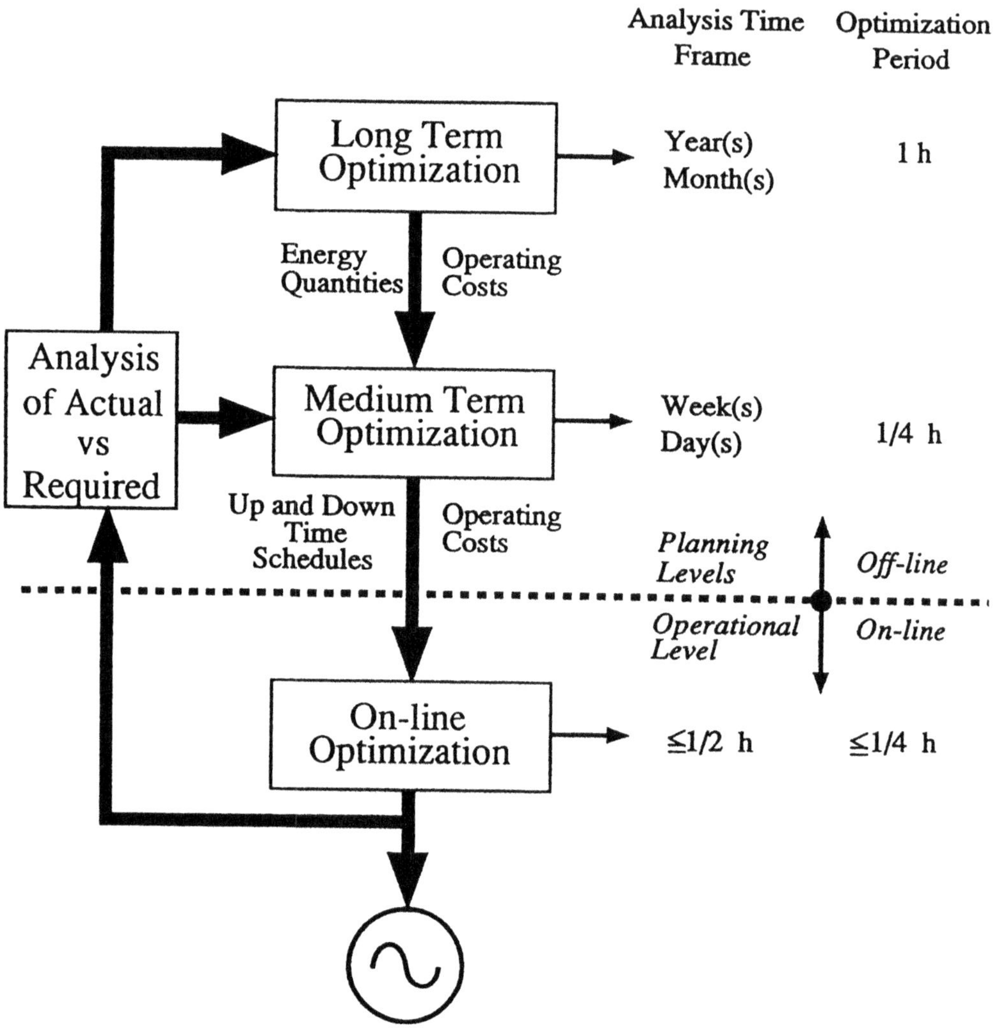

Figure 3 : Optimization Levels

The calendar year is used as a basis for long term optimization, because of the quantity constraints and in accordance with the financial and business considerations. Sequential optimization with respect to years can be used to examine a planning period of any duration. *Long term optimization* yields schedules for unit power outputs and contracted power on an hourly basis.

Instead of energy commitment, hourly schedules are produced in order to examine problems of the following type:

- Calculation of offers or purchases for electricity deliveries:
  determination of power station output which are required for special schedule deliveries or which are eliminated by purchases; determination of the corresponding cost or cost savings; this covers not only the fixed operating cost but also the start-up and shut-down cost for thermal units.
- Maintenance planning:
  determination of unit operating hours and start-ups in a particular time period.

The task of the second planning level, the *medium term optimization*, is the optimal utilization of the generation system for current and following weeks. This optimization level yields the commitment of the VEW units and of those units under contract to VEW as well as preliminary loading for these units.

More details are known about the expected load as well as about the available power for the weekly optimization process. The optimization results are therefore better than those of the long term optimization. The results are improved even further by switching to a 15 minutes time interval, providing a basis for short term optimization on the on-line computers.

Coupling medium term and long term optimization is necessary in order to ensure that long term strategies are accepted.

Short term optimization is continuously supplied with on-line data and meets final decisions concerning generation. It works in conjunction with load frequency control.

The following text considers long and medium term optimization which determine the committed units and further conditions on their utilization.

5. Annual optimization

5.1 Solution methodology

The first long term optimization package LFO-1 (Figure 4) was developed more than 20 years ago and used a heuristic method.

The optimization technique was based on increasing cost to obtain the optimal distribution of the energy between available generating units and electricity contracts. In a next step, an hourly unit commitment facility was

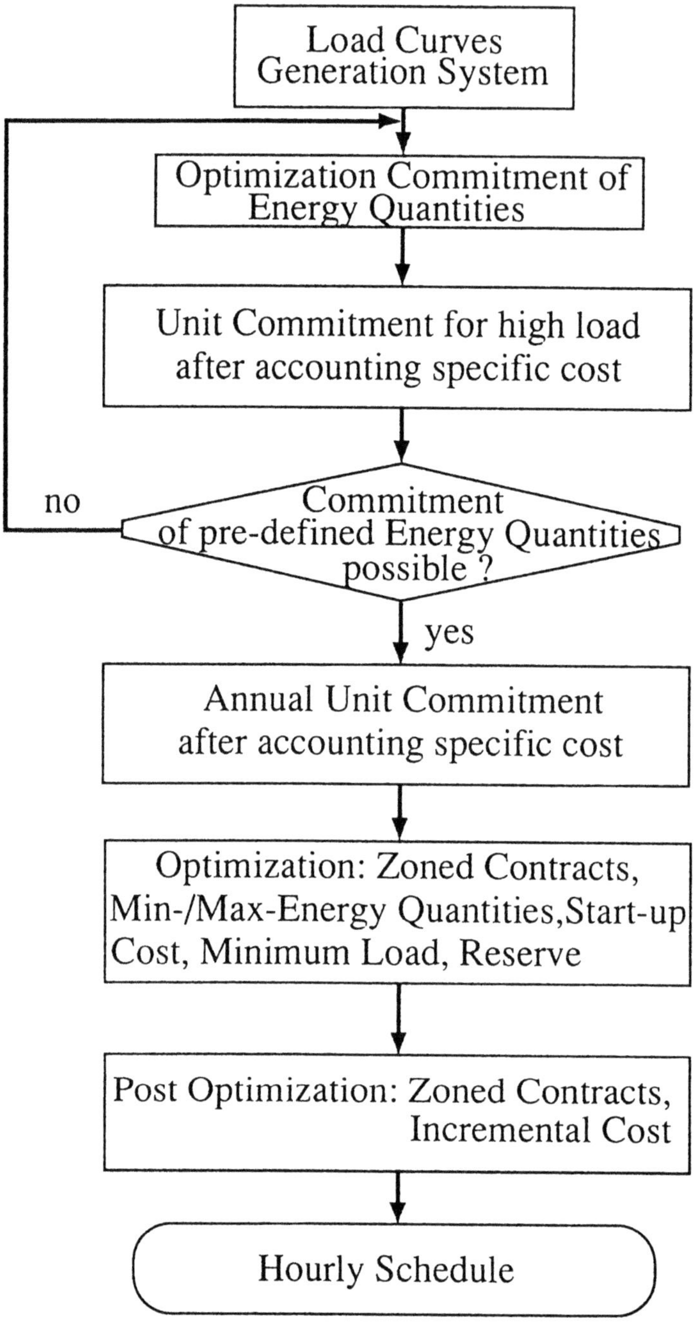

Figure 4: Long term optimization LFO-1

developed and progressively extended to include various constraints. By the help of a post optimization process it was able to provide usable results. This VEW program has been used regularly since 1972. It was repeatedly extended to cover the ever changing situation. However it became so complex and extensive that the heuristic technique was not really able to cope. In particular, constraints covering several generating units could not be introduced, the number of iterations became too large and the technique generally became too difficult to handle.

The limitations of LFO-1 led to the necessity to develop a better optimization package. The following requirements were put on the new package:

- the ability to handle many constraints in a consistent manner
- the use of mathematical procedures, instead of previously used heuristic methods to make sure that the solutions are optimal
- the use of short computing time.

VEW initiated the development of corresponding mathematical algorithms at several universities. A large number of comparisons and practical tests with real generation systems were carried out.

The long term optimization package LFO-2 operated according to the principle shown in Figure 5 [4]. The annual optimization procedure was divided into the following three stages :

- energy commitment planning
- weekly unit commitment and
- loading of units

The planning process for energy commitment calculates the optimal distribution of energy. This minimises the annual operating cost subject to the constraints which vary from week to week, and produces "shadow prices" for the subsequent optimization processes. The shadow prices are used for weekly unit commitment and loading processes in such a way that the results from the energy commitment planning process are implemented in the best possible way. Because of the complexity of the problem and the non-convexity of the objective function, Mixed Integer Programming was used. Simplifications in the model were carried out by a step-wise representation of the load curve, by approximating the spinning reserve and by neglecting short term constraints such as start-up cost and minimum up and down times.

Unit commitment with respect to one hour must not only take into consideration the short term constraints but also the long term energy constraints. The latter is reached by the help of the shadow prices. Because of

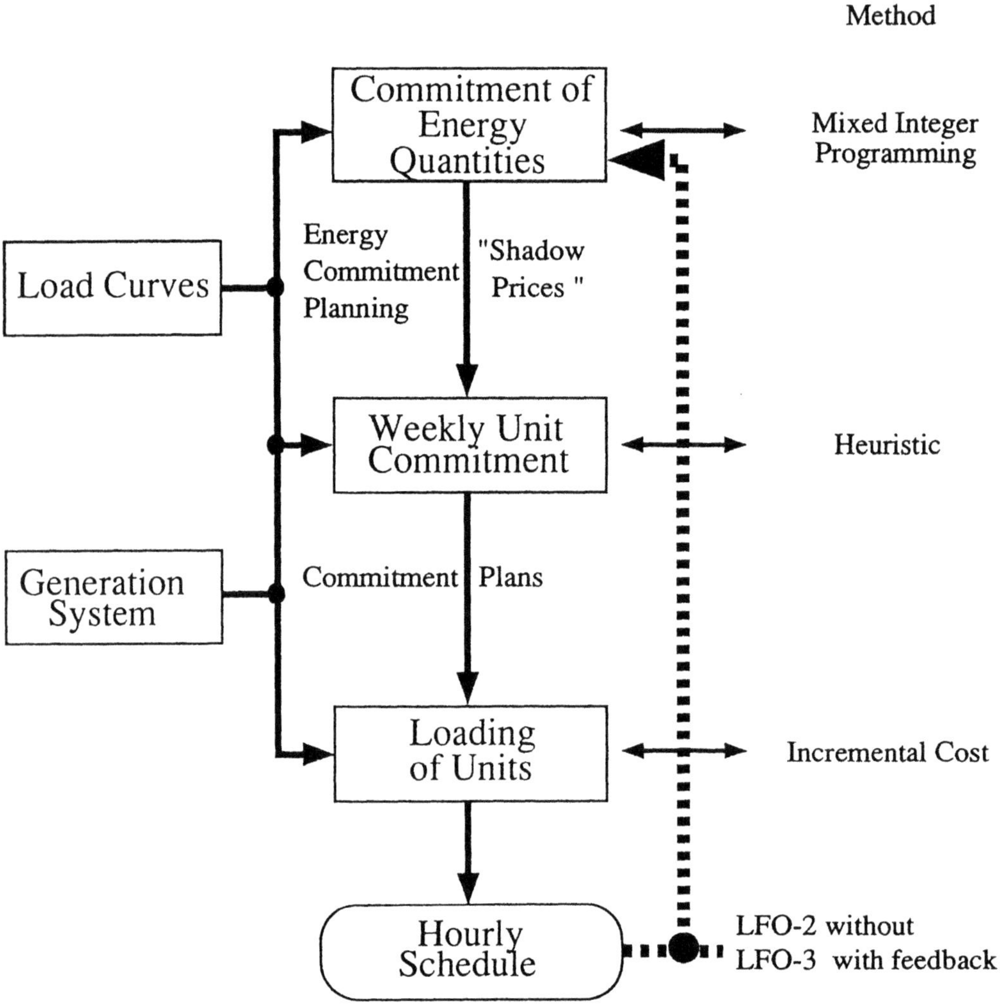

Figure 5: Hierarchical long term method LFO-2/-3

the many decisions necessary for unit commitment there was no mathematical algorithm available which was able to provide a result within reasonable computation time. For this reason a heuristic method was used to split the problem into two sub-tasks. First, a feasible commitment schedule is created according to cost merit order. This is then improved, mainly by

- continued operation of units during short shut-down periods in order to save start-up cost
- exchanging units with short operating periods
- reduction of excessive reserve.

The optimal loading of the committed units and electricity contracts is carried out via an economic dispatch process using constant, linear and non-linear incremental cost.

5.2 Comparisons

Numerous comparisons were carried out between LFO-1 and LFO-2.

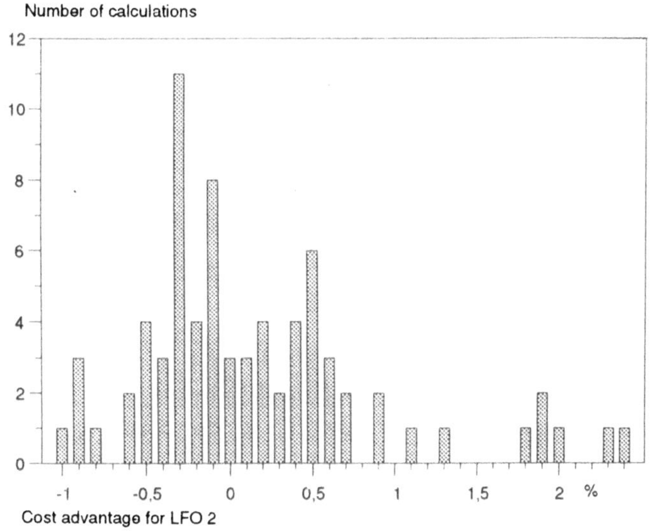

Figure 6: Cost comparison between LFO-2 and LFO-1 - 74 calculations

Among 74 cases (Figure 6) the least expensive solution was found

- by LFO-1 in 35 cases and
- by LFO-2 in 39 cases.

LFO-2 produced results which cost 0.17% more than LFO-1 (taking the average of all 74 cases).

The results of the different cases were analysed in detail in order to detect the less efficient parts of LFO-2. The experiences were used to improve the package via the following techniques :

1. Because of simplifications in the energy planning model the weekly unit commitment and the corresponding loading were not acceptable, and the required energy was not achieved. For this reason the energy planning has been repeated monthly. Energy deviations detected during the month are transferred to the rest of the year by the feedback shown in Figure 5. The so obtained schedule is now successively established.
2. In the energy planning process load and generation models were improved taking into account
   - separate working-day and week-end load curves
   - better representation of spinning reserve and
   - more details for natural gas and coal combinded-cycle units.

These improvements lead to the package LFO-3 which is used nowadays [5,6].

5.3 Today's solution

LFO-3 and LFO-1 were compared with respect to selected difficult planning cases. In 11 out of 12 cases LFO-3 produced between 0.06 and 0.38% less expensive solutions (see Figure 7a). In one case it produced a solution which was 0.13% worse. On an average LFO-3 was approximately 0.18% better. The corresponding computing times on a VAX 6410 are shown in Figure 7b. These times varied between 30 minutes and 107 minutes depending on the available scope for optimization of the individual cases. The average time was 52 minutes for cases where the electricity contracts had annual zones.

As a result of these comparisons LFO-3 replaced LFO-1 at VEW as the standard annual optimization algorithm.

In the mean time many annual optimization runs using LFO-3 have shown that the output is greatly improved with respect to solutions and computing time. Nevertheless the optimization results must be checked for plausibility. Sometimes a local minimum is found instead of a global minimum. In this case the optimization process is continued with a restricted degree of freedom in order to force a new and better solution. It is often possible to achieve this by preventing the shut down of heavily used units or by introducing fictitious

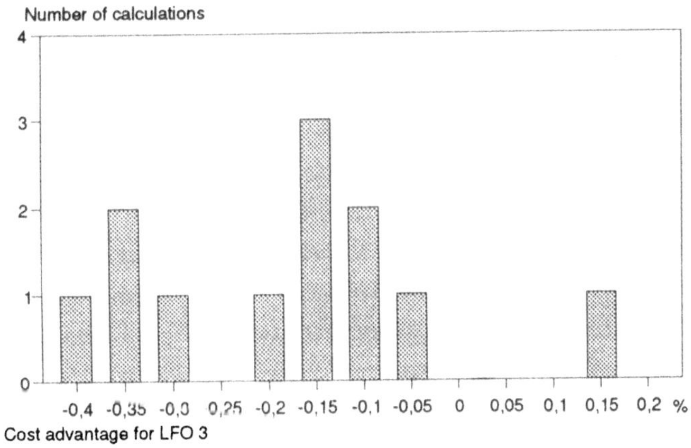

Figure 7a: Cost comparison between LFO-3 and LFO-1 (12 calculations)

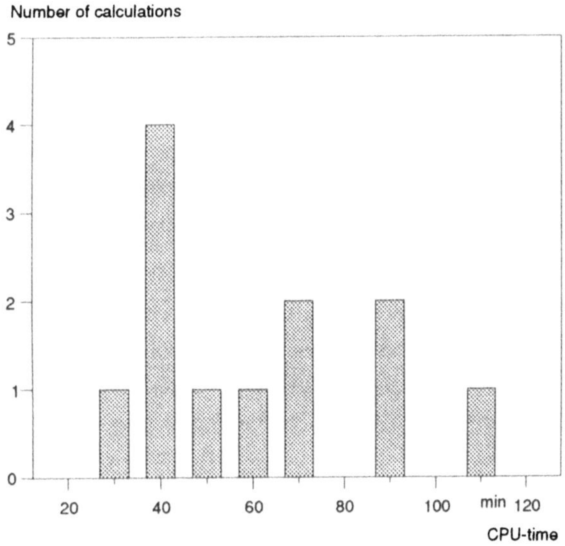

Figure 7b: Computation time-line for LFO-3 (12 calculations)

maintenance to influence the use of monthly zoned electricity contracts. By using these methods in various combinations, it was possible to improve the results compared to the base calculations in 12 out of 25 cases.

The overall cost curve is flat, giving rise to several different planning cases with about the same optimum cost. An overall saving of approximately

0.19% has been achieved, and this represents a saving of several million marks (DM) per year.

6. Medium term planning

The basis of the medium term optimization process are hourly schedules delivered by the long term optimization process. The medium term optimization task is to produce optimal 15 minute loading schedules for each generating unit and each electricity contract taking into account the energy constraints established by the long term schedules. At the same one has to pay attention to the following factors:

- unit loading rates
- reserve requirement for defined periods of the day and mobilisation-times within these periods
- power output during unit start-up
- operational requirements of waste-heat and Benson operation modes of the natural gas combinded-cycle units.

6.1 Previous experience with MFO-1

VEW has been using its own medium term optimization program since the beginning of the 70s to commit the individual generating units according to merit order. This is derived from the energy commitment planning provided by the long term optimization process. The load is then dispatched among the units according to the incremental cost and scheduled limits. These scheduled limits are used to reflect the previously defined commitment and to influence the schedules, so that the energy constraints are satisfied.

The computing (CPU) time on a Cyber 170/835 takes only a few seconds for the optimization of one week. This facility therefore represents a reliable instrument for the planning engineer to analyse generation planning cases and to react quickly and flexibly to short term input-changes; for example, such as daily trading in special purchases and deliveries of electricity, forced unit outages and unexpected load developments.

The disadvantage of the current VEW package is that not all of the factors needed for complete planning are taken into account in a consistent manner. For example, minimum up and down time for electricity contracts, necessary spinning reserve or start up cost are not considered. This means that several cases must be studied in order to obtain an optimal solution.

6.2 Recent experience with MFO-2

As a result of several year's cooperation with universities on the topic of optimization, an algorithm was developed which satisfied the needs of VEW and which was applicable to the long term planning process [7, 8]. Optimal unit commitment and optimal loading of units are calculated with the help of processes based on the results of long term optimization. The best medium term optimization results are obtained when the optimization period is large enough, so that the energy constraints associated with the long term process can be satisfied. As this requires more computing time, VEW usually carries out its medium term optimization with respect to one week (from Saturday to Friday).

The unit commitment problem can be formulated as Mixed Integer Program. To reduce computing time Lagrangian relaxation is used. Lagrangian relaxation allows for splitting the unit commitment problem into small subproblems which are solved effectively by Dynamic Programming [9].

6.3 Comparisons of MFO-1 and MFO-2

The coupling of long term optimization with medium term optimization has turned out to be very difficult. As expected inevitable deviations between monthly and weekly optimum were caused by:

- the relevant load forecast:
  without taking into account short term electricity deliveries, variations can occur up to +/- 6 % in the weekly energy levels and up to +/- 13% in overnight loading; this is due to dependencies on winter temperature.
- the utilization constraints for individual generating units and/or for electricity contracts:
  power variations can occur for some units because of disturbances or water cooling problems; sometimes units are fixed in a test procedure to perform activities such as optimization of the firing, calibration of measuring devices or acceptance testing.

In order to test coupling between long and medium term optimization under different operating conditions, MFO-2 was applied extensively for January and February with respect to the following cases:

1. annual planning updated monthly with the following coupling-strategies:
   Case A: energy requirements with fixed annual band with a tolerance of

+/- 1% of the desired value and the possibility of modifying this restriction for individual units.

Case B: energy requirements with a tolerance band adapted to the weekly situation

Case C: requirements based on weekly merit order of energy pricing derived by means of long term planning.

2. annual planning updated weekly and the coupling-startegy:

Case D: as in case A

6.3.1 Comparisons of cost and use of primary energy

Comparisons of the results of program MFO-1 are described in the following diagrams with respect to the utilization of primary energy (Figure 8a) and its cost (Figure 8b).

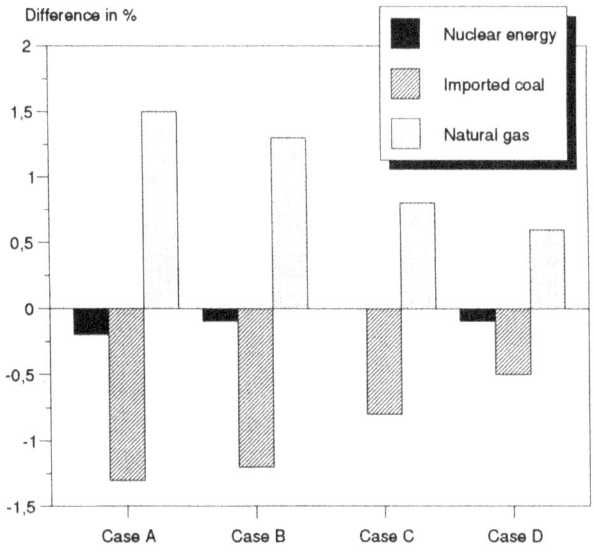

Figure 8a: Primary energy commitment (relative differences between MFO-2 and MFO-1)

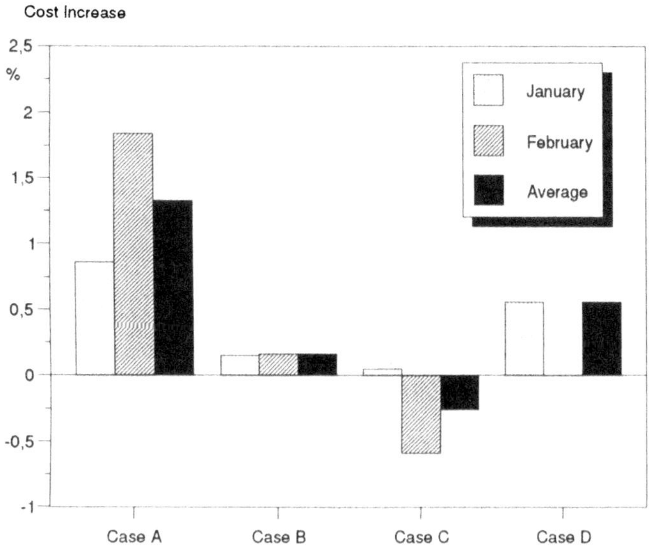

Figure 8b: Relative cost increase from MFO-1 to MFO-2

Case A:

- Nuclear energy is by far the least expensive and is not fully exploited. Furthermore the second least expensive energy source, imported coal, is used less and the more expensive natural gas is increased in comparison to the initial plan of MFO 1.
- The cost of the above poor utilization of primary energy resulted in an average cost increase of 1.33% within two months (of 0.86% in January and of 1.84% in February).
- A detailed analysis of this situation shows that coupling is too restrictive for achieving optimal utilization of generation.

Case B: Based on the experience of the first case coupling between medium and long term optimization was loosened, and energy requirements were adapted to the relevant weekly situation:

- Generation by means of imported coal and nuclear plants are below the corresponding maximum possible values. Whereas generation by means of natural gas is still too high.

- The cost exceeds the value of the control calculation by 0.15% in January and by 0.16% in February, only.
- After seeing the results it would have been possible to improve the energy commitment further and thereby reduce the incomplete utilization of nuclear energy in another iteration. Unfortunately, no additional information of the coupling-method was obtained.

Case C: Coupling takes place by means of a merit order of energy pricing, which is based essentially on the solution of long term optimization instead of using energy requirements:

- Nuclear generation is completely utilised and the use of natural gas is reduced by increased utilization of imported coal.
- Generation cost is considerably reduced as cheaper forms of primary energy are considered more. This increased the efficiency of natural gas units and lead to better utilization of electricity contracts. It is even 0.26% less than that of the control calculation.
- In general, this coupling strategy produced better results than the previously used medium term optimization package MFO-1.

Case D: Long term planning was updated weekly for January and medium term planning was carried out with using the coupling-strategy of case A, in order to prevent deviations in load forecasts and in unit availability from influencing the medium term optimization process. It had the following characteristics:

- Even with using information on load forecast and on unit availability at a high standard it was not possible to achieve the optimal situation. Again, insufficient use of nuclear generation and excessive use of natural gas lead to a reduction in imported coal.
- Generation cost is higher than that of control calculation by 0.56%.
- Improvements to the previous case cannot be achieved even when the energy requirements are updated. The coupling-strategy is too restrictive for obtaining the true medium term optimum and for taking the constraints into account.

6.3.2 *Typical daily schedules and computing times*

Figures 9a,b illustrate a weekly schedule for a 700 MW coal fired unit and for a 400 MW natural gas unit in cases A and C. In case A the natural gas unit is used excessively on working days, from Monday to Friday without stopping but reducing to Waste-Heat Operation overnight. In case C the unit is shutdown during specific periods which is not withstanding the start-up cost economically. The coal fired unit is well utilised during the night on working days and throughout Saturday and Sunday, more than in case A.

The following ranges of computing times correspond to MFO-2 for a 7 day calculation on a VAX-6410 :

Case A : 16 minutes +/- 4 minutes
Case B : 16 minutes +/- 4 minutes
Case C : 5 minutes
Case D : 40 minutes +/- 7 minutes.

7. Conclusions and consequences

The annual optimization package LFO-3 was used for a wide range of actual planning calculations within a period of 2 years. It clearly showed that many local optima near the global optimum can be expected for such a complex task.

Since there is no method available which guarantees that the global optimum is found, it seems necessary to restrict the decision space manually, for excluding the suspected local optima. A solution near the global optimum was obtained by means of the outlined approach. In fact, through restricting the decision space it was possible to reduce cost by 0.2%.

The accuracy of the solution corresponding to the basic method without restricting the solution space has turned out to be sufficient for long term planning calculations over one to two years, taking into account that input data is also not very accurate with respect to the underlying time period.

However a certain level of accuracy is desirable for studies of current and future years for operational planning purposes. It is necessary to obtain reliable data

- for fuel supply of each generating unit
- for electricity contracts and
- for monthly and weekly optimization.

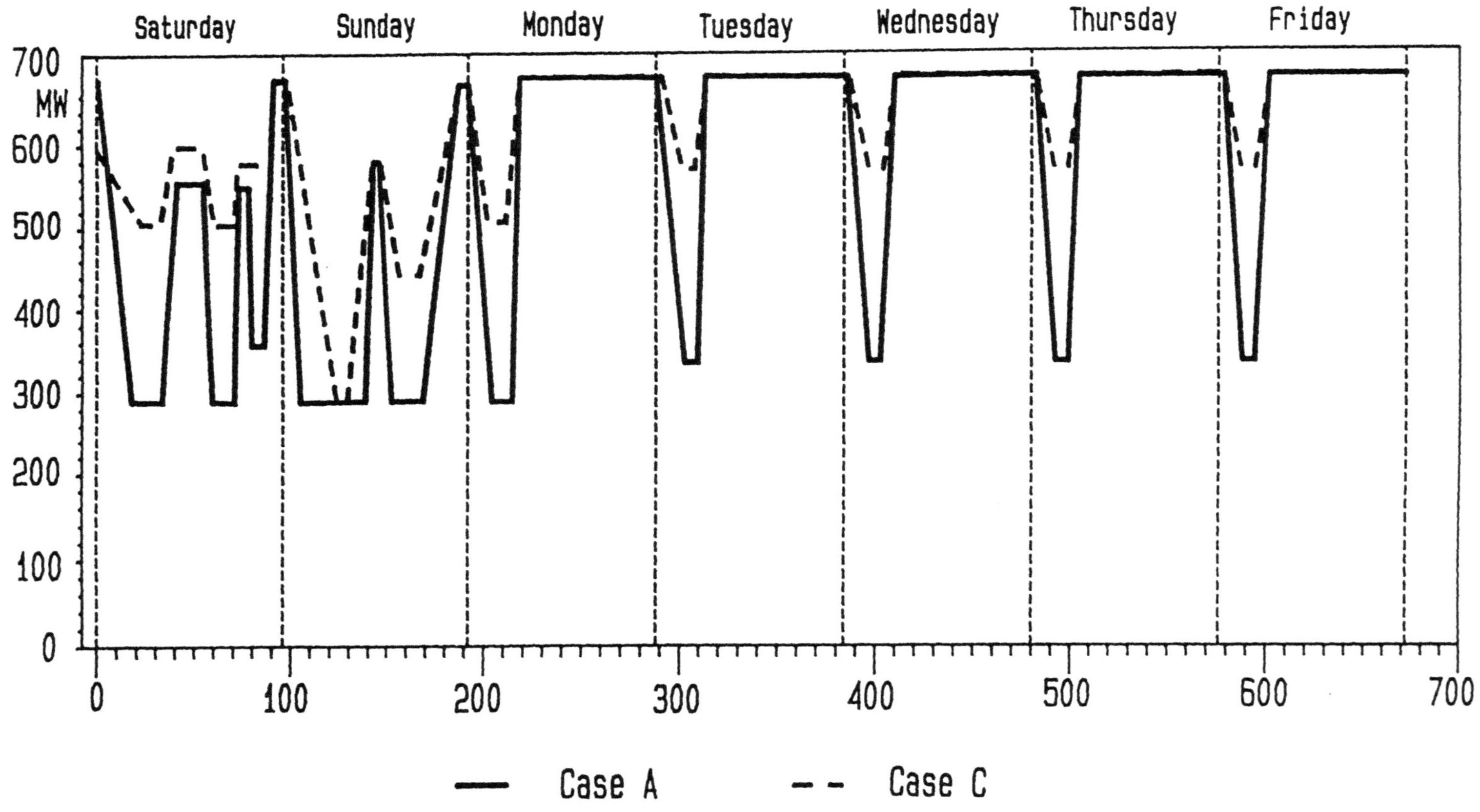

Figure 9a: 700 MW Coal Fired Unit

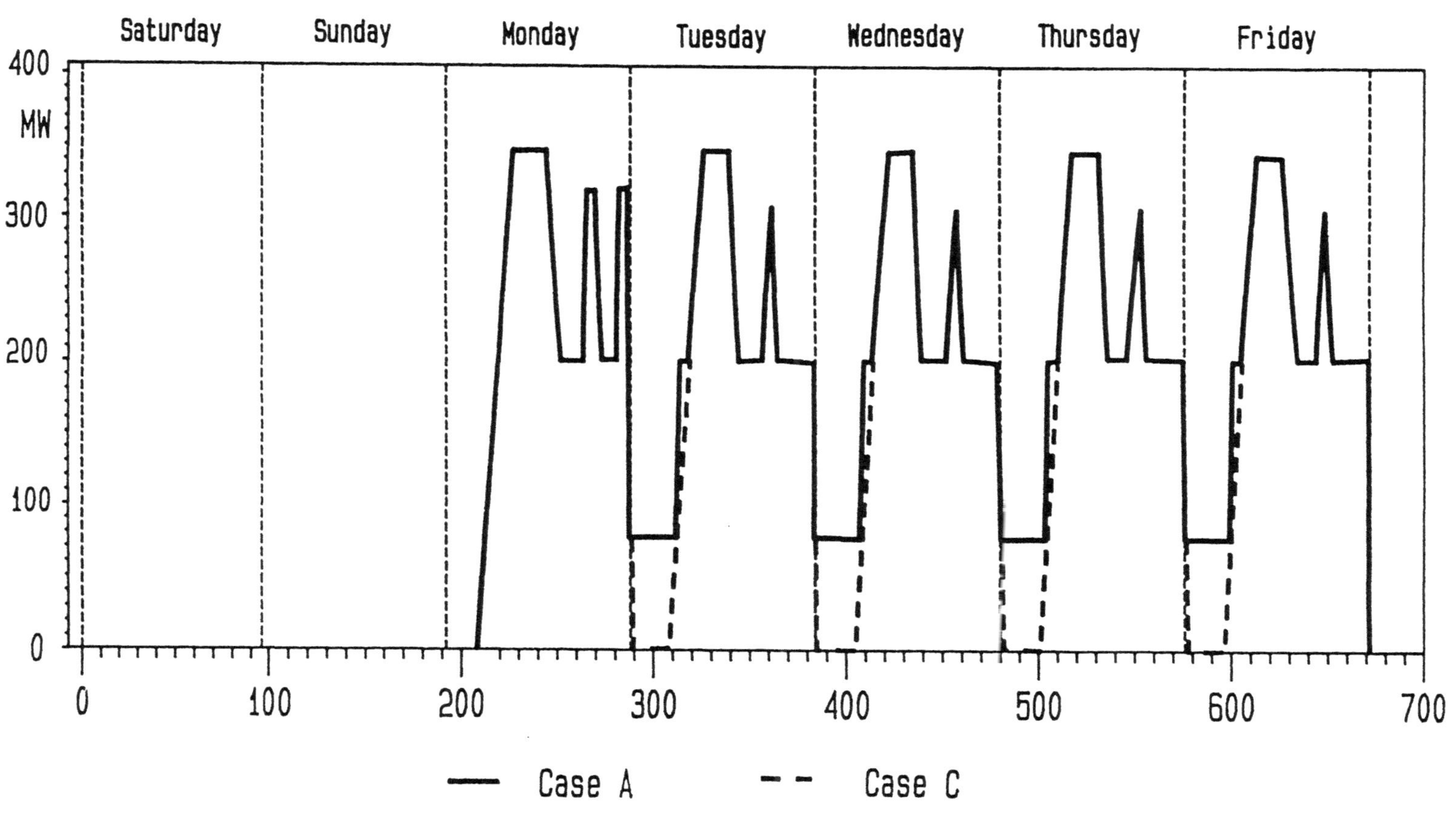

Figure 9b: 400 MW Natural Gas Unit

Concerning monthly or weekly optimization it can be very significant whether, for example, a particular zoned electricity contract is used, instead of imported coal during low load periods and natural gas during heavy load periods.

Above all, the goal of the annual optimization process is to find the commitment plan for the year's operation with the lowest generating cost. In contrast to the computing time requirements for long term planning, an increased amount of computing time is acceptable in order to reach the above goal. For example a run time of one night would be allowable for an optimization calculation on the available computer (VAX-6410).

In contrast, long term planning calculations for power station expansion with respect to periods of up to 20 years are calculated on an annual base. Several different variations are required to be completed in a shorter timeframe. Computing time for an annual optimization run should not last longer than one or two hours. Dependent on the situation to be optimized different programs could be used to cast the problem of long term optimization. Accuracy is of high priority for annual operational planning, whereas short computing time has a higher priority for long term planning calculations of power station construction and de-commissioning.

In order to improve the accuracy of the annual optimization for operational planning it seems to be necessary to improve the energy utilization planning, so that the optimal energy quantities of hourly schedules lead to optimal unit commitment and optimal loading of units.

For medium term optimization the problem of how to couple daily/weekly optimization and monthly/yearly optimization is not yet solved completely. Experiences have shown that weekly commitment planning can be improved by tricky coupling of medium and long term optimization. Further work is planned on this topic.

The target of this development is to allow for integrating the "expert knowledge" of experienced planning engineers into the coupling process, so that solutions compatible to both planning levels are found even when there are deviations in the constraints.

With the help of the participating universities, VEW has been able to make a significant step towards achieving the goal of using medium and long term optimization for determining the global optimum for generation utilization. Unfortunately, this goal has not been reached, yet. Hence, the planning engineer strives for an optimal solution through restricting the decision space. VEW's extensive experiences have shown that "optimum at the push of a button" is not yet available for complex systems, and the experienced human being is still an essential component in the solution process.

References

1. K. Linke: Kraftwerkseinsatzoptimierung; Energiewirtschaftliche Tagesfragen 11 (1985) 751 - 764
2. K. Linke: Kraftwerkseinsatzoptimierung im Verbundnetz; ETG-Fachbericht 26, VDE-Kongreß, Okt. 1988
3. K. Albers, P. Stelzner: Ein Verfahren zur Prognose von Lastganglinien für unterschiedliche Einsatzbereiche; Publication projected by Elektrizitätswirtschaft (1993)
4. T. Schroeder: Jährliche Kraftwerkseinsatzplanung; Energiewirtschaftliche Tagesfragen, 35. Jg., 11 (1985) 765-771
5. H. Wolter: Ankopplung der kurzfristigen Blockeinsatzplanung an die langfristige Energieeinsatzplanung; Institut für Elektrische Anlagen und Energiewirtschaft, RWTH Aachen, Jahresbericht 1989
6. H. Wolter: Kurzfristige Kraftwerkseinsatzplanung in thermischen Systemen mit langfristigen Nebenbedingungen; Dissertation RWTH Aachen 1990
7. E. Handschin, H. Slomski, E. Ortjohann, J. Voß: Longterm operation planning for thermal power systems; Proc. of the 9th PSCC, Lisbon, (1987) 41-47
8. E. Handschin, H. Slomski: Unit commitment in Thermal Power Systems with Long-Term Energy Constraints", PICA Conference Mai 1989, Seattle, 211-217, 1989
9. H. Slomski: Optimale Einsatzplanung thermischer Kraftwerke unter Berücksichtigung langfristiger Energiebedingungen; Dissertation Universität Dortmund 1990

MODELLING IN HYDRO-THERMAL OPTIMIZATION

A. Schadler, E. Steinbauer

STEWEAG, Graz, Austria

Abstract. The energy requirements of STEWEAG - the power and energy supply utility of the Austrian province of Styria - are met by water power, thermal power stations and by contracted power supply from the national utility (VG). The power system consists of thermal power stations with and without fuel contracts; combined heat and power stations, run-of-river plants, series of run-of-river-plants and water power stations with storage (year, week or day). The peak power demand is estimated and has to be contracted in advance. If it is exceeded, a rather expensive penalty must be paid. The contracted energy must be paid according to a strict tariff taking into account a rather high fixed premium related to power. Hence STEWEAG has been interested in the optimization of its energy resources and in determining optimal schedules and contracts. In a first approach which was oriented towards a horizon of a year, methods of calculus of variations have been used thereby accepting far reaching simplifications. Recently emphasis was put on daily, weekly optimization as well as on annual optimization for operations planning. For this purpose a software package based on mixed-integer programming and suitable hardware was used. Starting from an annual optimal schedule the weekly commitment schedule is optimized based on time steps of 2 to 4 hours. The main goal of the weekly optimization is to deal with the influence of the water reservoirs during the weekends and with the start-up and shut-down decisions. Embedded in the weekly optimization the daily optimization is calculated based on 48 time steps. The model expressed in tableau size is approximately 7000 rows by 10 000 columns for the week and 4000 rows by 5000 columns for the day. Details on modeling approaches, computational experiences and considerations on the economic gains will be given.

1. Problem definition

The Styrian Water Power and Electricity Company, STEWEAG, the power and energy supply utility of the Austrian province of Styria, has a very complex power system. This is due to the fact that nearly all types of power plants are available in the power system. Figure 1 shows a simplified model of the power system as it is used for daily and weekly production scheduling. In the upper part of the figure there is the hydro network with its reservoirs and plants situated at the river ENNS. It consists of two hydro storage plants with daily/weekly storage scycle (SÖLK and WAG) and one yearly cycle (SALZA). The water released from the upstream reservoirs SÖLK and SALZA reaches the WAG plateau after a delay of 11 hours, where it can be stored in the WAG reservoir or directly used for energy production. The pumps for storing the water are only necessary if the water level in the reservoir is higher than the water level of the inflow. Downstream to the WAG reservoir there is a chain of 3 hydro power plants which consist of main plants utilizing the water from bypass conduits and additional turbines utilizing the remaining water in the river (water release obligations). The optimization of this hydro system is further complicated by water supply contracts with the downstream utility (EKW). The water supply contract guarantees certain water discharge quantities over certain time periods.

For the load dispatcher it is difficult to operate this hydro system in an optimal manner, because he has to pay attention to different nominal discharges, different delay times and to different obligations for guaranteed water flows for each plant and for the neighbouring utility (EKW).

Beside the hydro system at the river ENNS there are other hydro storage systems located at the river MUR (BODENDORF reservoir with daily and weekly storage cycle and 2 plants) and TEIGITSCH (reservoir with yearly storage cycle and 3 plants).

The power demand for electric energy is further covered by thermal units and power exchange contracts with the national utility VERBUNDGESELLSCHAFT (VG). The eight large thermal units, which are shown in the lower part of figure 1, are fired by coal (VOITSBERG, FHK MELLACH), gas (NEUDORF, FHK GRAZ) and oil (PERNEGG, NEUDORF). Some of them are combined units for electric power generation and heating (FHK MELLACH, NWD1 and FHK2: condensing turbines, FHK1: extraction turbine).

For the thermal plants it is necessary to take into account minimum up and down times, rates for power changes and fuel contracts.

A special sort of thermal plants are substitution contracts which are operated like real thermal plants. Instead of operating own thermal plants it is possible to buy the corresponding power from the national utility (short term agreements).

The power exchange contract with the national utility is very sophisticated. In principle, the power exchange cost consists of a relative high power price and a time dependent energy price (low tariff, high tariff, low tariff for special days). The maximum power is contracted in advance. If it is exceeded during winter time a penalty must be paid which may prove very expensive. During summer no penalty has to be paid for limit violations, only the energy

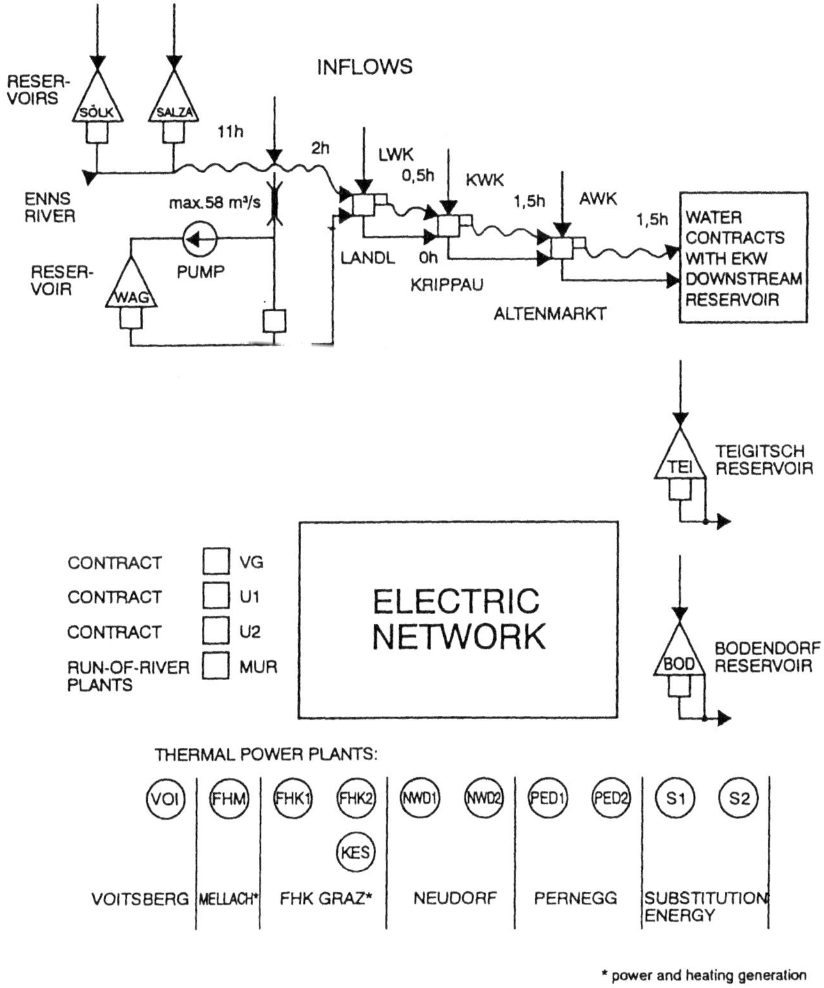

Figure 1: Model of the STEWEAG power system for daily and weekly optimization

prices increases.

Furtheron, there are time integral constraints for the imported quantities over the seasons (summer, winter). The limits themself are related to the energy production of STEWEAG. Due to the complexity the contract is built up by several simpler model contracts which are interconnected by logical and time integral conditions (VG,U1,U2,S1,S2).

STEWEAG's energy requirements are met, in similar proportions, by water power, by the thermal power stations and by contracted power supply from the national utility (VG) (Fig. 2). The problems of water power generation arise from the seasonal fluctuations of the storage inflows. Therefore it is not enough for the load dispatcher to study the average year; he must also be prepared for extremes of dry and wet years. The responsibilities of the load dispatcher may be summarized :

- determination of an operation plan for the year with regard to fuel stocks, management of the hydro reservoirs and fixing and ordering contracted power, for a variable hydraulic production;
- frequent adjustment of this plan depending on the actual hydraulic production and the actual energy consumption;
- determination of an operation plan for the week especially because of water management in the small hydro reservoirs; attention must be paid to this, as already mentioned, with regard to the series of power plants on the river Enns and all its special conditions;
- a schedule for the day in half-hourly steps taking into account the rest of the week.

STEWEAG started in the eighties together with IBM AUSTRIA the development of a program system based on IBM's standard mixed integer programming package (MPSX/MIP 370). In the first two project steps the software product KWOPT for the daily and weekly optimization was developed and tested (1982 - 1985), followed by the annual optimization product called JARO (1986 - 1988). A major breakthrough with respect to performance was achived in 1991 as more powerful optimization software tools became available, which support the RISC hardware of the IBM/R6000 computer family. The new optimization tools are called OSL (optimization subroutine library). Since we use the standard mathematical planning system format (MPS), it was easy to take advantage of the new facilities. By putting the model generator and the optimization software on a R6000/550 we were able to speed up the performance by a factor of 10 to 15 in relation to a IBM 4381/11.

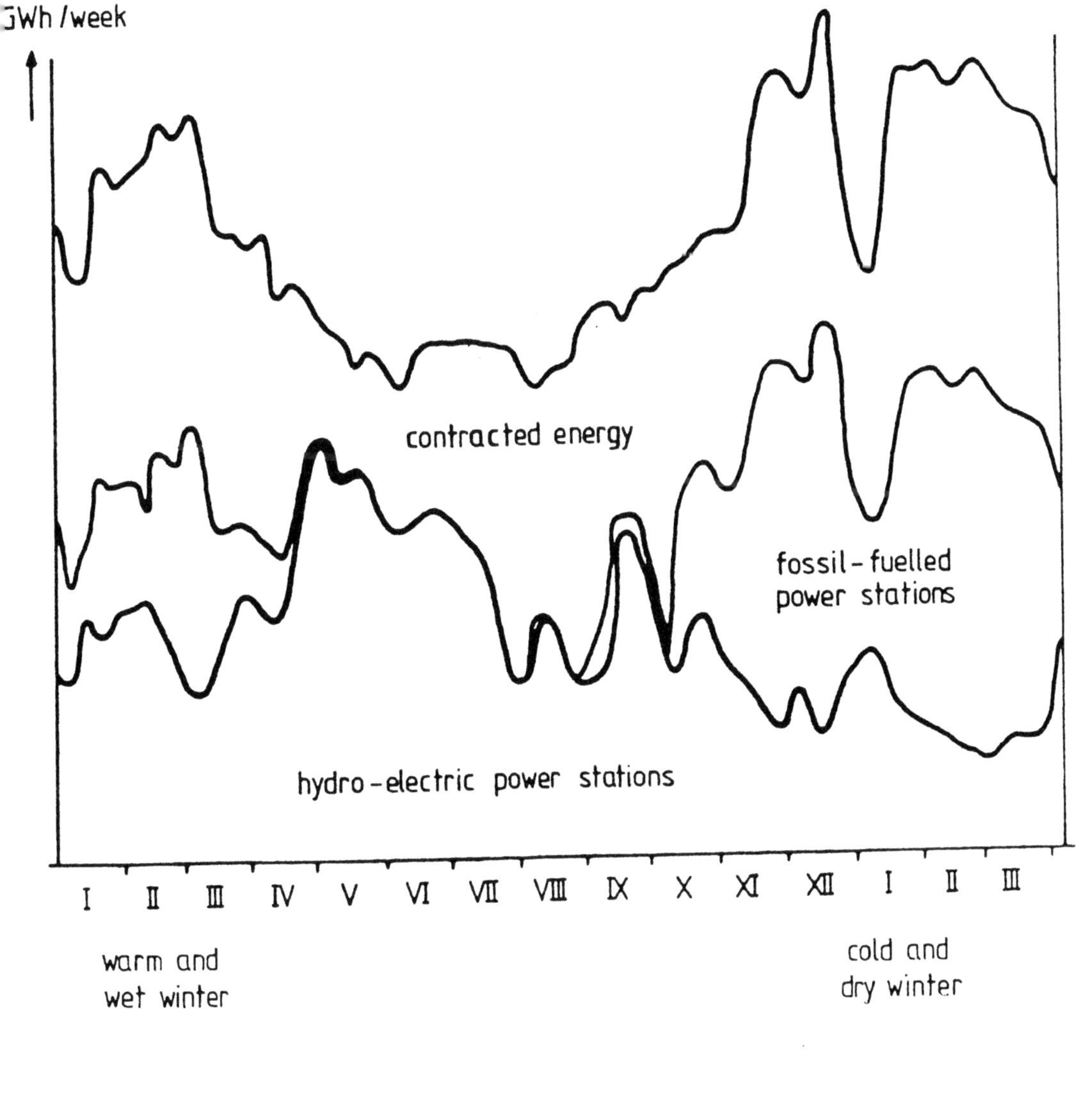

Figure 2: Energy, demand, production and contracted energy of STEWEAG (example with different winter characteristics)

The objectives for the new software package were from STEWEAG's point of view:

- to model the Steweag problems sufficiently accurate
- to get short execution times without using super computers with high cost
- to be able to vary conditions in a simple way
- to get a user-friendly menu driver application
- to present the obtained results graphically.

Fom IBM's point of view the aim was:

- to produce software which can be used universally (i.e.: in any utility, for any contracts and any energy system).

The basis of the long term and short term optimization packages JARO and KWOPT is the same. The main difference between them is that the annual optimization package JARO can handle flexible scenarios. The planning horizon for the annual optimization can be one or several years. Normally the planning period is split into subperiods like months, holiday periods, peaking weaks etc. (a maximum of 60 subperiods is possible). The subperiods are represented by characteristic weeks, which consist of 1-7 days. The time step length can be 0.5, 1, 2 or 4 hours. The step length for thermal plants can be chosen different to hydro plants. Week days can be joined together to "Super"-days using appropriate weighting factors. The annual optimization is executed for the normal, wet and dry year. For this reason there is no automatic data transfer between the annual optimization and the weekly optimization. The main objective of the weekly optimization is to deal with the start-up and shut-down decisions of the thermal plants taking into account the different load patterns of the days, the weekend influence of the water reservoirs and the fuel restrictions. Embedded in the weekly optimization the daily optimization is calculated on a 48 time step basis. The boundary constraints for the daily models are automatically transferred from the weekly model into the daily one.

2. Modelling

The packages KWOPT and JARO, now generally available, are universally applicable to electric power and heating systems. The model generator is able to handle things like:

- complex water reservoir systems including: reservoir networks, time delays for the water flow, water contracts to downstream plants;
- coupled energy systems with electric and heating power generation;
- sophisticated interchange contracts taking into account several types of contracts.

The mathematical model is generated by a model generator, consisting of modules for:

- storage systems
- hydro power plants
- thermal power plants
- heating plants
- combined heat and power production
- operational conditions
- power balance
- energy exchange contracts
- resource management constraints etc.

The following section shows the basic model as it is used for short term operation planning. For simplicity many constraints have been omitted and others have been simplified (e.g. heating requirements, weighting factors etc.). The presentation does not claim completeness with respect to the implemented software package KWOPT/JARO.

Minimize the *objective function*

$$\overbrace{\sum_{k,f,t} C_{k,f} Q_{k,f,t}}^{\text{fuel cost}} + \overbrace{\sum_{k,t} C^a_{k,t} Y^a_{k,t}}^{\text{start-up cost}} + \overbrace{\sum_{v,s,t} C_{v,s,t} P_{v,s,t}}^{\text{power exchange cost}} +$$

$$+ \overbrace{\sum_{k,t} C^r_k (\Delta P^p_{k,t} + \Delta P^n_{k,t})}^{\text{regulation cost}} + \overbrace{\sum_{k,s,t} C^p_{k,t} P_{k,s,t}}^{\text{penalty cost}}$$

The objective function contains the cost for primary energy, the start-up cost and the regulation cost. While the cost for primary energy - normally fuel prices and cost coming from energy exchange contracts - are always known exactly, the start-up cost is not well known in many cases, especially for hydro power plants. This is also true for regulation cost, which takes account of cost arising in certain power plants like thermal plants. In MIP- and LP-models, which use the Simplex method for optimization, all plants must have regulation cost, because the optimizer tends to swing otherwise (no smooth schedules). Furthermore, it has been necessary to introduce penalty cost which is not obvious at the first sight: In power systems which contain several storage power plants and contracts, there are generally many possible schedules which are equivalent with respect to cost, but not with respect to operation. That means the schedules are more or less extreme or practicable. No load dispatcher likes an optimal schedule that inherently risks a very expensive contractual violation, especially when such a situation can be avoided. It is necessary to get optimal and well balanced schedules. Unfortunately the criterion for this situation is not an exact one in our utility; it only says: "In case of equivalent schedules, the solution with the best balanced commitment for storage power plants and external power import is searched for." Because it is not completely possible to handle this problem by "hard" constraints like power price constraints for the contract or by constraints for the load change, a load dependent penalty for the storage power plants was introduced. This cost term is considerably smaller than the other cost terms; it results in a preferred commitment of the storage power plants during peak hours and avoids excessive power imports.

The *power balance* equation guarantees that the production of power follows the load demand curve. The power produced by thermal and hydro plants consists of a fixed amount representing the minimum power and additional power. The latter is given by piecewise linear segments. Since there is a forbidden zone between zero and minimum power, the generation curve is not continuous. Therefore switching state variables is necessary; these take the value 1 when the plant is operating and 0 when it is off (fixed charge problem).

$$\sum_k P_{k,t} = P_t^n - P_t^k$$

where $P_{k,t} = \underline{P}_{k,t} \cdot U_{k,t} + \sum_s P_{k,s,t}$ and $\underline{P}_{k,t} \leq P_{k,t} \leq \overline{P}_{k,t}$.

In connection with the power balance it is important that the commitment schedule allows sufficient reserve to ensure adequate reliability in the case of outages. The *spinning reserve* relation ensures that the committed plants are able to cover unexpected loads.

$$\sum_k (\overline{P}_{k,t} \cdot U_{k,t} - P_{k,t}) \geq P_t^r$$

Due to the necessary risk estimation this constraint is of difficulty for the load dispatcher. In practice reserve considerations are handled in a flexible way, especially if certain storage reserves or import contracts are available.

The *generation equation* shows the relationship between primary power input (fuel, water etc.) and electric power output.

$$\underline{Q}_{k,t} \cdot U_{k,t} + \sum_s \alpha_{k,s} \cdot P_{k,s,t} = \sum_f Q_{k,f,t}$$

Storage power plants may be handled like thermal plants if the influence of the storage level are negligible. If the storage level has a strong influence on the generated power, it is also possible to select a generation curve in dependence of the storage level during optimization. It should be noted, that model size and computation time increase considerably if generation curves are selected during optimization.

The *rate equation* limits the rate of power change for the generation units. This equation is especially needed for thermal units, as the generation cannot be increased, decreased respectively, above a certain rate.

$$0 \leq \Delta P_{k,t} \leq \Delta \overline{P}_k$$

$$\Delta P_{k,t} = \sum_s (P_{k,s,t} - P_{k,s,t-1})$$

A little more sophisticated are constraints which keep the generated power for a given time, before it can change again. By introducing the integer variables X^p it is possible to keep the power on the same level for a minimum time and to limit the number of power changes.

$$\sum_{\tau=t}^{\min(t+T_k^p, T)} X_{k,\tau}^p \leq 1$$

$$\sum_t X_{k,t}^p \leq N_k^p$$

If start-up cost must be included or if the number of start-ups should be limited it is necessary to introduce start-up variables.

$$Y^a_{k,t} = U_{k,t} - U_{k,t-1}$$

It is recommended to limit the start-ups as tight as possible.

$$\sum_t Y^a_k \leq N^a_k$$

In addition minimum up-time and down-time limits must be taken into account for thermal plants. This means that a generation unit must or must not be in operation for a specified time once it has been started or stopped.

$$\sum_{\tau=1}^{T^d_k} U_{k,t+\tau} \geq T^d_k \cdot Y^a_{k,t}$$

$$\sum_{\tau=1}^{T^s_k} U_{k,t+\tau} - T^s_k \cdot Y^a_{k,t} \leq T^d_k$$

Also logical operation conditions are given in the STEWEAG power system. (i.e.: operational states are connected by Boolean operators like AND, OR, NOT). In everyday operation especially the following AND and NOT conditions can often be found:

- $U_{k_1,t} - U_{k_2,t} \leq 0$ (a unit can be started only if another one has already been started - combined plants);
- $U_{k_1,t} + U_{k_2,t} \leq 1$ (a unit cannot be operated if another unit or substitution contract is active).

Furthermore, it is advantageous to introduce logical conditions if for example a unit has a much better efficiency than another one. This saves a lot of computing time in the MIP optimization phase.

Input restrictions are time integral constraints. They are used to limit the primary energy resources (oil, gas, coal etc.) for one or several units.

$$\sum_{k,t} Q_{k,f,t} \in \left[\underline{Q}_i, \overline{Q}_i\right]$$

Storage equations of hydro power plants also belong to this kind of constraints. They define the relationship between storage volume and the water inflows and outflows.

$$V_{r,t} - V_{r,t-1} = \left(Q^n_{r,t} + \sum_j \delta_{r,j} \cdot Q_{j,t-T_j}\right) \cdot \Delta t$$

$$V_{r,t} \in \left[\underline{V}_{r,t}, \overline{V}_{r,t}\right]$$

$$Q_{j,t} \in \left[\underline{Q}_{j,t}, \overline{Q}_{j,t}\right]$$

Hydro storage systems can be very complicated if many reservoirs are connected by the in-/outflows. Outflows from upstream reservoirs influence - often with time delay - the operation of downstream storage systems. If smaller reservoirs with daily or weekly storage cycles are given, the problem becomes "hard", because storage limitations lead to "BANG-BANG" phenomena which must be handled by the optimization algorithm without difficulty.

Output restrictions are also integral. They are used especially to limit energy imports and exports or the generation of several plants.

$$\sum_{k,t} P_{k,t} \in \left[\underline{E}_0, \overline{E}_0\right]$$

A similar equation is used in the STEWEAG model for integral water delivery contracts.

3. Optimization

In the following section the experiences with the optimization algorithm OSL (Optimization Subroutine library) are discussed. This software package solves the optimization problem in two steps:

1. solving the corresponding problem without integer conditions (LP relaxation);
2. solving the mixed integer problem using a branch and bound algorithm.

Under the assumption that the integer variables are bounded, it can be shown that the branch and bound algorithm solves the underlying MIP problem in a finite number of steps.

Solving the relaxed problem. The LP is solved with the revised simplex algorithm of OSL and causes no difficulties in everyday application. No program errors have been detected in the past. Data errors - that means inconsistent input data - are eliminated as far as it makes sense - before optimization starts. Further additional slack variables have been implemented in the model to avoid infeasibilities (for example a SLACK generator, slack variables for the storage volumes etc.). If nevertheless infeasibility occurs due to undetected logical contradictions in the input data, infeasibility will be detected by the model preprocessor or in the first stage of the optimiziation process (search for feasible solution). The model preprocessor, is even able to detect integer infeasibilities which normally can be detected in the branch and bound search rather late. The infeasible equations and variables are displayed together with a description, so that it is possible for the operator to detect and handle malicious data errors. It is a great advantage in the daily schedule production that the LP solution can be obtained quickly and reliably. The LP solution is often very useful for basic schedule consideration, where the violation of some integer variables does not matter.

Solving the MIP problem. Branch and bound methods which solve a relaxed LP in each node, are still very successful. Nevertheless mixed integer programming software packages like MPSX/MIP370 and OSL should not be treated as black boxes. Their efficiency can be considerably improved by exploiting the possibilities these packages offer to the user.

With MPSX/MIP we had to do a lot of work to speed up the search in order to get sufficiently good solution in reasonable computing times. We implemented and tested different things like integer starting solutions, iterative fixing of integer variables, special adjustments of the MIP procedures and special preprocessing techniques (see Steinbauer and Schadler 1990 [0]). OSL performs much better than MPSX/MIP and it has not been necessary to implement special strategies anymore. This is due to the fact that the MIP preprocessor can be called not only intially prior to the branch-bound processing starts but also during branching on 0-1 variables. This technique is referred to as *supernode processing*. The experience has shown that supernode processing has two main impacts:

- the number of nodes necessary for getting an integer solution is considerably reduced (in many cases an integer solution is obtained in node zero);
- the quality of the solution which can be obtained in a given time is much better when preprocessing is switched on.

The principle of the branch and bound method is a systematic search in the solution space through partitioning the convex feasible region of the LP into smaller and smaller subsets. A lower bound of the objective value is calculated for each subset by solving the corresponding LP problem. Any infeasible subset is dropped. At all stages of the process, the best known integer solution represents an upper bound on the optimal solution. Therefore any subset whose lower bound is greater than the upper bound is also dropped. If there is no subset whose lower bound is less or equal to the upper bound the global optimum for the best integer solution is proven and the search ends.

The performance of solving MIP problems can be improved if it is possible to reduce the feasible region. The MIP preprocessor, which is standard in OSL, tries to add constraints, to alter coefficients and looks closely at the interactions between integer variables. This analysis is called "supernode processing" as it analyzes many nodes of the branch and bound tree at once. In a *first* step matrix and bounds are analyzed to either tighten the bounds or declare the problem infeasible.
In a *second* step logical tests are performed to determine if a valid solution can be obtained. Each 0-1 variable is fixed in turn, first to zero and then to one, impacts on the remaining variables are registered and implication lists are generated.
In a *third* step cuts are added by examining the implication lists.

This analysis is repeated as long as the number of fixed variables or the improvement of the objective function is greater than a given threshold.

Finally it must be stated that a model preprocessor does not suggest better model formulations. Therefore it is important to consider all information that limits the integer variables, like limitations on start-up decisions.

4. Computational experience

In practice the operation planning strategy consists of yearly optimization, weekly optimization and daily optimization. Yearly optimization is based on characteristic weeks (usually 13 - 15) which consist of at least three days with 4 hours time steps (Saturday, Sunday, working day). The models have approximately 10000-15000 rows and 20000-30000 columns (of which 1000 are integer). The solution time depends on the data structure, the model size and the computer power. On an IBM R6000/550 computer (57 MIPS) it takes approximately less than 30 minutes to solve such models. Based on results of yearly optimization and on weekly short term forecast the schedule of a week is optimized periodically in time steps of 2 hours. The main goal of the weekly optimization is to consider the influence of water reservoirs and start-up and shut-down decisions during the weekend. The models for weekly optimization have approximately 7000 rows and 10 000 columns. The mean value for computing time is 7.1 minutes. Embedded in the weekly optimization the daily optimization is calculated on a 48 time step basis. These models have about 4000 rows and 5000 columns, the average value for the computing time is 2.8 minutes.

As the decisions of the load dispatcher influence the expense of all (fuel consumption and of contracted energy, as well as unused water and efficiency of thermal stations) an improvement of much less than half of a percent would be enough to justify the cost of hardware and software. Practical experiences show varying improvements, depending on the system of plant in operation. On average the required aim was surely achieved.

Appendix - Notation

Indices

t	time	k	plant	o	output
f	fuel	s	segment for plants	v	contract
r	storage	j	flow	i	input

Coefficients

$C_{k,f}$	primary energy cost	$C^a_{k,f}$	start up cost
$C_{v,s,t}$	power exchange cost	C^r_k	regulation cost
$\alpha_{k,s}$	generation coefficient	$C^p_{k,t}$	penalty cost
T^d_k	minimum up-time	T^s_k	minimum down-time
$\delta_{r,j}$	incidence coefficient -1,0,+1	N^a_k	number of start-ups
N^p_k	number of power changes	P^r_t	spinning reserve
$\underline{P}_{k,t}, \overline{P}_{k,t}$	min/max active power	$\underline{Q}_{j,t}, \overline{Q}_{j,t}$	min/max primary energy for flow
$\underline{V}_{r,t}, \overline{V}_{r,t}$	min/max storage volume	$\underline{Q}_{k,t}, \overline{Q}_{k,t}$	min/max primary energy for plant
$\underline{\Delta} P_k, \overline{\Delta} P_k$	min/max rate of power change	P^n_t	network load
$\underline{E}_i, \overline{E}_i$	min/max secondary energy	$Q^n_{r,t}$	natural water inflow
P^k_t	optimization invariant generation (run of river plants)	$\underline{Q}_i, \overline{Q}_i$	min/max available primary energy
T^p_k	time between consecutive power changes	T_j	time delay for water flows

Variables

$Q_{k,f,t}$	primary energy flow	$V_{r,t}$	storage volume
$Q_{j,t}$	primary energy for hydro systems	$X^p_{k,t}$	power change variable (integer)
$Y^a_{k,t}$	start-up variable	$P_{k,s,t}$	segmented power variable
$\Delta P^p_{k,t}$	positive rate of power change	$\Delta P^n_{k,t}$	negative rate of power change
$U_{k,t}$	switching state variable		

References

0. E.Steinbauer, A.Schadler: Optimizing the operation planning of the STEWEAG Power System, Proceedings of the 10, PSCC, Graz (1990) 1099-1106
1. E.Steinbauer: Kraftwerke an der steirischen Enns - Einsatz und Betriebsführung der Kraftwerkskette", ÖZE 26 (1973) 185-191
2. M.Muschick und A.Schadler: Operation Planning of Power Systems; in: System Modelling and Optimization, Proceedings of 12th IFIP-Conference (1985)
3. E.Steinbauer: Experience of the STEWEAG control engineer in optimizing the operation planning of the STEWEAG power system - development and installation of a new software product; UNIPEDE symposium on the application of advanced software technologies in electricity supply undertakings, (Juni 1-2, 1989)
4. E.Steinbauer: Planungs- und Betriebsoptimierung in der Elektrizitätswirtschaft mit Hilfe von Rechenanlagen; ÖZE 26 (1973) 221-229
5. P.G. Harhammer: Wirtschaftliche Lastaufteilung auf Basis der Gemischt-Ganzzahligen Planungsrechnung; ÖZE 29 (1976) 87-94
6. E.Steinbauer, M.Muschick, A.Sillaber, P.Harhammer, H.Strobl und H.Haschka: Kraftwerkseinsatzoptimierung; ÖZE 38 (1985) 1-7
7. IBM-Programminformation: Programmsystem KWOPT, IBM- Österreich (1987)
8. A.Brearley, G.Mitra and H.P.Williams: Analysis of mathematical programming problems prior to applying the simplex method; Math. Progr. 8 (1975) 54-83
9. H.P.Crowder, E.L.Johnson, M.Padberg: Solving Large scale Zero-One Linear Programming Problems; Oper. Res. 31/5 (1983) 803-834
10. Mathematical Programming System Extended/370 Version 2, Program Reference Manual
11. Optimization Subroutine Library, Guide and Reference

Chapter III

OPTIMAL POWER FLOW

POWER SYSTEM MODELS, OBJECTIVES AND CONSTRAINTS IN OPTIMAL POWER FLOW CALCULATIONS

Rainer Bacher

Swiss Federal Institute of Technology (ETH)
CH-8092 Zürich, Switzerland

Abstract. The principal goal of the Optimal Power Flow (OPF) program is to provide the electric utility with suggestions (setpoints) to optimize the current power system state online with repect to various objectives. Typical objectives are minimization of the total generation cost, minimization of the total (or regional) active power network losses, maximization of the degree of security of a network or even a combination of some of them. The achievement of these goals is important to utilities, since often they are obliged by law to operate the network with consumption of minimal resoures and a maximum degree of security.

The OPF problem consists of three parts: The set of equality constraints representing the power system model for static computations, the set of inequality constraints representing real-world and practical operational constraints whose violation is not acceptable in the power system or only acceptable during a given short period, and the objective function.

The main problem in the OPF formulation is the overwhelming size of the problem, especially if contingency constraints representing the possible power system element outages are included. A detailed derivation of the type and the number of equality and inequality constraints is given in the paper. Also, in the case where contingency constraints are included in the OPF problem, the underlying coupling between the contingency network and the actual network is discussed.

The typical OPF objective functions including important practical considerations are discussed and derived to give the reader a complete picture about the OPF problem formulation.

1 Introduction

The electric power system is defined as the system with the goal to generate and to transmit the electric power user via the transmission system to the electric power end users. In order to achieve a practically useful power system engineers have come up with principal components and characteristics of the power system: These components are pieces of hardware like high voltage transmission lines, underground cables, high voltage transformers, nuclear, hydro or hydro-thermal power stations, compensators, etc.

Electric power can be provided in direct or alternating current form, however, in the beginning of the 20th century alternating current has been determined as the more economical and technically feasible solution. An electric power system where the sources are alternating currents is principally of a dynamic nature, i.e. all current and dependent quantities like voltage and electric power itself are varying over time. A physical law exists among these three quantities:

$$Power(t) = Voltage(t) * Current(t) \tag{1}$$

In (1) the variable (t) refers to time.

From a control point of view it is the electric power consumption of end users which is hard to control: Today in most civilized countries the end user consumes the electric power with almost no restrictions. Fortunately, due to a regular daily behavior of the businesses and households electric power consumption follows more or less regular patterns. Due to the regular behavior it is possible to use some kind of prediction method to get good approximative values for power consumption at any future time.

Electric power is of such a nature that whatever electric power is needed must be generated by generators at any time t. This generated power includes the power losses in the resistances of the electrical transmission network.

Thus the approximate total sum of generated power is always defined for any given total end user power consumption. Since the total generation can be produced by a high number of generators, a power system could physically be operated in a huge number of states. However, many of them are not tolerable due to physical or operational constraints. Examples are extreme material usage due to e.g. excessive power flow through a cable, excessive voltage at an electrical busbar, etc.

Quality standards have quickly been identified and standards have been formulated which are valid world-wide with minor variations. These standards say that the power system operation must be **safe, reliable and economical**. Safety relates mainly to life threatening aspects for people involved in the usage of the electric power. Reliability standards determine the total time per year during which an electric power end user can be without electric power. Today, in the western world, the standard for electric power end user availability at any time during the year is very high at almost 100%. Since the generation and transport of power is very costly and since this high cost must mainly be paid by the electric power consumers the importance of economical aspects related to all aspects of power system operation are obvious.

This very complex dynamic behavior, the wide geographical distribution of the power system and also the high cost and economical risk of planning, building and maintaining power system components and to achieve the above mentioned power system goals have led to a mostly regional organization of the power system. The power system apparatus and power system operation for these regional areas ('the power system control areas') are coordinated by electric utilities which are mostly privately held or mixed privately held / state owned companies in the western countries.

With the size of the interconnected power systems getting larger and larger in the early 50's manual ad hoc operation of the power system even with local control mechanisms has become harder and harder and high level, often hierarchical control systems have been introduced into the power system to have an automatic or semi-automatic operation of parts of the power system satisfying new standards.

Since almost all major power system control areas are interconnected with mainly high voltage transmission lines operating rule standards have been established in Western Europe (UCPTE) and also in the U.S.A. around 1960. The standards are mainly related to the quality of the power system frequency (Europe: 50 Hz, U.S.A.: 60 Hz) and also to the control of power or energy contract based inter-utility transfers.

This automatic high level control mechanism is called Load Frequency Control (LFC) and has quickly been established as a standard control mechanism to hold both the frequency and also the inter-utility power transfers within given tolerances. LFC establishes some standards for the cooperation of even foreign electric utilities but still allows each utility to operate and maintain its power system part autonomously.

The LFC concept, respecting autonomous power system area control is accepted today by all participating utilities. It is within this framework where the individual utility must satisfy the goals of the power system: Safe, reliable and economic operation.

The above keyword 'safety' meaning, has been discussed before. 'Reliability' and 'Econonomy' can be realized or enforced by computer controlled operation. It is obvious that the result of some optimization could bring the utility nearer to achieve the above goals. The Optimal Power Flow (OPF) is one very important computer tool to help the operator achieve this reliable and economic power system operation for which the utility has autonomous responsibility. However, the OPF is only one of several tools within a real-time based hierarchical computer assisted control system as will be briefly discussed in the next section.

Much literature can be found about the characteristics and modelling of the power system. A good overview of the data acquisition and the generation control system is given in [1] and [2]. These two papers give the reader a brief overview of how real-time data is transferred from the power system to the computer where it is both displayed and used as input for many kinds of algorithms.

2 The role of the optimal power flow (OPF) computation within the overall power system control

Due to the complexity and very high degree of freedom of power system operation hierarchical models have been established which are valid for different time frames. Usually, the output of the higher level model based computations can be taken as input for the next lower level model, etc.

Within each model type different optimization problems can be formulated. The distinction from model type to model type is both the time frame for which the models are valid and also the resulting solution algorithms.

By decoupling the power system into hierarchical models for different time frames global mathematical optimality is lost. However, practical experience has shown that the chosen time based hierarchical characterization leads to quite a practicable approach which is probably not too far from a global optimum.

From an optimization point of view two different steady state power system models exist:

Category 1: In this category the models are derived in such a way to be applicable to power system simulation from approx. 24 hours to 10 years from actual time. Algorithms based on category 1 models are executed in a real-time environment at a rate of about one hour or slower. Within this time frame the goal is to determine the least expensive subset of power generators for each discrete time step based on uncertain and predicted total system load determined by some method which constitutes the only hard equality constraint per discrete time step. Many other inequality type constraints for the individual generation units are incorporated. The objective function is usually to minimize the sum of the cost of all generators for all discrete time steps.

The main mathematical solution problem comes from the mixed discrete, continuous variable problem formulation and also from the fact that inequality constraints for variable changes from time step at time t to those from time step $t + \Delta t$ exist.

The problem area of category 1 is often called Optimal Unit Commitment (UC). Different problem statements, mainly depending on the chosen time horizon and the discrete time steps are possible, however, all have in common, that the power transmission system, i.e. the transmission lines, the transformers, etc. are **not modelled**. In summary, all computations within this category 1 have in common

- that the total end user power consumption is predicted by some method and is given from now until some given future time in discrete intervals.
- that the power system active power transmission losses for each network state in the future are zero or simply predicted by some heuristic approaches and
- that no power transmission based power system limits are violated.

Category 2: Model validity: 5 minutes to 1 hour from actual time into the future. All dynamic quantities are assumed to be in sinusoidal wave form with constant frequency and constant wave peak amplitude (a network state called 'steady state') which allows the application of complex numbers to the solution of the problem. The goal is to determine the optimal scheduled values or control set points for a set of generators and other non-generator controls such that the power system with dynamically varying end user power consumption is continuously operated in or towards a reliable and economic network state.

The main algorithmic problem is given by the need to incorporate the model of the power transmission system with elements like transmission lines, underground cables, transformers, shunts and associated operational constraints. Also, simple models for the generators and loads must be used. This is in contrast to the models used in category 1 where the generation and load are the only power system elements and variables respectively.

The aspect of load behavior uncertainty in algorithms of this category 2 is much smaller than in the algorithms of category 1, due to the relatively short period in the future for which the load must be predicted. Today, due to the complexity given by the need to model the operational constraints of the power transmission system and also due to the additional complexity of the immediate application of the optimization result as control means, many algorithms of this category 2 neglect the varying load behavior at all. In order to compensate for this, a different, less accurate power system model which incorporates the changing load behavior is used in some utilities and applied in real-time more often, e.g. at intervals of 1 minute or even faster.

The OPF which is the main theme of this paper, today must be put within category 2. It is the function in an Energy Management System that schedules the power system controls in some optimal way being at the same time constrained by the power flow network model and power system operating limits.

The modelling challenges are mainly coming from two problem areas:

- Due to the necessity to model the transmission system and its operating constraints for sinusoidal waves with constant frequency and amplitude (which means usage of complex variables) the dimension of the mathematical formulation is enormous. This and the non-linearity of the underlying power system model results in an optimization problem statement whose solution is not at all straightforward.

- The fact that the optimization output should be applicable to the controllers of the power system in real-time mandates very fast and robust algorithms solving a problem based on models which represent the real power system behavior as closely as possible.

In this paper an OPF problem statement is derived independent on the actual computer solution and the fact that the result must be transferred in real-time to the controllers. Thus emphasis is given to the facts that the OPF formulation

- is derived in a form to get a model of very high quality of the power system for a steady state power system;

- is precise in mathematical terms;

- satisfies most constraints given from an operator controlled real-time application of the OPF output.

This is in constrast to the actual solution processes for the OPF problem formulation which are not presented in this paper. However, it must always be kept in mind, that the solution process itself can lead to necessary simplifications of the original problem formulation. Also it is very hard if not impossible to prove that a closed form mathematical solution process with an optimal solution exists for the given OPF problem formulation. It is known that at least for certain defined subproblems given in the following sections, optimal solutions and clear, straighforward solution algorithms exist (see paper by H. Glavitsch).

During the past 20 years much literature has been written about OPF problem formulations and related solutions algorithms. In the appendix literature is cited which is mainly related to the modelling of the power system in connection with the OPF (see [3] ... [7]). Most of these papers have detailed literature references which are not repeated in this paper.

3 The power flow model as equality constraint set of the OPF

3.1 Basic model assumptions

3.1.1 Power system loads

The OPF is based on a power system model for category 2, see the preceding section. The end user power consumption is varying permanently over time and this has to be (or should be) considered in any power system model, especially when applying the model based algorithmic output back into the power system via control mechanisms. As mentioned in the preceding section the validity for the OPF model can be found in the 5 minute to approx. 1 hour time frame. The change of load within this time frame can result in different power system states at time t and time t+5 Min., t+15 Min. or t+1 hour which again could lead to quite different optimal controller settings. Two possibilities exist to handle this situation, knowing that the OPF is executed only once during the chosen time interval:

- Either the utility moves the controller setpoints only once at every discrete time interval, i.e. after the OPF algorithm output is obtained, or

- an additional mechanism with the incorporation of measured or accurately predicted end user power consumption is used to adapt or recompute the optimal controller settings at faster intervals than the main OPF execution rates.

The first approach leads to a non-optimal or only near-optimal network operation. The second approach seems to be better from an optimal operation point of view, however, a more complicated, probably hierarchical algorithmic usage is needed.

In conclusion electric end user power consumption is discretized at pretermined intervals. The individual end user power consumption at a discrete time step is called a *load* in OPF models and computations.

3.1.2 Power system model: Steady state and symmetric power system operation

All electric quantities power system like current, power and voltages are quantities varying over time. For the power system model it is assumed that

- All currents, powers and voltages are quantities with sinusoidal wave form with constant amplitude. This behavior is called steady state power system operation and leads to the very important fact that the power system states can be modelled with complex variables. The complex variables for voltages, currents and powers are transformations of the corresponding steady state power system quantities.

- The power flow model of the three-phase power system assumes so-called (phase-) symmetric power system operation. Without going into details this assumption allows the modelling of the power transmission system with electric two-ports.

Note that both above assumptions are only models of the real power system. Thus model errors (errors with respect to the real-world power system) are introduced. However, practical experience shows that this assumption is valid for a wide variety of cases within category 2.

3.1.3 Power system model: Generation, load and the transmission system

The main components of the power system which must be modelled under the assumption of known or predicted loads at geographical locations and under steady state and symmetric power system operation are the following:

Overhead transmission lines, underground cables, transformers, shunt elements: Each of these passive power system transmission elements is modelled as a two-port mathematical element, situated between electrical nodes i and j (shunt elements are only associated with one electrical node i). Thus the power system model is composed of many passive elements placed between nodes i and j ($i \neq j$) yielding a network of branch elements. The special properties of this passive network are summarized as follows:

An **electrical node** (called from now on 'node') is connected via passive elements only to about 2 or 3 other nodes (on the average). The resulting connectivity matrix is very sparse: Matrix element (i,j) is non-zero only if there is a connection between i and j.

Since the power system is divided into power system control areas, each operated by an electric utility, each utility can model its own control area with high accuracy.

Often the individual utilities split the model of their control area into submodels to reduce the size and the complexity of the power system model. This splitting is possible especially in the lower high voltage levels (e.g. below 110 kV or 60 kV), due to the organization of the electric transmission system: Often the lower voltage network parts are only connected at one point via a transformer to the highest voltage level network parts. Thus a split into parts at these transformers is possible and does not lead to very high modelling inaccuracies.

At the highest voltage level, however, other circumstances exist: Here, power system control areas of different utilities are interconnected. The modelling problem is obvious: Due to the high degree of connectivity of highest voltage networks, each individual utility should also model at least parts of the highest voltage levels and associated network elements of neighboring utilities. However, especially in the network parts representing the highest voltage levels, ownership and mostly profit-based economic reasons often prevent one utility to know the exact data of the network components of the neighboring utilities.

Thus assumptions about passive network elements of mainly the highest voltage levels of the neighboring or even more distant electric utilities must be made by each individual utility. It must be emphasized that modelling only the data of the highest voltage levels of the own control area usually leads to very inaccurate simulation results which might not be useful in practice. This modelling problem has been recognized by the utilities and data exchange for the most sensitive power system elements and their status (on-off) has been established. Thus for the power system model of the transmission network operated by a utility, it can be assumed that the model comprises at least parts of the neighboring utilities. Some theoretical approaches which determine some criteria for the neecessary size of neighboring areas exist today, although not in perfect form. Often they are based on heuristic assumptions.

For power system models (category 2) it must be assumed that the chosen network comprises an expanded geographical area and that the most important power system elements of the own area and neighboring areas are modelled with high accuracy. In this text, the term 'network' refers to this expanded power system network, at least for highest voltage levels networks.

Depending on the size of the utility and the data-exchange related cooperation between neighboring utilities, networks can have varying sizes, starting at about 100 nodes up to 5000 or more nodes with about 150 to more than 10000 passive network elements, each modelled by two-ports, derived in the next subsection.

Generators: They are modelled individually, i.e. at their geographical location at a node of the electrical network. For OPF models it can be assumed that it is known which ones of the individual, geographically distributed generators are 'on' and which ones are 'off'. Only those generators which have a status of 'on' can deliver power into the network and are important in the category 2 models. Certain new modelling aspects are introduced per generator as compared to the category 1 model, such as constraints on the upper and lower reactive generator power (to be defined later in this paper) and upper and lower voltage magnitude levels.

Loads: It is assumed that loads are modelled individually at their geographical location at a node of the modelled network. If passive elements for subareas of the power system control area are not modelled or if they are split apart to be treated in separate models (discussion see above), the individual loads of such an area are assumed to be collected together at one precisely known node. Also, it is assumed that the power values for all the loads are known, either because they are measured precisely or because they are predicted by some method.

DC-Lines: DC-Lines and associated control equipment represent important power system elements in certain power system control areas. A detailed model description is not given in this paper. However, a simple model for a DC line with given MW flow at each end is a generator with given MW. The generator model for OPF computations is described in detail in this paper.

3.2 Power flow: Mathematical model of the power system for steady state simulation

3.2.1 Passive power system elements

Branches

Branches are passive network elements which can all be modelled by the same type of two-port equation, given below. Branch elements always refer to two different nodes i and j:

For each branch-element a relationship between the current, the voltage and the line parameters, all in complex quantities, exists:

$$\begin{bmatrix} \underline{I}_{el-i-j_i} \\ \underline{I}_{el-i-j_j} \end{bmatrix} = \begin{bmatrix} \underline{y}_{ii}^{el-i-j} & \underline{y}_{ij}^{el-i-j} \\ \underline{y}_{ji}^{el-i-j} & \underline{y}_{jj}^{el-i-j} \end{bmatrix} \begin{bmatrix} \underline{V}_i \\ \underline{V}_j \end{bmatrix} \tag{2}$$

(2) is the two-port equation for a branch in the model for category 2. The meaning of the variables is summarized in the appendix of this paper.

Branch elements can be subdivided into two major subcategories: **Lines and transformers**:

Lines can be categorized into two main classes:

- Overhead transmission lines and
- underground cables.

The four complex matrix terms of (2) for a line are computed with the line parameters (variables with capital Y) as follows:

$$\begin{aligned} \underline{y}_{ii}^{el-i-j:Line} &= \underline{Y}_{iio}^{el-i-j} + \underline{Y}_{ij}^{el-i-j} \\ \underline{y}_{ij}^{el-i-j:Line} &= -\underline{Y}_{ij}^{el-i-j} \\ \underline{y}_{ji}^{el-i-j:Line} &= -\underline{Y}_{ij}^{el-i-j} \\ \underline{y}_{jj}^{el-i-j:Line} &= \underline{Y}_{jjo}^{el-i-j} + \underline{Y}_{ij}^{el-i-j} \end{aligned} \tag{3}$$

Transformers are passive network elements which allow the transformation of one voltage to another (or some even to two or even three other voltage levels). Transformers represent a (rather small) subset of the branch-elements defined with (2).

The four complex matrix terms of (2) for a transformer are computed as follows:

$$
\begin{array}{ll}
\underline{y}_{ii}^{el-i-j:Trafo} & = \underline{Y}_{iio}^{el-i-j} + \underline{Y}_{ij}^{el-i-j} \\
\underline{y}_{ij}^{el-i-j:Trafo} & = -\underline{t}^{el-i-j}\underline{Y}_{ij}^{el-i-j} \\
\underline{y}_{ji}^{el-i-j:Trafo} & = -(\underline{t}^{el-i-j})^{*}\underline{Y}_{ij}^{el-i-j} \\
\underline{y}_{jj}^{el-i-j:Trafo} & = (|\underline{t}^{el-i-j}|)^{2}(\underline{Y}_{jjo}^{el-i-j} + \underline{Y}_{ij}^{el-i-j})
\end{array}
\qquad (4)
$$

The elements denoted by capital Y in (3) and (4) can be assumed to be numerically known at precise values for each line or transformer from node i to another node j (although this is not perfectly true due to data uncertainties in the range of plus or minus 10 %). This data uncertainty comes from the fact that the variables are dependent on many effects which are hard to quantify precisely. For example, a complex geometry of the transmission line tower, earth resistance dependency on ground condition which again is weather dependent, geometric construction of the transformer, etc.

The complex transformer-related variable $\underline{t}^{el-i-j}$ (t: tap) represents the measure for varying the transformer two-port parameters which in the power system allows to have a variable voltage relationship of side i to side j of the transformer.

In practice, $\underline{t}$ represents a discrete control of the power system. However, in OPF algorithms, $\underline{t}$ is taken as a continuous variable within upper and lower bounds and is set to the next practically possible discrete step after the optimization.

Shunts

Shunts are passive network elements which can all be modelled by the same type of equation, given below. Shunt-elements always refer two one node i:

For each shunt-element a relationship between the current, the voltage and the shunt parameters, all in complex quantities, exists:

$$
\underline{I}_{el-i-o} \;] = [\; \underline{y}_{ii}^{el-i-o} \;] \;\; \underline{V}_i \;] \qquad (5)
$$

(5) is the equation for a shunt in the model for category 2.

The complex network element of (5) for a shunt is computed as follows:

$$
\underline{y}_{ii}^{el-i-o:Shunt} = s^{el-i-o}\underline{Y}_{iio}^{el-i-o} \qquad (6)
$$

The element with capital Y in (6) can be assumed to be numerically known at precise values for each shunt at node i. However, also here, data uncertainties exist for the same reasons as given for branch elements.

This data uncertainty, however, is **not considered** in most OPF models. Thus all variables written with capital Y can be assumed to be precisely known.

The real variable s^{el-i-o} represents the measure for varying the shunt admittance. Practically this must be seen as a measure indicating how many individual shunts of a shunt bank at a node i must be switched in and how many are in an 'out'-status. Thus, in practice, s represents a discrete control of the power system. However, in OPF algorithms, s is taken as a continuous variable within upper and lower bounds and is set to the next practically possible discrete step only after the optimization.

3.2.2 Kirchhoff-law: All currents at a node must add up to zero

A fundamental law - one of the Kirchhoff laws - says that all currents flowing into an electrical node must sum up to exactly zero. This together with conventions on the direction of currents is represented in Fig. 1.

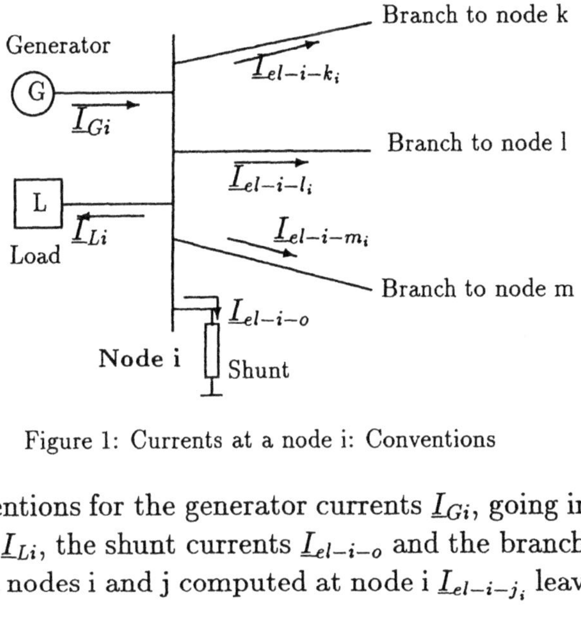

Figure 1: Currents at a node i: Conventions

Note the conventions for the generator currents $\underline{I}_{Gi}$, going into the node, the load currents $\underline{I}_{Li}$, the shunt currents $\underline{I}_{el-i-o}$ and the branch currents for a branch between nodes i and j computed at node i $\underline{I}_{el-i-j_i}$ leaving the node i.

For these currents the Kirchhoff Law is as follows:

$$\underline{I}_{Gi} - \underline{I}_{Li} - \sum_{j=1}^{N} \underline{I}_{el-i-j_i} - \underline{I}_{el-i-o} = 0 \ ; \mathrm{i} = 1 \ ... \ \mathrm{N} \tag{7}$$

In both (7) and Fig. 1 and also throughout the rest of this paper the following assumptions are made:

- N represents the total number of electrical nodes
- All generators at a node i are summarized in one generator ($\underline{I}_{Gi}$).
- All loads at a node i are summarized in one load ($\underline{I}_{Li}$).
- All shunts at a node i are summarized in one shunt ($\underline{I}_{el-i-o}$).
- All parallel branches from node i to node j are summarized into 1 branch element with one two-port equation ($\underline{I}_{el-i-j_i}$ and $\underline{I}_{el-i-j_j}$).
- If there is no generator at a node i then $\underline{I}_{Gi} = 0$.
- If there is no load at a node i then $\underline{I}_{Li} = 0$.
- If there is no shunt at a node i then $\underline{I}_{el-i-o} = 0$.
- If there is no branch-type connection between nodes i and j then $\underline{I}_{el-i-j_i} = 0$ and $\underline{I}_{el-i-j_j} = 0$.

With (2), (5) and (7) a mathematical relationship between the nodal voltage related variables $\underline{V}$ (node-related) and the element related current variable $\underline{I}_{Gi}$ (generator-related), $\underline{I}_{Li}$ (load-related), $\underline{I}_{el-i-o}$ (shunt-related) and $\underline{I}_{el-i-j_i}$ (branch-related) is given.

3.2.3 Power - voltage - current - relationship

For every network element (generator, load, branch, shunt) relationship between the complex power ($\underline{S}$), the complex element-related voltage ($\underline{V}$) and the complex element-related current ($\underline{I}$) is valid:

<u>Generators</u>

$$\underline{S}_{Gi} = \underline{V}_i \underline{I}^*_{Gi} \quad ; \text{i} = 1 \dots \text{N} \tag{8}$$

<u>Loads</u>

$$\underline{S}_{Li} = \underline{V}_i \underline{I}^*_{Li} \quad ; \text{i} = 1 \dots \text{N} \tag{9}$$

<u>Branch-Elements</u>

$$\underline{S}_{el-i-j_i} = \underline{V}_i \underline{I}^*_{el-i-j_i} \quad ; \text{i-j} = \text{all branch elements} \tag{10}$$

$$\underline{S}_{el-i-j_j} = \underline{V}_j \underline{I}^*_{el-i-j_j} \quad ; \text{i-j} = \text{all branch elements} \tag{11}$$

<u>Shunts</u>

$$\underline{S}_{el-i-o} = \underline{V}_i \underline{I}^*_{el-i-o} \quad ; \text{i} = 1 \dots \text{N} \tag{12}$$

I.e. the complex power of an element is always the product of the complex voltage times the conjugate complex current of the corresponding element.

3.3 Mathematical formulation of the various equality constraint sets

3.3.1 General characteristics

The equations (2) to (12) written for all corresponding network elements, represent the set of equality constraints for the power flow model. Its characteristics are:

- All variables are complex with the exception of the shunt taps.
- There is an enormous number of variables and equality constraints which define non-linear relations among the variables.
- The connectivity of the power system elements (the branch elements) determines the number of terms in the equality constraints, especially in (7).

3.3.2 Classification of the variables and equality constraints

Since this set of equations represents a subset of the constraint set a classification of the already used variables in subsets is advantageous:

The variables of the OPF problem will be called $\underline{x}$ from now on. Similarly behaving variables will receive the same index and belong to the same group.

Subsets of variables can be characterized as follows (Sets of variables are put into a bracket like { .. }).

- $\{\underline{x}_{A1}\} = \{\underline{Y}_{ii}^{el-i-j}, \underline{Y}_{ij}^{el-i-j}, \underline{Y}_{ji}^{el-i-j}, \underline{Y}_{jj}^{el-i-j}, \underline{Y}_{iio}^{el-i-o}\}$ (numerically given parameters of all passive network elements)
- $\{\underline{x}_{A2}\} = \{\underline{t}^{el-i-j}, s^{el-i-o}\}$ (transformer and shunt related variables)
- $\{\underline{x}_{B}\} = \{\underline{I}_{el-i-j_i}, \underline{I}_{el-i-j_j}, \underline{I}_{el-i-o}\}$ (branch and shunt related current variables)
- $\{\underline{x}_{C}\} = \{\underline{I}_{Gi}, \underline{I}_{Li}\}$ (generator and load related current variables)
- $\{\underline{x}_{D}\} = \{\underline{V}_{i}\}$ (nodal voltage variables)
- $\{\underline{x}_{E}\} = \{\underline{S}_{Gi}, \underline{S}_{Li}\}$ (generator and load related complex power variables)
- $\{\underline{x}_{F}\} = \{\underline{S}_{el-i-j_i}, \underline{S}_{el-i-j_j}, \underline{S}_{el-i-o}\}$ (branch and shunt related complex power variables)

Note that all of the above variable sets are complex, except s^{el-i-o} related to shunt elements.

Some of the variable sets created above can be further categorized: There is one variable set whose elements can be assumed to be numerically known at fixed, given values:

- $\{\underline{x}_{A1}\}$: Numerically given network parameters of the passive elements.

With this variable classification the equality constraints (2) to (12) can also be classified into four categories:

(2) and (5) are characterized as follows:

$$\mathbf{g}_B = \underline{\mathbf{x}}_B - \mathbf{f_B}(\underline{x}_{A2}, \underline{x}_D) = \mathbf{0} \tag{13}$$

(7) is characterized as follows:

$$\mathbf{g}_A = \mathbf{f_A}(\underline{x}_B, \underline{x}_C) = \mathbf{0} \tag{14}$$

(8) and (9) are characterized as follows:

$$\mathbf{g}_E = \underline{\mathbf{x}}_\mathbf{E} - \mathbf{f_E}(\underline{x}_C, \underline{x}_D) = \mathbf{0} \tag{15}$$

(10), (11) and (12) are characterized as follows:

$$\mathbf{g}_F = \underline{\mathbf{x}}_\mathbf{F} - \mathbf{f_F}(\underline{x}_B, \underline{x}_D) = \mathbf{0} \tag{16}$$

(13) .. (16) represent the sets of equality constraints which must be satisfied for any numeric power flow solution. They show certain properties:

- Each equality constraint set has its own properties. These properties have consequences in the OPF solution process: The Jacobian submatrices, representing the first partial derivatives of the equations with respect to all variables will have special and different properties like sparse matrices, block-diagonal matrices, symmetric matrices, etc..

- The sets of equality constraints $\mathbf{g}_E$ and $\mathbf{g}_F$ could be eliminated from a mathematical solution process if the set of free variables $\{\underline{x}_E\}$ and $\{\underline{x}_F\}$ do not appear in any other equality **and/or** inequality constraint set. Note that as shown later in this paper, the latter is the case, thus $\{\underline{x}_E\}$ and $\{\underline{x}_F\}$ cannot be eliminated from the equality constraint set.

- $\{\underline{x}_B\}$ of the equation set $\mathbf{g}_B$ could immediately be replaced into $\mathbf{g}_A$ and $\mathbf{g}_F$, thus eliminating the variables $\{\underline{x}_B\}$ in the equality constraints and reducing the number of equality constraints. However, for certain mathematical formulations and because of OPF solution based reasons, it is better not to eliminate the variables at this stage.

(13) .. (16) can be summarized into one compact equality constraint set as follows:

$$\mathbf{g}(\underline{\mathbf{x}}) = \mathbf{0} \tag{17}$$

with $\underline{\mathbf{x}}^T = (\underline{\mathbf{x}}_{A2}^T, \underline{\mathbf{x}}_B^T, \underline{\mathbf{x}}_C^T, \underline{\mathbf{x}}_D^T, \underline{\mathbf{x}}_E^T, \underline{\mathbf{x}}_F^T)$ and $\mathbf{g} = (\mathbf{g}_A, \mathbf{g}_B, \mathbf{g}_E, \mathbf{g}_F)$.

4 Mathematical formulation of operational constraints

4.1 Introduction

In the preceding section a power system model and needed equality constraints have been formulated. Satisfying these equality constraints with any numerical set of variables means that the physical characteristics of the power system for a model of category 2 are satisfied. The problem is that many of these physically possible states do not make operational sense or are not operationally possible. Thus in order to model the power system behavior more realistically additional constraints have to be formulated. Different types of operational constraints can be formulated:

- Physical damage to network equipment must be prevented since power system equipment is often very expensive and hard to repair.
- Laws dictate mandatory standards to be satisfied by all utilities, e.g. the voltage amplitude at a node must be within certain upper and lower limits, or: a power end user must have power available at almost 100% during the year, or: if any network element is unvoluntarily outaged the power system must be brought back to an acceptable network state within a given time period.
- Physically given limits for power system sources (any generator has its upper power limit).
- Power consumption at certain discrete time steps force the power to flow from the generators via the transmission system to predefined geographical places, i.e. the power consumed by loads at certain nodes at certain times is given and must be considered.
- Contracts among utilites determine precisely at what times how much power must be imported or exported from one utility to another. This imposes limitations on the power system operation and also on its model.

- Operational limits being computed by other power system problem analysis areas like network stability determine e.g. the maximum allowable complex voltage angle shift from node i to another node j.
- Human operators cannot implement more than say 10 % of all possible controls manually within a give short time period.

These examples show that a huge number of operational constraints exist which must be translated into mathematical constraint types.

These mathematical constraints are derived in the following subsections. Three main constraint groups are identified: The *transmission constraints*, representing all operational limits of the actual network. The *contingency constraints* are related to all operational aspects if any network element is outaged as compared to the actual network and its associated network state. In the third constraint group the *operational policy based constraints* are formulated. They represent e.g. limits of human operator based system control.

The correct mathematical representation of the three constraint groups is emphasized in the following subsections, although this might lead to problems which are not solvable today with classical optimization tools.

4.2 Transmission constraints

Transmission constraints are always related to the actual network, i.e. to a network with given branch connectivity.

4.2.1 Given complex loads

As already discussed in a preceding section individual loads cannot be influenced by operating policies and must be satisfied at any time by the corresponding generation. Thus for any given discrete time step where the individual load values $\underline{S}^0_{Li}$ are either measured or predicted with some method, the following must be valid:

$$\underline{S}_{Li} = \underline{S}^o_{Li} \quad ; \mathrm{i} = 1 \; ... \; \mathrm{N} \tag{18}$$

An upper index ^{o} means a numerically given value (complex if underlined). This equality constraint set deals with a subset of the variable set $\mathbf{x}_E$.

4.2.2 Branch current magnitude - maximum limit

The maximum current magnitude values for transmission branches, i.e. lines and transformers, are given due to limitation of the branch material. Excessive currents would damage the transmission elements.

$$\begin{aligned} |\underline{I}_{el-i-j_i}| &\leq I_{el-i-j}^{max} \quad ; \text{i-j: all branches} \\ |\underline{I}_{el-i-j_j}| &\leq I_{el-i-j}^{max} \quad ; \text{i-j: all branches} \end{aligned} \tag{19}$$

An upper index ^{max} means a numerically given maximum limit value. An upper index ^{min} means a numerically given minimum limit value. The symbol $|(.)|$ refers to the absolute value of the variable in (.). This inequality constraint set deals with the variable set $\mathbf{x}_B$.

4.2.3 Branch MVA-power - maximum limit

The same reason as the one for branch maximum current limit, discussed before, is valid.

$$\begin{aligned} |\underline{S}_{el-i-j_i}| &\leq S_{el-i-j}^{max} \quad ; \text{i-j: all branches} \\ |\underline{S}_{el-i-j_j}| &\leq S_{el-i-j}^{max} \quad ; \text{i-j: all branches} \end{aligned} \tag{20}$$

This inequality constraint set deals with the variable set $\mathbf{x}_F$.

Either (19) or (20) or both must be formulated for each branch of the network.

4.2.4 Lower and upper nodal voltage magnitude limits

These limits are often given by very strict standards. Too high or too low voltages could cause problems with respect to end user power apparatus damage or instability in the power system. This could lead to unwanted and economically expensive partial unavailability of power for end users.

$$V_i^{min} \leq |\underline{V}_i| \leq V_i^{max} \quad ; \text{i} = 1 \text{ .. N} \tag{21}$$

This inequality constraint set deals with the variable set $\mathbf{x}_D$.

(21) is valid for every node of the network. There are nodes (often a subset of the generator nodes) where the upper and lower voltage limits are identical, i.e. the voltage magnitude of this node is numerically given.

4.2.5 Lower and upper generator active power limits

The active power of a generator i is defined to be the real part of the complex variable $\underline{S}_{Gi}$. This quantity is physically limited in each generator.

$$P_{Gi}^{min} \leq Real(\underline{S}_{Gi}) \leq P_{Gi}^{max} \quad ; \text{i} = 1 \text{ .. N} \tag{22}$$

This inequality constraint set deals with the variable set $\mathbf{x}_E$.

(22) must be formulated for every generator. Often the lower limit is zero.

4.2.6 Lower and upper generator reactive power limits

The reactive power of a generator i is defined to be the imaginary part of the complex variable $\underline{S}_{Gi}$. It is an important measure of voltage magnitude quality, e.g. a low voltage indicates a local shortage of reactive power.

$$Q_{Gi}^{min} \leq Imag(\underline{S}_{Gi}) \leq Q_{Gi}^{max} \quad ; \text{i} = 1 \text{ .. N} \tag{23}$$

This inequality constraint set deals with the variable set $\mathbf{x}_E$.

(23) must be formulated for every generator. The upper and lower reactive power limits are often not given as numeric values but as functions of the active generator power:

$$\begin{aligned} Q_{Gi}^{min} &= f_{min_i}(Real(\underline{S}_{Gi})) \\ Q_{Gi}^{max} &= f_{max_i}(Real(\underline{S}_{Gi})) \end{aligned} \tag{24}$$

4.2.7 Upper and lower transformer tap magnitude limit

These limits come from the fact that the range of a tap is limited by the physical construction of each transformer. Thus clear limits exist for each transformer.

$$t_{el-i-j}^{min} \leq |\underline{t}_{el-i-j}| \leq t_{el-i-j}^{max} \quad ; \text{i-j: all transformer branches} \tag{25}$$

This inequality constraint set deals with a subset of the variable set $\mathbf{x}_{A2}$.

4.2.8 Upper and lower transformer tap angle limit

The same reasons as discussed for the transformer tap magnitude limits are valid here.

$$\begin{aligned} \delta t_{el-i-j}^{min} &\leq \angle(\underline{t}_{el-i-j}) \leq \delta t_{el-i-j}^{max} \\ &\quad ; \text{i-j: all transformer branches} \end{aligned} \tag{26}$$

Note that the symbol $\angle$ refers to the angle of the following complex variable. This inequality constraint set deals with a subset of the variable set $\mathbf{x}_{A2}$.

In the power system mainly two types of transformers can be found with variable tap position:

- In-phase tap changing transformer: The transformer can vary or regulate the voltage magnitude
- Phase shifting transformer: The transformer can vary or regulate the voltage angle.

For an in-phase tap changing transformer usually a constraint of type (25), for a phase shifting transformer a constraint of type (26) must be formulated.

4.2.9 Upper and lower shunt tap limit

This limit has to be understood as follows: In practice a limited number of discrete shunt elements is available at some electrical nodes. These individual shunt elements can either be switched in or out. Thus the upper shunt tap limit corresponds to the state where all shunts are switched in and the lower limit to the lowest number of possible switched-in shunts.

$$s_{el-i-o}^{min} \leq s^{el-i-o} \leq s_{el-i-o}^{max} \text{ ; i-o: all shunt elements} \tag{27}$$

This inequality constraint set deals with a subset of the variable set $\mathbf{x}_{A2}$.

4.2.10 Upper and lower limits on branch voltage angle

This type of constraint must be formulated if some stability based criterion is formulated for the branch angle. Violating some of the constraints can cause severe dynamic stability problems which could lead to power outage and severe economic penalties.

$$\begin{aligned} \delta_{el-i-j}^{min} &\leq |\angle(\underline{V}_i) - \angle(\underline{V}_j)| \leq \delta_{el-i-j}^{max} \\ &\text{; i-j: all critical branch elements} \end{aligned} \tag{28}$$

This inequality constraint set deals with a subset of the variable set $\mathbf{x}_D$ and must be formulated for every branch determined by some stability criterion.

4.2.11 Minimum generator active power spinning reserve

$$P_{reserve}^{min} \leq \sum_{i=1}^{N}(P_{Gi}^{max} - Real(\underline{S}_{Gi})) \tag{29}$$

This inequality constraint is necessary in order to force each power system control area to have a certain amount of total active generator power available for unforseen cases. An example is the generator outage in a neighbouring power system control area: Here the power system control areas with intact power generation automatically help to provide the power for a certain time after the outage.

4.2.12 Upper and lower limits on total active power of a given set of branches

$$\begin{array}{l} P_{BS_k}^{min} \leq |\sum_{(i-j)\in\{BS_k\}} Real(\underline{S}_{el-i-j_i})| \leq P_{BS_k}^{max} \\ \quad ;\ BS_k\text{: set of branches with limited total active power} \end{array} \tag{30}$$

This inequality constraint set deals with a subset of the variable set $\mathbf{x}_D$ and must be formulated for every branch set for which a constraint on the sum of the active power is valid. Often the upper and lower limits for this total branch set flow is identical and numerically given. This is the case if the network comprises the network data of more than one utility and if the utilities have contracts for power transfer from one into the other power control area.

4.2.13 Control variable time-related movement limits

In a power system the operator can only control certain quantities. These quantities are called controls and represent a physical apparatus to implement any operationally feasible and acceptable network state. Each control has an associated control variable. The most important control variables are the following:

- Generator active power control P_{Gi}: This control variable is the real part of the variable $\underline{S}_{Gi}$.

$$P_{Gi} = Real(\underline{S}_{Gi}) \quad ;\ \text{i: all generators} \tag{31}$$

- Generator reactive power control Q_{Gi}: This control variable is the imaginary part of the variable $\underline{S}_{Gi}$.

$$Q_{Gi} = Imag(\underline{S}_{Gi}) \quad ;\ \text{i: all generators} \tag{32}$$

- Generator voltage magnitude control V_{Gi}: This control variable is the voltage magnitude $|\underline{V}|_i$ only of the generator nodes.

$$V_{Gi} = |\underline{V}_i| \quad ; \text{ i: all generators} \tag{33}$$

- Phase shifter transformer tap position control δt_{el-i-j}: This control variable is the angle of the complex tap of a transformer $\underline{t}_{el-i-j}$.

$$\delta t_{el-i-j} = \angle(\underline{t}^{el-i-j}) \quad ; \text{ i-j: all phase shifter transformers} \tag{34}$$

- In-phase transformer tap position control t_{el-i-j}: This control variable is the magnitude of the transformer tap $\underline{t}_{el-i-j}$.

$$t_{el-i-j} = |\underline{t}^{el-i-j}| \quad ; \text{ i-j: all in-phase transformers} \tag{35}$$

- Shunt value control: This setpoint is usually the real-number shunt tap variable s_{el-i-o} itself.

$$s_{el-i-o} = s^{el-i-o} \quad ; \text{ i: all shunts} \tag{36}$$

- Active power interchange transaction control: A utility connected to neighbouring utilities can buy or sell active power via the tie-lines which connect the areas. The control variable is called $P_{interchange-k-l}$ (The index k refers to area k and the index l to a neighbouring area l):

$$\begin{aligned} P_{interchange-k-l} &= \textstyle\sum_{(i-j)\in\{k-l\}} Real(\underline{S}_{el-i-j_i}) \\ &\quad ; \text{ k-l: branches connecting utility l to utility k} \end{aligned} \tag{37}$$

The above mentioned variables on the left hand side of the equality constraints are called control variables and represent variables with real values which can be directly influenced by the power system operator.

The mathematical representation of control variables as used in this paper is $\mathbf{u}$. They can always be derived from variables of the vector $\mathbf{x}$ as shown with (31) .. (37) and are summarized as follows:

$$\mathbf{u} = \mathbf{g}_u(\underline{\mathbf{x}}_{\mathbf{A2}}, \underline{\mathbf{x}}_{\mathbf{E}}, \underline{\mathbf{x}}_{\mathbf{D}}) = \mathbf{g}_u(\underline{\mathbf{x}}) \tag{38}$$

It is important to understand that the power system cannot realize the state of the desired values at the time t when the control variables and the associated physical controls are changed by the operator: Due to the dynamic nature of the power system there is always a time delay from changing the control setpoint values to the time when the power system actually shows the desired values. Also, it does not make sense to change the controls to values which cannot be realized in the power system at all or only after quite a long time.

Thus there is an additional set of inequalities, representing the maximum difference between actual power system state and the state where the power system can be moved from this state within the time frame for which the OPF result is valid. This time frame corresponds often to the discrete time interval in which one OPF calculation is done, see sections 2 and 3 of this paper.

These additional lower and upper control variable constraints are formulated as follows in general form:

$$\mathbf{u}^{min} \leq \mathbf{u} \leq \mathbf{u}^{max} \tag{39}$$

In (39), the limit vectors are derived both from the values of the actual network state and the ability to move controls within a given time period. They can be assumed to be numerically given.

4.2.14 Summary: Transmission constraints

Using the variable characterization of the preceding section the additional equality (18) and inequality constraints (19) .. (30) can be setup as follows:

(18)	$\{\mathbf{g}_{E1}(\underline{\mathbf{x}}_{E1}) = 0\}$
(38)	$\{\mathbf{g}_u(\underline{\mathbf{x}}_{\mathbf{A2}}, \underline{\mathbf{x}}_{\mathbf{E}}, \underline{\mathbf{x}}_{\mathbf{D}}) - \mathbf{u} = \mathbf{0}\}$
(19)	$\{\mathbf{h}_B(\underline{\mathbf{x}}_B) \leq 0\}$
(20), (22), (23), (30)	$\{\mathbf{h}_E(\underline{\mathbf{x}}_E) \leq 0\}$
(21), (28)	$\{\mathbf{h}_D(\underline{\mathbf{x}}_D) \leq 0\}$
(25), (26), (27)	$\{\mathbf{h}_A(\underline{\mathbf{x}}_{A2}) \leq 0\}$
(39)	$\{\mathbf{h}_u(\mathbf{u}) \leq \mathbf{0}\}$

All of the above equality constraints and those of (17) and all inequality constraints above can be summarized in compact form as follows:

$$\mathbf{g}(\underline{\mathbf{x}}, \mathbf{u}) = 0 \quad (40)$$

$$\mathbf{h}(\underline{\mathbf{x}}, \mathbf{u}) \leq 0 \quad (41)$$

Both (40) and also (41) are functions of mixed complex and real variables. However it is always possible to formulate the problem with real variables by using the simple rectangular or polar coordinate formulation.

From a mathematical point of view these equality and inequality constraints are relatively simple functions of related complex OPF variables. At this point the choice of the variable coordinate system for the OPF variables $\underline{\mathbf{x}}$ has a direct influence on the complexity of the real-variable formulation of the constraints:

When choosing the **polar coordinate system** then all angle related inequality constraints ($\angle(.)$) (like (26), (28), (34)) and all magnitude related functions ($|(.)|$) (like (19), (20), (21), (25), (27), (30), (33), (35) and (37)) are very simple 1:1 functions (meaning $x_{magnitude_i}$ or x_{angle_i}) of the corresponding polar magnitude and polar angle related variables.

These functions are more complicated if using the **rectangular coordinate system**: Here the functions for magnitude related functions take the form $\sqrt{(x_{real_i} + x_{imag_i})}$ and for angle related functions $\arctan \frac{x_{imag_i}}{x_{real_i}}$.

For functions where the real or imaginary part of a complex variable is desired (like (22), (23), (31) and (32)), the polar coordinate formulation is more complex and takes the form $cos(x_{angle_i}) \cdot x_{magnitude_i}$ for 'Real'- type functions and for 'Imag' - type functions $sin(x_{angle_i}) \cdot x_{magnitude_i}$.

In rectangular coordinates the corresponding functions are simple: For 'Real'- type functions x_{real_i} and for 'Imag' - type functions x_{imag_i}.

(40) and (41) represent the set of equality and inequality constraints. Both together are called the **transmission constraints** within the complete OPF formulation.

Note that for each discrete time step for which an OPF is valid the transmission constraints can be slightly different with respect to the symbolic formulation, but significantly different with respect to the numerically given limit and load values.

4.3 Contingency constraints

In the preceding subsection all equality and inequality constraints refer to the actual network state. This means that the OPF transmission constraints are based on a given network structure with known branch, shunt, and generator statuses. It has been assumed that all these elements are in a 'switched-in' status. The network where all elements are in this 'switched-in' status is called the 'actual network'.

In addition to satisfying the transmission constraints for this 'actual network', law or economic reasons force utilities not only to satisfy all transmission constraints of the 'actual network', but also to satisfy the so-called 'n-1-security' based constraints which can be defined as follows:

The state of the 'actual network' is defined as 'n-1-secure' with post-contingency rescheduling at any time T if

1. all operational constraints of the 'actual network' are satisfied, i.e. the **transmission constraints and if**

2. any one of the branches, shunts or generators is outaged (representing a **contingency**), the new network state must be such that within a given short time period τ_{short}, the 'new actual network' can be shifted to a transmission constrained state by control means. Since this control variable movement occurs immediately after the contingency happens it is called **post-contingency control movement.**

Figure 2 shows typical network state changes caused by the outage on an element:

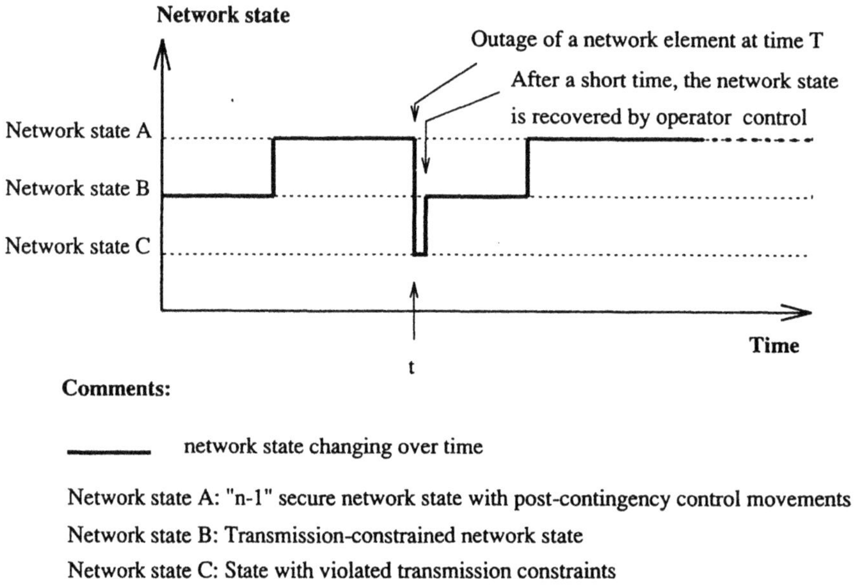

Figure 2: Network state changes due to the outage of a network element at time t

Many utilities, e.g. at the east coast in the U.S.A., choose the time for recovering the network state after the contingency has occured to a transmission constrained network state, i.e. τ_{short} as zero. This means that the utility wants to operate its actual network in such a state that if any contingency occurs the new network state after the contingency is immediately such that no transmission constraints are violated.

It is important to note that it can be assumed that the control variables $\mathbf{u}$ do not change from the network state immediately before to immediately after the contingency occurs. Only the OPF variables $\underline{\mathbf{x}}$ change immediately due to the physical behavior of the power system. Thus a contingency can immediately lead to operational constraint violations which must be corrected by moving the control variables $\mathbf{u}$ within a short time τ_{short}.

A dynamic simulation not being the objective, the recovery time effect is translated into a maximum control variable move after the outage occurs (the so-called post-contingeny control variable movement):

It is known how long it takes to move a certain power system control by a certain amount, e.g. the active power of a generator from a generation of 100 MW to 130 MW. Thus it is assumed that the maximum possible and operationally acceptable movements of all control variables are known. The resulting new inequality constraint set representing the maximum allowable upward and downward movements of all control variables per outage is as follows:

First the maximum movements for all control variables are computed numerically. Note that different values can be chosen per contingency i:

$$\mathbf{\Delta u}^{(i)^{max}} = \tau_{short} \cdot \Delta\mathbf{u}/\mathbf{sec.}^{(i)^{max}} \tag{42}$$

The scalar value τ_{short} is given (in seconds) and the vector $\Delta\mathbf{u}/\mathbf{sec.}^{(i)^{max}}$ represents the numerically given maximum possible movements per second for each control variable.

$$\begin{array}{c} \mathbf{u} - \mathbf{\Delta u}^{(i)^{max}} \leq \mathbf{u}^{(\mathrm{i})} \leq \mathbf{u} + \mathbf{\Delta u}^{(i)^{max}} \\ \text{i: all possible contingencies} \end{array} \tag{43}$$

In (43) the vector $\mathbf{u}^{(\mathrm{i})}$ refers to the control variable set valid at time $t + \tau_{short}$, assuming that the outage of network element i has occured at time t. I.e. for every possibly outage network element i such a vector is created, representing the new state at time $t + \tau_{short}$. The size of the vector $\mathbf{u}^{(\mathrm{i})}$ per outage element i is the same as the size of $\mathbf{u}$, with the exception of a control variable related to an outaged network element.

Since the new network state after the outage must satisfy the transmission constraints the model must satisfy the same equality and inequality constraints like in the 'actual network' state (see (40) and (41)). Thus for each outage case i the following equality and inequality constraints have to be formulated with the new outage state related variable set as defined above:

$$\mathbf{g}^{(i)}(\underline{\mathbf{x}}^{(i)}, \mathbf{u}^{(i)}) = \mathbf{0} \quad ; \text{i: all possible contingencies} \tag{44}$$

$$\mathbf{h}^{(i)}(\underline{\mathbf{x}}^{(i)}, \mathbf{u}^{(i)}) \leq \mathbf{0} \quad ; \text{i: all possible contingencies} \tag{45}$$

In summary, by formulating contingency constraints the following points are significant:

- The number of variables is increased to (n = number of possible contingencies) n times the number of variables of the 'actual network', i.e. the problem is tremendously increased.
- The absolute values for the difference of control variables of the 'actual network' and corresponding control variables of the outage cases must be less or equal to a numerically given value.
- A network state valid after the contingency has occured, must satisfy all operational equality and inequality constraints. The difference in the symbolic formulation to the actual network equations are as follows:
  - A new set of variables is used for each contingency.
  - The outaged network element must not be modelled in the equality constraints
  - If there is an inequality constraint for the outaged element this inequality constraint must not be formulated for the contingency network state.

4.4 Operational policy based constraints

4.4.1 Introduction

The constraints formulated in the preceding sections represent important restrictions on the power system model used in the OPF calculation. Without including these constraints the model will result in non-practical results.

One of the mathematically challenging problems is the fact that a power system operator cannot change too many controls during a given time period. This is the case if the result of the OPF optimization is not transferred automatically to the power system control. The automatic control meachanism is called closed loop control. The mathematical formulation for this problem is given in the following subsection.

4.4.2 Limited number of controller movements

In a preceding subsection the control variables $\mathbf{u}$ have been defined as functions of the OPF variables $\mathbf{x}$. Control variable movements are limited by the maximum physically possible change during a given time period.

In the case of non-closed loop OPF operation further constraints must be formulated, since the total number of control variable movements must be limited to a certain numerically given value:
The problem can be stated in words as follows:
Given: The actual power system state is given in such a way that all equality constraints (2) .. (12) are satisfied. The numeric values of the control variables for this state called 'Base Case'(BC) state are represented with the vector $\mathbf{u}_{BC}$. They are computed with the equations (31) .. (37).
Goal: Do not move more than a given number of control variables to some other network state which is operationally more convenient. This new state and the associated control variables are called $\mathbf{u}$. A typical value for this number of maximum movable control variables is e.g. 10 % of all control variables.

This problem of limiting the number of control variable movements is given not only from the base case to an optimized base case network state, but also for each outage case i:

Especially when a contingency occurs the operator does not want to deal with too many control variable movements which are needed to bring the network state back to an acceptable new transmission constrained network state after the outage has occured. By the value τ_{short}, representing the time during which the power system must have been moved back to a transmission contrained state, the operator is under heavy pressure and is interested to have a limited number of movable control variables.

Outage related control variables movements refer the movement starting point to the state represented by the vector $\mathbf{u}$. Note that the number of movable control variables for each individual outage case can differ from the number of control variable movements for the base case.

The control variable movements for the individual outage cases $\mathbf{u}^{(i)}$ are derived from the outage-related variables $\underline{\mathbf{x}}^{(i)}$ in analogy to (38) by simple transformations of the complex outage case variables $\underline{\mathbf{x}}^{(i)}$ to real outage case related control variables $\mathbf{u}^{(i)}$.

$$\mathbf{u}^{(i)} = \mathbf{g}_u^{(i)}(\underline{\mathbf{x}}^{(i)}) \quad ; \text{ i: all possible contingencies} \tag{46}$$

The problem is that it is not known beforehand which subset of the control variables to move, both base case and also outage case related. This means that the status of each control variable must be a variable and this is formulated mathematically as follows:

Each of the control variables can have a status of moved (1) or not moved (0). This status is represented by the variable vector $\mathbf{w}$ which refers to the control variables of the base case. For outage case i the corresponding status variable vectors are called $\mathbf{w}^{(i)}$.

$$\mathbf{w}, \mathbf{w}^{(i)} = \begin{bmatrix} \mathbf{0} \\ \mathbf{1} \end{bmatrix} \quad ; \text{ i: all possible contingencies} \tag{47}$$

(47) shows that for each element of the vectors $\mathbf{w}$ and $\mathbf{w}^{(i)}$ the value can only be either 0 or 1.

With these additional, discrete variable type constraints the control variables $\mathbf{u}$ and $\mathbf{u}^{(i)}$ must satisfy the following conditions:

$$\mathbf{diag}(\mathbf{u} - \mathbf{u}_{BC}) \cdot (\mathbf{w} - \mathbf{1}) = \mathbf{0} \tag{48}$$

and for the outage cases:

$$\mathbf{diag}(\mathbf{u}^{(i)} - \mathbf{u}) \cdot \left(\mathbf{w}^{(i)} - \mathbf{1}\right) = \mathbf{0} \quad ; \text{ i: all possible contingencies} \tag{49}$$

(47) together with (48) and (49) guarantee that the individual control variables are either moved or not moved at all.

The number of control variable movements is limited with the following simple inequality constraints representing the sum all individual variables of the vectors $\mathbf{w}$ and $\mathbf{w}^{(i)}$:

For the base case:

$$\mathbf{w}^T \cdot \mathbf{1} \leq CV^{max} \tag{50}$$

CV^{max} stands for the maximum number of base case related control variable movements (a scalar, numerically given value).

For the individual outage cases:

$$\mathbf{w}^{(\mathrm{i})^T} \cdot \mathbf{1} \leq CV^{(i)^{max}} \quad ; \text{ i: all possible contingencies} \tag{51}$$

$CV^{(i)^{max}}$ stands for the maximum number of outage case i related control variable movements (i.e. per outage element i, a scalar, numerically given value).

(48) .. (51) represent the mathematical formulation for the operational problem of the limited number of control variable movements.

4.5 Overview: Network type, network state and constraint set related term definitions

Important goals of the power system control are the reliable and economic operation. These goals are translated into OPF problem parts: Reliability is translated into the operational constraints, economy into the objective function (discussed later in this paper). These operational constraints can be grouped and important terms have been introduced in this section. They are summarized in Tab. 1.

5 OPF objectives and objective functions

5.1 Introduction

An important part in any mathematical optimization problem is the objective function which allows to make a distinct and often unique, optimal selection out of the solution region defined by the equality and inequality constraints. Also, the objective function is needed to drive a mathematical optimization process towards an optimal solution.

As already discussed before utilities can have different goals or power system operation objectives. Some goals are clearly defined like minimum active power losses in the resistive parts of the transmission system branches or minimum total cost for the active power generation of the generators. These objectives are of economical nature and can thus easily be justified.

On the other side less clear goals exist which can be of the following types and often depend on operational utility policies: If the power system is monitored and if the power system is found to be in a state where either transmission or contingency constraints are violated, operate the system in such a way that the violations are eliminated as quickly as possibly.

It is obvious and mathematicians can prove that objectives often are exchangeable with equivalent inequality constraints, i.e. putting harder

Term	Description
Actual network	Network with given branch, shunt and generator statuses. I.e. it is known which of these elements are switched in and which ones are switched out.
Contingency i network	This is the network where one element i is outaged as compared to the 'actual network'. If the outage is lasting for a long time this new network will be a new 'actual network'.
Base case network state	This is any network state (i.e. voltages, currents, powers) computed or measured at an 'actual network'. Only the power flow equality constraints of the 'actual network' are satisfied. There can be violations on all kinds of operational constraints. It is the most general state from which an OPF optimization can be started. One of the goals of the operator is to shift the network from this network state via the 'transmission constrained network state' to the 'n-1-secure network state'.
Transmission constrained network state	An 'actual network' is in this state if all transmission constraints are satisfied (some of the contingency constraints are violated).
n-1-secure network state	This is a computed state of the 'actual network' where all operational constraints including those for contingencies are satisfied.

Table 1: State and network naming definitions

inequality constraints limits can also limit the value of an objective function or the other way around: Formulating inequality constraints as part of the objective function can give valid solutions by enforcing so-called soft limits.

This paper concentrates on the formulation and not on the solution of the OPF problem by giving equality and inequality constraints. In this section possible objectives and the mathematical objective function formulations are discussed. However, one must always keep in mind that the possibility to partially exchange inequality constraints and objective function formulations is given. The solution process actually decides how to combine the constraint and objective function formulations.

5.2 Mathematical formulation of various OPF objective functions

5.2.1 Objective: Minimum active power losses

The active power losses (called 'losses' from now on) represent a quantity whose minimization can easily be justified by an electric utility if the effort to actually operate in a minimum loss mode is not too expensive. Losses are generated in the resistances of the transmission lines and are a measure of the difference of the generated total active power to the total active load at any time.

Two different loss-related cases exist: If the network model comprises only network parts of the own utility controlled area the losses are computed as follows:

Loss case A:

$$\text{Minimize } P_{Loss} = \sum_{i=1}^{N} (Real(\underline{S}_{Gi}) - Real(\underline{S}_{Li})) \tag{52}$$

Thus in this case, the losses are a simple function of elements of the vector $\underline{\mathbf{x}}_E$ representing the generator and load related nodal complex powers.

Loss case B:

Here the modelled network also comprises network parts of the neighbored utilities. Reasons for this have been discussed in previous section and are often the case if the highest voltage levels of the network must be modelled. For this case B the losses are the sum of the active powers of each branch of the utility controlled area. For each branch the active power flow from node i to node j must be added to the branch flow from node j to node i. This results in the active power losses per branch. Thus the losses for a defined area 'a' are computed as follows:

$$\text{Minimize } P_{Loss}^{\text{area 'a'}} = \sum_{i-j \in \text{area 'a'}} \left(Real(\underline{S}_{el-i-j_i}) + Real(\underline{S}_{el-i-j_j}) \right) \quad (53)$$

Thus in this case, the losses are a simple function of elements of the vector $\underline{\mathbf{x}}_F$ representing the complex power of branches.

5.2.2 Objective: Minimum total active power operating cost

Each utility has control over generation of different kind, e.g. hydro, hydro-thermal or thermal power generation. Each type of generation has cost associated with it: Hydro power is, if available, usually the cheapest power (water does not cost much in mountain areas) and thermal power generation is more expensive (often oil, gas or nuclear material is used as primary energy resource. These resources have to be bought at market prices).

In addition to own power resources the utility can buy or sell power from or to neighbouring utilitities. Sometimes it can be cheaper to buy power than to produce it within the power system control area.

It is in the interest of the utilities and also of the paying power consumers to minimize the cost associated with active power generation. It can be assumed that the cost of each generator can be represented as a distinct curve of relating cost to the active power which the generator delivers. Usually this curve is given for the full range of the operating capability of the generator. The general type of a cost curve can be written as follows:

$$C_{Gi} = c_i(Real(\underline{S}_{Gi})) \quad (54)$$

where c_i is a general function of the active power of the generator i. A typical cost curve is shown in Fig. 3.

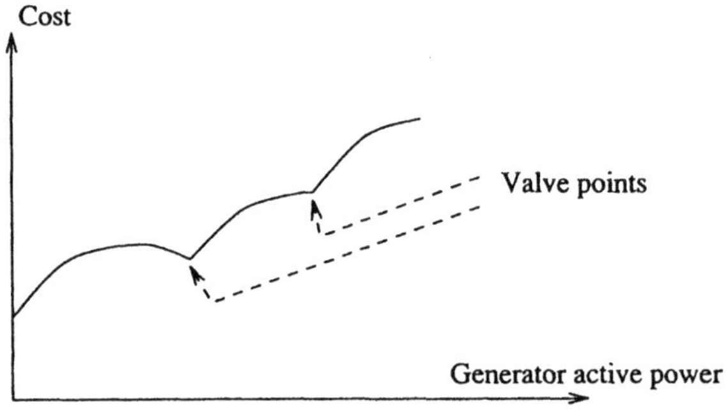

Figure 3: Typical true cost curve for hydro-thermal power generation

As seen in Fig. 3 the general cost curve type is very complex: Non-convex cost curves are possible. Also, it is important to note that the cost curves can be assumed to be separable with respect to the active power generation of the generatorsm, i.e. the cost of each generator is only dependent on the cost of its own active power generation and not on the cost of another generator power.

The cost curve associated with buying or selling power is not as clear. Sometimes it is step-wise linear, sometimes linear over the full range. Note that selling power leads to a cost curve with negative cost values.

The problem is the shape of the cost curves. Optimization algorithm usually cannot deal with cost curve shapes as shown in Fig. 3. Thus, the cost curves are usually modified and a convex, smooth cost curve as shown in Fig. 4 can be assumed to be the cost curve model for the OPF.

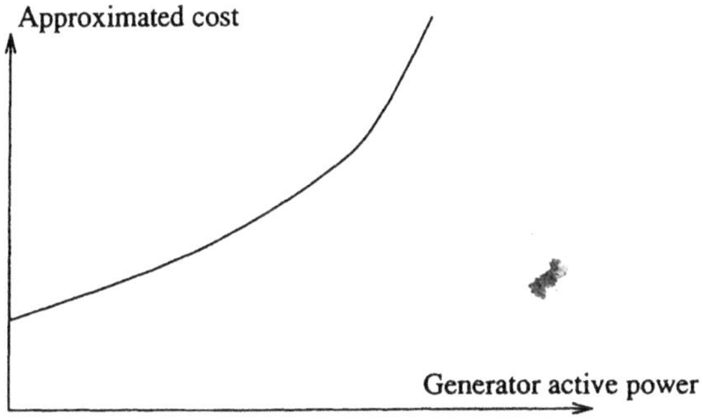

Figure 4: Corrected smooth and convex power generation cost curve for OPF model

Often these cost curves are assumed to be piecewise quadratic with smooth transition at the cost curve break points. Also, exponential cost curve representations are seen in OPF models. Usually the approximated smooth and convex cost curves (one per active power generation) is given as the exact reference along which the OPF optimum solution must be found. Whether this makes sense or not is not the issue of this paper. However, it should be kept in mind that the cost curve issue is open and needs to be resolved the more accurate the remaining power system model is chosen. An inaccurate cost curve model can prevent an accurate optimization result, even if the remaining model parts are model with very high accuracy or in other words, the optimization result is not more accurate than the chosen accuracy of the model components.

The total cost of all generators and transaction interchange active power can be represented as objective function as follows:

$$\begin{aligned}\text{Minimize } C &= \sum_{i=1}^{N} c_{approximated_i}(Real(\underline{S}_{Gi})) \\ &+ \sum_{l \in k-l} c_{int-approximated_{k-l}}(P_{interchange-k-l}) \\ &\text{k-l: all areas l connected to area k}\end{aligned} \quad (55)$$

5.2.3 Objective: Fast transition from a violated to a non-violated network state

For many reasons a power system can suddenly move into a state where important constraints are violated. If this happens a very important objective of a utility can be, to move the power system as quickly as possible back to a network state without violating constraints, including transmission, contingency and utility policy constraints.

The problem to get a mathematical objective function is the term 'quick' which is time related. The OPF, however, does not have time as variable in the formulation. Thus another measure for the quick operation must be found. One measure can be to determine a feasible state (a state where no inequality constraints are violated) with a minimum number of control variables movements. Provided a solution for the minimum number of control variables can be found in reasonable time, this is probably one of the fastest ways for the utility to achieve a feasible operation. Thus the objective function would be:

$$\text{Minimize } F = \mathbf{w}^T \cdot \mathbf{1} \quad (56)$$

There might be other objective function formulations for this special objective. The main problem is to translate the operating philosophies into the objective function. Once it is given the mathematical process tries to solve the problem as accurately as possible. The optimization result could be such that although satisfying the mathematical model, it does not conform to the often rule-based operational policies of the utility. This will, especially in the first implementation phase of the OPF, lead to some type of iterative process for defining the objective function and the resulting mathematical optimum. At the end of this iterative process the optimization result must both satisfy the mathematical optimum criteria and also the operational, often rule based philosophies of the utility.

6 The complete OPF formulation

6.1 Mathematical OPF formulation - summary

Description	Mathematical formulation	Equations
Actual network state	$\mathbf{g}(\underline{\mathbf{x}}, \mathbf{u}) = \mathbf{0}$	(40)
Contingency i network state	$\mathbf{g}^{(\mathbf{i})}(\underline{\mathbf{x}}^{(i)}, \mathbf{u}^{(i)}) = \mathbf{0}$	(44) ∀ i
Control variable movement status	$\mathbf{w}, \mathbf{w}^{(i)} = \left[\mathbf{0}^T \;\; \mathbf{1}^T\right]^T$	(47) ∀ i
Actual network control variable movement	$\mathbf{diag}(\mathbf{u} - \mathbf{u}_{BC}) \cdot (\mathbf{w} - \mathbf{1}) = \mathbf{0}$	(48)
Post-contingency case i variable movement	$\mathbf{diag}(\mathbf{u}^{(i)} - \mathbf{u}) \cdot (\mathbf{w}^{(\mathbf{i})} - \mathbf{1}) = \mathbf{0}$	(49) ∀ i

Table 2: Equality constraints

Description	Mathematical formulation	Equations
Actual network state	$\mathbf{h}(\underline{\mathbf{x}}, \mathbf{u}) \leq \mathbf{0}$	(41)
Contingency i network state	$\mathbf{h}^{(\mathbf{i})}(\underline{\mathbf{x}}^{(i)}, \mathbf{u}^{(i)}) \leq \mathbf{0}$	(45) ∀ i
Post-contingency i control movement	$\mathbf{u} - \boldsymbol{\Delta}\mathbf{u}^{(i)^{max}} \leq \mathbf{u}^{(\mathbf{i})} \leq \mathbf{u} + \boldsymbol{\Delta}\mathbf{u}^{(i)^{max}}$	(43) ∀ i
Actual network control movement status	$\mathbf{w}^T \cdot \mathbf{1} \leq CV^{max}$	(50)
Contingency case i control movement status	$\mathbf{w}^{(\mathbf{i})^T} \cdot \mathbf{1} \leq CV^{(i)^{max}}$	(51) ∀ i

Table 3: Inequality constraints

Description	Equations
Active power losses	(52), (53)
Total active power operating cost	(55)
Fast transition from a violated to a non-violated network state	(56)

Table 4: Typical OPF objective functions

The OPF formulation can be split in three main parts: The equality constraints, the inequality constraints and the objective function. Each of these three parts has its own properties, as discussed in the preceding sections. The three tables Tab. 2 ... Tab. 4 summarize the OPF problem.

Note that only one of the possible objective functions together with all equality and inequality constraints can be chosen and can be solved for one optimal solution variable set.

Here another problem with the present OPF formulations and solutions becomes obvious: Although a utility would like to satisfy more than one objective at the same time, the classical mathematical optimization algorithms allow only one objective function at a time. The problem of getting a solution which considers more than one objective function (a multi-objective function problem) at a time exists and should be addressed in the future to make the OPF result more practically acceptable and applicable.

6.2 Infeasibility of OPF problem formulation

Due to its practical nature and due to many existing constraints the problem can be formulated in such a way that no mathematical solution exists. This is often true if many contingency constraints are to be satisfied.

This mathematical infeasibility problem, however, is in contrast to the practical demand to get an answer under any circumstances. From an operators point of view it is desirable to get a solution which represents a network state which is 'as near as possible' to an feasible network state and which has an objective function value 'as low as possible' (in the case of minimization). Obviously mathematically some of the original inequality constraint must be violated in this solution point. The problem consists now of setting up a relaxed constraint set and/or moving some of the constraints into the objective function in the form of penalties.

Note that the operator should only be involved on a 'high level' with this problem, i.e. no mathematical relaxation algorithm question can be asked.

However, often it is possible to get some kind of priority schemes related to the importance of the operational constraints.

Today much heuristics must be used to transform a mathematically infeasible OPF problem into an optimization problem which has a feasible solution but with relaxed inequality constraints. A systematic analysis of this problem must be addressed in the future to get practically acceptable OPF solutions.

6.3 Sensitivity of the OPF solution with respect to parameter changes

When using the OPF with any of the above mentioned objectives and the corresponding mathematical objective functions, the OPF formulation is today such that parameters like the network data, the inequality constraint limit values and the measured or predicted load values are numerically precisely given. The operator, however, has a different view of these parameters, especially with respect to the limit values of the inequality constraints: They do not represent precisely given values but are often derived based on experience and on heuristic assumptions. The operator would like to know how much the OPF result varies when the parameters are slightly varied. The idea is to give the operator a feeling for example how much the violation of a maximum branch MVA flow reduces the chosen objective function. From such a sensitivity analysis the operator could derive important operating policies and it would provide deeper insight into the OPF optimum.

Mathematically the problem is often formulated under the assumption that the binding inequality constraint set of the OPF solution remains identical even with a slight change of the parameters, mentioned before. Thus the problem is as follows: Assume that the OPF result (solution process to obtain an OPF result see paper by H. Glavitsch) has been obtained by some solution method. The resulting, numerically given values for this optimum are as follows (denoted with index ^{opt}):

$$\begin{array}{ll} \mathbf{u}^{opt}, \underline{\mathbf{x}}^{opt} & \text{(optimum state for base case variables)} \\ \mathbf{u}^{(i)^{opt}}, \underline{\mathbf{x}}^{(i)^{opt}} & \text{(optimum state for outage case i related variables)} \\ \mathbf{w}^{opt} & \text{(optimum base case control variables statuses)} \\ \mathbf{w}^{(i)^{opt}} & \text{(opt. contr. var. movement status for cont. i)} \end{array} \quad (57)$$

The problem is now a parametric equality constrained optimization problem and can be formulated as follows, assuming that all variables of the

OPF problem are summarized into $\mathbf{X}$, all equality constraints into $\mathbf{G}(\mathbf{X})$, all binding inequality constraints into $\mathbf{H}_{binding}(\mathbf{X})$ and the objective function is $F(\mathbf{X})$:

$$\begin{array}{ll} \text{Minimize} & F(\mathbf{X}^{opt} + \delta\mathbf{x}) \\ \text{subject to} & \mathbf{G}(\mathbf{X}^{opt} + \delta\mathbf{x}) = \delta\mathbf{g} \\ \text{and} & \mathbf{H}_{\mathbf{binding}}(\mathbf{X}^{opt} + \delta\mathbf{x}) = \delta\mathbf{h} \end{array} \tag{58}$$

In (58) the values of the vectors $\delta\mathbf{g}$ and $\delta\mathbf{h}$ are slight variations to the right hand sides of the equality and binding inequality constraints. The problem is to get a solution of the type

$$\delta\mathbf{x} = \text{function of } (\mathbf{X}^{opt}, \delta\mathbf{g}, \delta\mathbf{h}) \tag{59}$$

where the term 'function of' can in the simplest case be a numeric matrix.

6.4 Discussion and future outlook

The OPF formulation as shown in section 6.1 has several characteristics which are summarized in the following points.

- The OPF formulation comprises a huge number of variables and equality / inequality constraints. Typical OPF problem sizes are summarized in the following table:

Number of nodes	Number of actual network related constraints		Number of contingency network related constraints	
	equality	inequality	equality	inequality
80	160	640	32000	120000
500	1000	4000	$1.4 \cdot 10^6$	$5 \cdot 10^6$
1000	2000	8000	$2.8 \cdot 10^6$	$20 \cdot 10^6$
2000	4000	16000	$5.6 \cdot 10^6$	$40 \cdot 10^6$

  Note that in this table the equality constraints are assumed to be formulated in the most compact form (this includes elimination of variables).

- The variable sets of the actual and contingency case related networks are almost perfectly decoupled. Each variable set is related to a different network with distinct operational constraints. The coupling between the variable sets is 'simple' in that only maximum difference values between a subset of the variable sets (the control variables) for each different contingency case to the base case must be satisfied.

- The variables are either real, complex or discrete. Either the polar or rectangular coordinate system must be used to represent the complex variables in real variable form. The polar form introduces sin and cos - type functions into the OPF equations. The rectangular form is responsible for square or square root - type and also arctan - type equations.
- Today the complete OPF formulation as given before cannot be solved in closed form. Typically, steps are undertaken to make the OPF problem solvable with classical optimization methods:
  - The discrete variables of the vector $\mathbf{w}$, $\mathbf{w}^{(i)}$ are assumed to be known, i.e. the set of moveable control variables is known. Doing this leads to an optimization problem with continuous variables only.
  - The maximum possible moves of the control variables within the time τ_{short} after the contingency occurs is often assumed to be zero, meaning that the control variable values do not change from the n-1-secure state to any post contingency state. This leads to an optimization problem with much less control variables.
  - The contingency based equations $\mathbf{g}^{(i)}$ and $\mathbf{h}^{(i)}$ are often linearized around the n-1 secure state variables. The equations resulting from this linearization are seen as the exact equation set which the OPF optimum has to satisfy.

Today (1992), the OPF problem as stated in this paper is not solvable in real-time with computers presently available in Energy Management Centers. Faster computers, new optimization algorithms and the feedback obtained from practical OPF usage will change the OPF formulation and its real-time usage in the near future. Modelling the OPF as realistically as possible together with a robust, fast OPF execution will lead to high acceptance of this powerful power system control tool.

For the author of this paper, it is clear that the OPF will be more and more important for the electric power industry due to its potential in increasing both power system reliability and also power system economy without much investment in power system hardware.

A Appendix

A.1 Symbols

The following notations are used throughout this text:

- Symbols representing complex variables are <u>underlined</u>.
- Matrices are shown in capital boldface letters.
- Vectors are shown in small boldface letters.

*	Conjugate complex
^{opt}	Associated variable is optimum variable
^{T}	Transposed
^{low}	Low limit
^{high}	Upper (high) limit
Δ	Change
$\lvert(.)\rvert$	Absolute value of a variable (.) (variable can be complex)
$\angle(.)$	Angle value of a complex variable (.)
$Real(.)$	Real part of a complex variable (.)
$Imag(.)$	Imaginary part of a complex variable (.)
(i)	Related to contingency case i
$diag(.)$	Diagonal matrix

N	Total number of electrical nodes
n	Total number of network elements
$\underline{I}_{el-i-j_i}$	Complex current of element from node i to node j, computed at node i
$\underline{y}_{ii}^{el-i-j}$	Primitive **Y**-matrix element (i,i) of branch element between nodes i and j
$\underline{y}_{ij}^{el-i-j}$	Primitive **Y**-matrix element (i,j) of branch element between nodes i and j
$\underline{y}_{ji}^{el-i-j}$	Primitive **Y**-matrix element (j,i) of branch element between nodes i and j
$\underline{y}_{jj}^{el-i-j}$	Primitive **Y**-matrix element (j,j) of branch element between nodes i and j
$\underline{V}_i$	Complex voltage at node i
$\underline{Y}_{ii}^{el-i-j}$	Two-port admittance element (i,i) of branch between nodes i and j
$\underline{Y}_{ij}^{el-i-j}$	Two-port admittance element (i,j) of branch between nodes i and j
$\underline{Y}_{ji}^{el-i-j}$	Two-port admittance element (j,i) of branch between nodes i and j
$\underline{Y}_{jj}^{el-i-j}$	Two-port admittance element (j,j) of branch between nodes i and j
$\underline{t}^{el-i-j}$	Complex tap of transformer branch between nodes i and j
$\underline{I}_{el-i-o}$	Complex current of shunt element at node i
$\underline{y}_{ii}^{el-i-o}$	Admittance element of shunt at node i
s^{el-i-o}	Shunt tap value (corresponding to the number of switched in shunts) at node i
$\underline{Y}_{iio}^{el-i-o}$	Admittance element of total shunt at node i

$\underline{I}_{Gi}$	Complex generator current at node i
$\underline{I}_{Li}$	Complex load current at node i
$\underline{S}_{Gi}$	Complex power of generator at node i
$\underline{S}_{Li}$	Complex power of load at node i
$\underline{S}_{el-i-j_i}$	Complex power of branch from node i to node j computed at node i
$\underline{S}_{el-i-j_j}$	Complex power of branch from node i to node j computed at node j
$\underline{S}_{el-i-o}$	Complex power of shunt element at node i
$\{\underline{x}_{A1}\}$	$= \{\underline{Y}_{ii}^{el-i-j}, \underline{Y}_{ij}^{el-i-j}, \underline{Y}_{ji}^{el-i-j}, \underline{Y}_{jj}^{el-i-j}, \underline{Y}_{iio}^{el-i-o}\}$ (numerically given parameters of all passive network elements)
$\{\underline{x}_{A2}\}$	$= \{\underline{t}^{el-i-j}, s^{el-i-o}\}$ (transformer and shunt related variables)
$\{\underline{x}_{B}\}$	$= \{\underline{I}_{el-i-j_i}, \underline{I}_{el-i-j_j}, \underline{I}_{el-i-o}\}$ (branch and shunt related current variables)
$\{\underline{x}_{C}\}$	$= \{\underline{I}_{Gi}, \underline{I}_{Li}\}$ (generator and load related current variables)
$\{\underline{x}_{D}\}$	$= \{\underline{V}_{i}\}$ (nodal voltage variables)
$\{\underline{x}_{E}\}$	$= \{\underline{S}_{Gi}, \underline{S}_{Li}\}$ (generator and load related complex power variables)
$\{\underline{x}_{F}\}$	$= \{\underline{S}_{el-i-j_i}, \underline{S}_{el-i-j_j}, \underline{S}_{el-i-o}\}$ (branch and shunt related complex power variables)
$\mathbf{g}_A$	Equation set mainly related to variables $\underline{\mathbf{x}}_{A2}$
$\mathbf{g}_B$	Equation set mainly related to variables $\underline{\mathbf{x}}_{B}$
$\mathbf{g}_E$	Equation set mainly related to variables $\underline{\mathbf{x}}_{E}$
$\mathbf{g}_F$	Equation set mainly related to variables $\underline{\mathbf{x}}_{F}$
$\underline{\mathbf{x}}$	Variable set which include $\underline{\mathbf{x}}_{A2}$, $\underline{\mathbf{x}}_{B}$, $\underline{\mathbf{x}}_{E}$ and $\underline{\mathbf{x}}_{F}$
$\mathbf{g}(\underline{\mathbf{x}})$	Equation set which includes $\mathbf{g}_A$, $\mathbf{g}_B$, $\mathbf{g}_E$, $\mathbf{g}_F$

$\underline{S}^o_{Li}$	Given complex power value of load at node i
I^{max}_{el-i-j}	Maximum current of branch between nodes i and j
S^{max}_{el-i-j}	Maximum MVA-power of branch between nodes i and j
V_i^{min}	Minimum voltage magnitude at node i
V_i^{max}	Maximum voltage magnitude at node i
P^{min}_{Gi}	Minimum active power of generator at node i
P^{max}_{Gi}	Maximum active power of generator at node i
Q^{min}_{Gi}	Minimum reactive power of generator at node i
Q^{max}_{Gi}	Maximum reactive power of generator at node i
t^{min}_{el-i-j}	Minimum tap position of in-phase transformer between nodes i and j
t^{max}_{el-i-j}	Maximum tap position of in-phase transformer between nodes i and j
δt^{min}_{el-i-j}	Minimum tap position of phase shifter tap transformer between nodes i and j
δt^{max}_{el-i-j}	Maximum tap position of phase shifter tap transformer between nodes i and j
s^{min}_{el-i-o}	Minimum tap for shunt bank at node i
s^{max}_{el-i-o}	Maximum tap for shunt bank at node i
δ^{min}_{el-i-j}	Minimum angle value between nodes i and j
δ^{max}_{el-i-j}	Maximum angle value between nodes i and j
$P^{min}_{reserve}$	Minimum active generator power spinning reserve
$P^{min}_{BS_k}$	Minimum active power value for transfer at branch set k
$P^{max}_{BS_k}$	Maximum active power value for transfer at branch set k
P_{Gi}	Active power of generator at node i (a control variable)
Q_{Gi}	Reactive power of generator at node i (a control variable)
V_{Gi}	Voltage magnitude of generator at node i (a control variable)
δt_{el-i-j}	Phase shift transformer tap magnitude of transformer between nodes i and j (a control variable)
t_{el-i-j}	In-phase transformer tap magnitude of transformer between nodes i and j (a control variable)
s_{el-i-o}	Shunt tap (a control variable)
$P_{interchange-k-l}$	Active power transaction interchange between area k and area l (a control variable)

$\mathbf{u}$	Control variable vector of the actual network
$\mathbf{u}^{(i)}$	Control variable vector of contingency case i
$\mathbf{x}$	OPF state variable vector of the actual network
$\mathbf{x}^{(i)}$	OPF state variable vector of contingency case i
$\mathbf{u}^{min}$	Control variable minimum values
$\mathbf{u}^{max}$	Control variable maximum values
$\mathbf{g}(\underline{\mathbf{x}}, \mathbf{u})$	OPF equality constraint set represented in OPF variables and control variables (actual network)
$\mathbf{h}(\underline{\mathbf{x}}, \mathbf{u})$	OPF inequality constraint set represented in OPF variables and control variables (actual network)
$\Delta\mathbf{u}^{max}$	Maximum move of the control variables from base case state to some other (optimized) state (actual network)
$\Delta\mathbf{u}^{(i)^{max}}$	Maximum move of the control variables from (optimized) base case state to some state after contingency i occurs (contingency case i)
τ_{short}	Short time (in seconds) during which the operator has to move the control variables to some optimal state
$\Delta\mathbf{u}/\mathbf{sec.}^{max}$	Maximum move of the control variables per second (actual network)
$\Delta\mathbf{u}/\mathbf{sec.}^{(i)^{max}}$	Maximum move of the control variables per second (contingency i network)
$\mathbf{g}^{(i)}(\underline{\mathbf{x}}^{(i)}, \mathbf{u}^{(i)})$	OPF equality constraint set represented in OPF variables and control variables (contingency case i network)
$\mathbf{h}^{(i)}(\underline{\mathbf{x}}^{(i)}, \mathbf{u}^{(i)})$	OPF inequality constraint set represented in OPF variables and control variables (contingency case i network)
$\mathbf{g}_u^{(i)}(\underline{\mathbf{x}}^{(i)})$	Functions relating the OPF variables $\mathbf{x}^{(i)}$ to the control variable $\mathbf{u}^{(i)}$ of the contingency case i
$\mathbf{w}$	OPF control variable movement status variables (actual network)
$\mathbf{w}^{(i)}$	OPF control variable movement status variables (contingency i network)
$\mathbf{u}_{BC}$	Given base case (BC) control variable values at any time t describing a network state with possible violated operational constraints
CV^{max}	Maximum number of moveable control variables (actual network)
$CV^{(i)^{max}}$	Maximum number of moveable control variables (contingency i network)

P_{Loss}	Loss function of the whole network
$P_{Loss}^{area\ 'a'}$	Loss function of a part of the network
C_{Gi}	Cost function of the active power of the generator at node i
$c_{approximated_i}$	Approximated, smooth active power cost curve for generator i
$c_{int-approx._{k-l}}$	Approximated, smooth active transaction interchange power cost curve for between area k and area l
C	Total cost of all generators
F	Special objective function related to the quick move of control variables

References

[1] D.J. Gaushell, H.T. Darlington; *Supervisory Control and Data Acquisition*, Proceedings of the IEEE, Special issue on computers in power system operations, (1987), Vol. 75, No. 12, pp. 1645 - 1658

[2] T.M. Athay; *Generation Scheduling and Control*, Proceedings of the IEEE, Special issue on computers in power system operations, (1987), Vol. 75, No. 12, pp. 1592-1606

[3] B. Stott, O. Alsac, A.J. Monticelli; *Security Analysis and Optimization*, Proceedings of the IEEE, Special issue on computers in power system operations, (1987), Vol. 75, No. 12, pp. 1623 - 1644

[4] N. Balu, et. al; *On-Line Power System Security Analysis*, Proceedings of the IEEE, Vol. 80, No. 2, February 1992, pp. 262 - 280

[5] H. Glavitsch, R. Bacher; *Optimal Power Flow Algorithms*, Control and Dynamic Systems, Vol. 41, Academic Press, Inc. (1991), pp. 135 - 205

[6] H.Happ; *Optimal Power Dispatch. A Comprehensive Survey*, IEEE Transactions on Power Apparatus and Systems, Vol. 96. (1977), pp. 841-853

[7] J. Carpentier; *Optimal Power Flows;* Electrical Power & Energy Systems, Butterworths, Vol 1, No.1, April 1979

USE OF LINEAR AND QUADRATIC PROGRAMMING TECHNIQUES IN EXPLOITING THE NONLINEAR FEATURES OF THE OPTIMAL POWER FLOW

Hans Glavitsch

Swiss Federal Institute of Technology (ETH)
CH-8092 Zürich, Switzerland

Abstract. The ordinary power flow and the optimal power flow due to the nature of the problem have several nonlinear features. The important ones are related to the nodal power specifications and to the formulation of the objective function. Since there are many inequality constraints there is a strong tendency to linearize the problem and to apply LP and QP-techniques in order to achieve an efficient solution process. Any linearization, however, will increase the computational effort because of an increase in the number of iterations. Quadratic approximations adapt in a better way, however, they require more effort in the solution process. In application programs special approaches have been taken to exploit the well-known features of LP and QP. A special subject within the optimal power flow is the treatment of the objective function. When separability of the nonlinear conditions is given the process of linearization can be adapted in such as way that segments of varying sizes can be used to account for the prevailing accuracy. The LP-solution process is confined to a small number of segments thereby keeping the number of variables small. As the accuracy increases the size of the segments is reduced. An integrated LP method with a control process for the selection of segments is possible.

The second approach takes advantage of the basic property of a quadratic programming problem that it can be reduced to two linear problems, one solving the unconstrained optimality conditions and the other the inequality constraints. The latter is an LP problem. As it turns out the reduction constitutes the dual quadratic problem to the original formulation. In order to achieve satisfactory performance special measures have to be taken, e.g. to exploit the sparsity of the problem. As far as the nonlinearities of the load flow problem itself is concerned a practical approach is the linearization applied in each iteration. As a consequence a quadratic problem results which is solved by linear means. Formulations of the problem vary with the tratement of the sparsity of the system. Examples of solved problems will be given.

1 Introduction

In [1] and [2] the detailed development has brought to light the variety of specialties of the power system model, the objectives of control, practices of power system operation and its security all in view of its optimization. Thereby optimization has to be understood as a process for finding a best solution to a well-posed problem or for solving a problem which - without having an objective - would be non-determined.

Optimization in power systems follows a certain historical path as the determination of minimum cost for the dispatch of generation was the original problem at the outset. Economic dispatch [3], as it was called, had the distribution of generated thermal power at lowest cost as its main objective. When the calculation of the power flow was becoming a realtime computer process loss minimization has been attached or evolved as an objective of its own. More recently security of power system operation is moving into the center of analytical and computational efforts whereby the objectives of cost and losses are somewhat fading. New terms like transmission constrained, contingency constrained optimal power flow or post-contingency rescheduling have been created [4]. The power system model, i.e. the power flow with all constraints imposed by operational considerations has obtained central importance.

The optimal power flow program (OPF) or an optimal power flow package in its most general form lends itself for purposes of power system planning, operations planning or realtime operation. The emphasis of the application varies accordingly. Whereas in planning considerations of energy savings and cost maybe the most important objectives it is losses and the security of operation which become dominant in operations planning and actual operation. As the recent trends show there is an increasing interest in finding feasible solutions because of the large number of constraints which are imposed on a particular state of the system due to security requirements. This is particularly true for short time optimization, i.e. for half hour to half hour time periods. Losses and operating costs become of interest for medium term optimization and of course for the planning of daily schedules (unit commitment).

The applications have progressed to the point where an optimal power flow program is part of the Energy Management System (EMS) such that an optimum schedule or state is computed fully automatically starting from a measured set of input data which consists of loads, flows and breaker positions (topology). On the other hand an optimal power flow is becoming the standard tool for the operator which enables him to assess reserve margins, reactive requirements, stability margins, loss increments and cost information for the purposes of operations planning.

In general it should be stated that in the optimal power flow problem it is power and security of supply which is of interest in contrast to unit commitment where energy and cost are the main objectives.

Besides arriving at a well defined problem and finding an appropriate method of solution it is the performance of a particular program which is a central issue today. Although the computer hardware is offering a tremendous speed the overall performance of a particular program relies heavily on the analytical details and methodological specialties. Speed becomes a particular issue when a large size network is involved or when many load cases have to be checked. From the experience of calculations of the ordinary load flow the user is interested in obtaining a quasi-linear relation between network size and computing time. Thereby it is interesting to note that the choice of method becomes determinant for this desired computational behavior. Hence it will be the prime objective of this paper to outline the current methods of optimization as they seem most suitable for the solution of the optimal power flow problem and to underline their advantages and disadvantages.

2 Characteristics of the nonlinearities

2.1 Overview

As already outlined in [1] there are three domains upon which the power flow optimization resides. It is the model, the objective and the constraints. In all three domains there are nonlinearities which have an important bearing on the solution process and on the approach in general. Thereby the objective has a dominant influence in as far as it becomes determinant for the solution method. The nonlinearities of the model and of the constraints cannot be accommodated in closed form within a particular method according to the present state of the art. Hence, it has become common practice to develop incremental forms of the model and of the constraints which are treated as linearized forms within the solution process. The incremental forms have all the advantages of linear systems such that known and standard methods can be applied. The consequence of such a model is the need to work in an iterative fashion, i.e. apply the basic method repetitively.

2.2 Objectives

Objective functions derived from the practical requirements are usually continuous but not necessarily convex. The best example of a non-convex function as appearing in power systems is the cost curve of an individual generator with valve points, see [1]. Very often however, non-convexity is ignored and the curve is replaced by a continuous, not necessarily differentiable but convex function. This function can be represented by a piece-wise quadratic curve or combination of curves. An example of such a representation is given in Fig. 1 together with the derivative of the cost curves which is called the incremental cost curve. The segmentation of the cost curve can be increased as will be seen when it comes to the application of a particular optimization method.

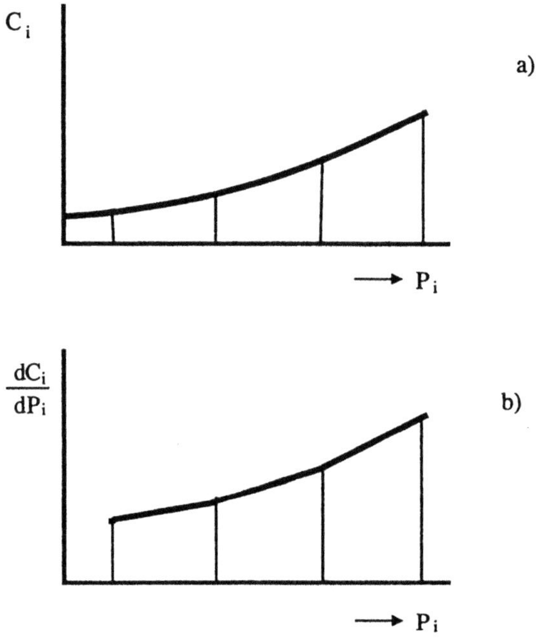

Figure 1: Representation of cost curves by a piece-wise quadratic function a) Cost curve b) Incremental cost curve

At this point it is to be mentioned and emphasized that curves representing cost of generated power in generating units have the unique property that the functions are separable. This means that the cost of a particular generator is dependent on its own generated power only, i.e.

$$C_i = a_{i0} + a_{i1}P_i + a_{i2}P_i^2 \tag{1}$$

There are no cross products or other mixed influences. The total cost of a park of generators is then the sum of all individual generator costs as shown below.

$$C = \sum_i C_i = \sum_i (a_{i0} + a_{i1}P_i + a_{i2}P_i^2) \tag{2}$$

Another way of representing cost of generation is by piece-wise linear functions. This may seem very crude at first sight but in more detail it has to be realized that todays methods strive for a high accuracy. Hence any approximation will be extended to give the desired accuracy at the end. Therefore a piece-wise linear approximation can be chosen with very small segments if needed. An example of a piece-wise linear cost curve is given in Fig. 2.

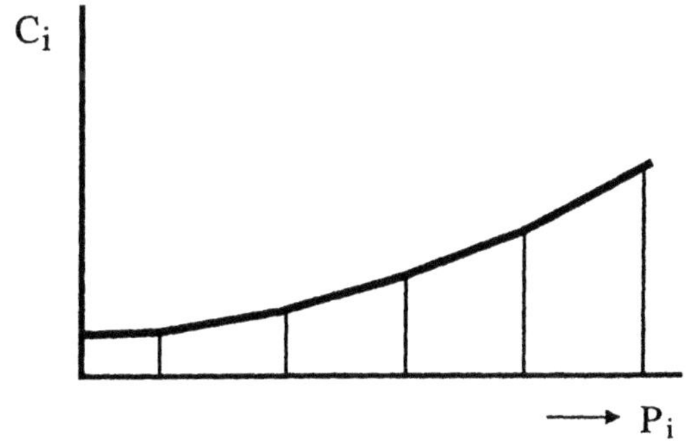

Figure 2: Piece-wise linear representation of a cost curve

What has to be mentioned also is that cost curves are given in terms of the active power generated which is a control variable u , see [1]. They are variables which can be immediately manipulated in contrast to state or dependent variable which vary in function of the rest of the variables. Another type of objective function which has a quite different character is the function representing active losses of the power system. In this case one is in the ideal situation that the function is ideally quadratic in terms of the nodal voltages or in terms of the line currents. The exact formulation, however, applies for state variables in rectangular coordinates only. The loss formula can be derived from a singular line element which extends to the whole network or any part of it by taking an appropriate sum of line elements. The elementary loss formula for a line element from node i to node j is

$$\left.\begin{array}{ll} G_{ij} & = \frac{R_{ij}}{R_{ij}^2+X_{ij}^2} \\ L_{ij} & = G_{ij}\left((E_i - E_j)^2 + (F_i - F_j)^2\right) \end{array}\right\} \tag{3}$$

which extends to any combination of elements by applying

$$L = \sum_m L_{ij} = \sum_m G_{ij}\left((E_i - E_j)^2 + (F_i - F_j)^2\right) \quad \text{m line elements} \tag{4}$$

The general representation is given by a quadratic form which is easily amenable to the optimization process or method, i.e

$$L = \sum_m \left[\begin{array}{cccc} E_i & E_j & F_i & F_j \end{array}\right] \left[\begin{array}{cc|cc} G_{ij} & & -G_{ij} & \\ & G_{ij} & & -G_{ij} \\ \hline -G_{ij} & & G_{ij} & \\ & -G_{ij} & & G_{ij} \end{array}\right] \cdot \left[\begin{array}{c} E_i \\ E_j \\ F_i \\ F_j \end{array}\right] \tag{5}$$

Note that this representation is valid for dependent variables only. In general terms any other objective may be represented by linear or piece-wise linear functions as well as quadratic or piece-wise quadratic functions. Closed forms or forms in terms of absolute variables, either dependent or control variables will be rare. However, incremental forms or approximations as just mentioned are quite common. As an example the deviation of voltages from an ideal voltage profile will be given. An incremental form is appropriate here.

$$F = \sum_i \left((E_{i0} - E_i)^2 + (F_{i0} - F_i)^2\right) = \sum_i (\Delta E_i^2 + \Delta F_i^2) \tag{6}$$

In this expression the deviation of nodal voltages from predetermined voltages is given in a square formulation whereby the individual deviations are increments of the nodal voltages. The overall objective function is again a scalar. The objective is a quadratic form without any linear expressions attached as known from loss formulae. According to this example various other objectives are possible. The general approach is to approximate them by linear or quadratic expressions of increments of dependent variables or control variables.

2.3 Constraints

As derived in [1] constraints are given in terms of equalities and inequalities which in general are nonlinear. The best example are the load flow equations. Inequalities may apply to dependent or to control variables. In case the inequality does not apply to the state or dependent variable a functional relationship exists which in most cases will be nonlinear. These functional constraints cannot - according to present knowledge - be accommodated within the standard optimization methods. Hence they are always a subject of irritation. For illustration, the nodal power at a load node is given as an example for an equality constraint. The rectangular coordinate system is chosen.

$$\left.\begin{array}{lll} -P_{Li} & -\Sigma_{j=1}^{N}\left(E_i(E_j g_{ij} - F_j b_{ij}) + F_i(F_j g_{ij} + E_j b_{ij})\right) & = 0 \\ -Q_{Li} & -\Sigma_{j=1}^{N}\left(F_i(E_j g_{ij} - F_j b_{ij}) - E_i(F_j g_{ij} + E_j b_{ij})\right) & = 0 \end{array}\right\} \tag{7}$$

Inequality constraints may apply to control variables directly or in functional form, i.e. depending on other variables. A typical example is the limitation of a voltage magnitude as given below.

$$\left.\begin{array}{lll} |V_i| & \leq & V_{imax} \\ & & \\ |V_i| & = & \sqrt{E_i^2 + F_i^2} \end{array}\right\} \tag{8}$$

There a quadratic form is valid for the components of the voltage in rectangular form. Another example is the limitation of a line current. Here the expression is more complex.

$$\sqrt{\frac{(E_i - E_j)^2 + (F_i - F_j)^2}{R_{ij}^2 + X_{ij}^2}} \leq I_{ijmax} \tag{9}$$

Line flows which are the transmitted powers over a particular transmission line lead to similar forms. The general experience with inequality constraints is that these expressions in a first approximation are linear with a quadratic term superimposed.

3 Incremental forms of system relations

3.1 Ordinary power flow

As outlined above the power flow itself is a nonlinear relation due to the nodal equality constraints which can be active and reactive power or voltage magnitude, each being either derived from a product voltage times current or given by the square-root of two components of the nodal voltage. Since the power flow relations quite often appear as equality constraints in the formulation of an OPF it is necessary to linearize the equations as it is known from the load flow solution. The nonlinear load flow equations

$$\left.\begin{aligned} P_i &= \sum_{j=1}^{N} \left(E_i(E_j g_{ij} - F_j b_{ij}) + F_i(F_j g_{ij} + E_j b_{ij})\right) \\ Q_i &= \sum_{j=1}^{N} \left(F_i(E_j g_{ij} - F_j b_{ij}) - E_i(F_j g_{ij} + E_j b_{ij})\right) \end{aligned}\right\} \tag{10}$$

are replaced by the relations among increments of active, reactive power and voltage magnitudes on one hand and increments of components of nodal voltages. The classical formulation is

$$\begin{bmatrix} \Delta P_i \\ \\ \Delta Q_i \\ \\ \Delta V_i \end{bmatrix} = \begin{bmatrix} \frac{\partial P_i}{\partial E_j} & \frac{\partial P_i}{\partial F_j} \\ \\ \frac{\partial Q_i}{\partial E_j} & \frac{\partial Q_i}{\partial F_j} \\ \\ \frac{\partial |V_i|}{\partial E_j} & \frac{\partial |V_i|}{\partial F_j} \end{bmatrix} \begin{bmatrix} \Delta E_j \\ \\ \Delta F_j \end{bmatrix} = J \begin{bmatrix} \Delta E_j \\ \\ \Delta F_j \end{bmatrix} \tag{11}$$

The matrix J in this relation depends on the operating point, i.e. the point around which the linearization has been performed. It is a functional matrix and is called the Jacobian or Jacobian matrix which is sparse due to the topology of the network. Within the Jacobian submatrices may be distinguished which show similar structures, however, there are no symmetries. At this point it is to be remarked that some of the quantities of the left hand side are control variables and some are constant. The constant quantities are the load quantities which are given. In the iterative solution process these incremental quantities have to become zero whereas the control variables, e.g. generator powers will assume certain values. For the accommodation of these relations within an iterative process, e.g. Newton-Raphson iteration, it is to be recommended to re-arrange the variables such that the unknown quantities appear on the right-hand side of the system of equations. The left-hand side still has increments of the same quantities but at this point these increments are mismatches.

$$\begin{bmatrix} \Delta P_i \\ \\ \Delta Q_i \\ \\ \Delta V_i \end{bmatrix} = J \begin{bmatrix} \Delta E_i \\ \\ \Delta F_i \end{bmatrix} - \begin{bmatrix} \Delta P_{iG} \\ \\ \Delta Q_{iG} \\ \\ \Delta V_{iG} \end{bmatrix} \tag{12}$$

All mismatches have to become sufficiently small when the solution has been reached. This linearized relation may be discussed and treated in its own rights since there are dependent variables which may be of no interest or no concern for the optimal solution. The dependent variables can be eliminated such that the number of relations is reduced. A simple process will illustrate this reduction. Assume that a power flow relation in linearized form is given whereby for simplicity increments of power will be assumed on the left hand side.

$$\begin{bmatrix} \Delta P_{n-1} \\ \Delta Q_{n-1} \end{bmatrix} = J_{2n-2} \cdot \begin{bmatrix} \Delta E_{n-1} \\ \Delta F_{n-1} \end{bmatrix} \tag{13}$$

The length of the vector on the left-hand side is taken as 2n-2 as it is the case for the ordinary power flow. For this condition the Jacobian can be inverted and the increments of the voltages can be expressed by the nodal power as given below.

$$\begin{bmatrix} \Delta E_{n-1} \\ \Delta F_{n-1} \end{bmatrix} = J_{2n-2}^{-1} \cdot \begin{bmatrix} \Delta P_{n-1} \\ \Delta Q_{n-1} \end{bmatrix} \tag{14}$$

The active power of the slack which is not part of the above relation can be written separately which would be just another row of the Jacobian. The slack node is one of the n nodes of the network which is designated to absorb the balance of active and reactive powers and which has a fixed voltage magnitude at phase angle zero.

$$\Delta P_n = \begin{bmatrix} \frac{\partial P_n}{\partial E_{n-1}} & \frac{\partial P_n}{\partial F_{n-1}} \end{bmatrix} \begin{bmatrix} \Delta E_{n-1} \\ \Delta F_{n-1} \end{bmatrix} = p_n^T \begin{bmatrix} \Delta E_{n-1} \\ \Delta F_{n-1} \end{bmatrix} \tag{15}$$

The vector of dependent variables in (15) is the same as the one in (14) and thus it can be expressed by (14) whereby all dependent variables disappear. The result is a scalar relation among the active and reactive nodal powers on one hand and the slack power on the other hand.

$$\begin{aligned} \Delta P_n &= \alpha_{P_1}\Delta P_1 + \alpha_{P_2}\Delta P_2 + ...\alpha_{P_{n-1}}\Delta P_{n-1} + \\ &\alpha_{Q_1}\Delta Q_1 + \alpha_{Q_2}\Delta Q_2 ... \alpha_{Q_{n-1}}\Delta Q_{n-1} \end{aligned} \tag{16}$$

This is an incremental power flow which establishes a sensitivity relation between the slack power and all the other nodal powers. Of course, this incremental power flow can be generalized by allowing any variable in the left-hand side vector of (13). Thus a voltage magnitude can be included showing its effect on the slack power. It should made clear that the incremental power flow replaces the equality constraint in an optimization process which at the outset was a matrix relation, e.g. (11). For the actual calculation of the coefficients of (16) a very efficient algorithm is available which uses the triangular factorization of the Jacobian and which requires one backward substitution only. The method is given in the appendix A1 which generates the coefficients [1] based on a solved load flow. As for the size of the coefficients a consideration of losses and loss increments will be helpful. The loss increment is the sum of all active nodal powers which can be added to (16) whereby the following relation results.

$$\Delta L = \sum_n \Delta P_i \tag{17}$$

If the sensitivity of the losses with respect to the other powers is considered it will be apparent that for the loss minimum without any other constraints being active the coefficients of (16) must fulfill the conditions

$$\alpha_{P_i} + 1 = 0; \qquad \alpha_{Q_i} = 0 \tag{18}$$

which means that all coefficients related to active powers are equal to minus one and all coefficients related to reactive powers are zero (also valid for voltage magnitudes). By simple reasoning it is obvious that in the general case the former coefficients will be near minus one and the latter will be small, i.e not larger than 0.1 in p.u..

3.2 Objective function

3.2.1 Cost function

For the further development two forms of the cost function have to be considered, namely a quadratic or piece-wise quadratic function and a piece-wise linear function. As has been outlined before generating costs are separable functions such that control variables of a particular generator enter the cost function of this generator only. If the generating cost of such a generator is taken as being quadratic assuming its validity over a certain range, e.g.

$$C_i = a_{i0} + a_{i1} P_i + a_{i2} P_i^2 \tag{19}$$

the control variables which is generated power can be assumed to be composed of a base value and an increment

$$P_i = P_{i0} + \Delta P_i \text{ or } u_i = u_{io} + \Delta u_i \tag{20}$$

which can be substituted in (19) where base values and increments are separated.

[1] Note that the sign of the α_{Pi}'s and α_{Qi}'s is inverted against the derivation in Appendix A1

$$C_i = a_{i0} + a_{i1}P_{i0} + a_{i2}P_{i0}^2 + (a_{i1} + 2a_{i2}P_{i0})\Delta P_i + a_{i2}\Delta P_i^2 \tag{21}$$

Thereby an expression is obtained which has exactly the same structure as the original cost function the difference being that the absolute value and the linear terms are changed and the quadratic terms are replaced by increments. The advantage of such an expression is that cost functions can be modelled around a near optimum solution expecting that the increments will be small which will tend to zero when the modelling process is repeated in subsequent iterations. The idea of modelling around a near optimum solution can also be applied to piece-wise linear cost functions whereby the base value is placed at the breakpoint of two consecutive segments. This will be a quite useful setup for the application linear programming methods in solving a problem with nonlinear cost functions. A plot of a piece-wise linear, separable cost function is taken as shown in Fig. 3 whereby the base value is placed in A.

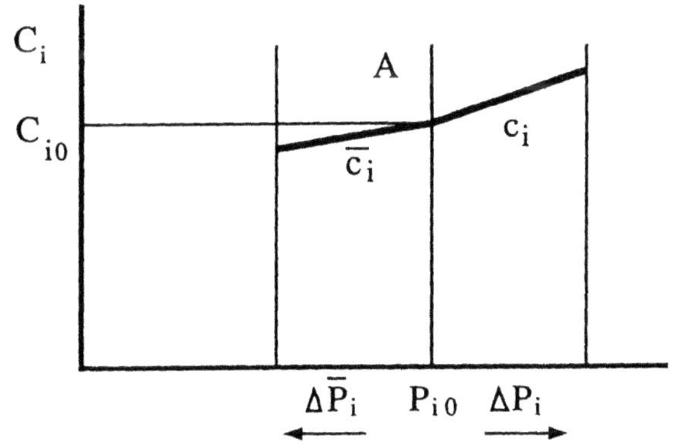

Figure 3: Plot of a piece-wise linear cost function

The control variable is represented by a base value and two non-negative increments

$$C_i = C_{i0} + c_i\Delta P_i - \overline{c_i}\Delta\overline{P_i} \tag{22}$$

By a concept known from linear programming the cost function in the vicinity of the base value can be represented by the following expression

$$C = \sum C_i = \sum(C_{i0} + c_i\Delta P_i - \overline{c_i}\Delta\overline{P_i}) \tag{23}$$

Again the cost is represented by a constant value and an increment which in this case is generated by two control variables which for the purpose of linear programming must be non-negative. Special measures will be taken in the application of the solution process to decrease the size of the increment as required for achieving a desired accuracy.

3.2.2 Loss formula

Network losses either for the whole of a network or for certain sections and lines are non-separable functions of dependent or independent variables. In case of a formulation where voltages or currents are used the loss function is an exact quadratic as already explained in chapter 2.2. Thus the incremental form is an exact quadratic too. Since the principle of substituting the variable by base value and increments has already been shown above the result of such a substitution is given here.

$$\Delta L = \begin{bmatrix} \Delta\mathbf{E} & \Delta\mathbf{F} \end{bmatrix} Q \begin{bmatrix} \Delta\mathbf{E} \\ \Delta\mathbf{F} \end{bmatrix} + p^T \begin{bmatrix} \Delta\mathbf{E} \\ \Delta\mathbf{F} \end{bmatrix} \quad (24)$$

There is no basic change in the functional relationship between losses and dependent variables. The advantage lies again in the iterative solution process where by subsequent modelling and linearization the increments will decrease to zero.

The situation is quite different when losses are to be modelled by control variables which is only feasible by means of increments. The functional relationship is not exactly quadratic any more as a simple consideration will show. Take the linearized relationship derived in the section on power flow and equality constraints, e.g. (11) for which dependent variables can be expressed by control variables. Substituting such a relation in the loss formula produces a quadratic which, however, has certain deficiencies since in the derivation starting from (11) the linearization is an approximation and the error is carried over in the final loss formula.

Such an error can be avoided by using a rigorous derivation as given in the appendix A2. According to this derivation the loss formula [5] expressed by increments of control variables is

$$\Delta L = \alpha_P^T \Delta\mathbf{P} + \alpha_Q^T \Delta\mathbf{Q} + \begin{bmatrix} \Delta\mathbf{P}^T & \Delta\mathbf{Q}^T \end{bmatrix} Q \begin{bmatrix} \Delta\mathbf{P} \\ \Delta\mathbf{Q} \end{bmatrix} \quad (25)$$

By comparison with the incremental power flow above it will be seen that the expression (25) can be used to form an extended incremental power flow. Instead of losses the expression can be applied to the slack power whereby the linear part obtained is equivalent to the linear incremental power flow. The quadratic extension can be taken as either incremental losses or the incremental slack power.

If for particular applications any other variable needs to be expressed in a quadratic form or with a quadratic extension the process given in the appendix will be useful. So losses or any other nonlinear quantity can be expressed in incremental form by a linear and a quadratic term with a very high degree of precision.

3.3 Constraints

At this point inequality constraints are of interest as these are required to include any limitations on dependent and independent variables. Due to facilities of the solution algorithms incremental forms including linear terms are required only. Hence inequalities being valid for either dependent or independent variables have to be expressed as outlined in the following

$$\mathbf{A}\begin{bmatrix}\Delta\mathbf{u}\\ \Delta\mathbf{x}\end{bmatrix} \leq \mathbf{b}_2 \tag{26}$$

They appear in the form that an increment of $\mathbf{u}$ or $\mathbf{x}$ has to be smaller than a certain value. In this form constraints can be attached to the problem formulations of the optimal power flow using standard solution methods.

4 Solution concepts and considerations on the choice of a method

4.1 Various points of emphasis

It is to be emphasized that the development of the problem formulation is oriented towards a convex problem, preferably a convex and differentiable problem. This is given by modelling the objective functions either in piece-wise linear or piece-wise quadratic form. The equality and inequality constraints originally are nonlinear but by converting them to incremental forms they are to be considered as linear. Under these presumptions optimality conditions [6] can be set up in any case, i.e. using a Lagrangian the Kuhn-Tucker conditions can be applied. The existence of optimality conditions, however, does not automatically provide a solution algorithm, the most serious problem being the inequality constraints, in particular when they appear in large numbers.

In the development of efficient algorithms for the solution of the OPF-problem it is observed that there is a tendency to employ linear programming methods [7] because LP is very strong in handling inequality constraints. Quadratic features as they show up in this problem area have to be modelled by piece-wise linear approximations. In order to achieve the desired accuracy it will be necessary to improve the approximations as the solution process goes on. This additional effort will require a correspondingly heavy computational effort which affects the overall computation time. The user has to decide if it will be worth the ease of using LP when certain restrictions or drawbacks have to be encountered in other aspects of the problem. As it turns out there are problem formulations in the OPF area which are quite suitable for the application of LP, in particular when the objective function is separable. A good example is economic dispatching [8].

Another aspect of choosing the right method is sparsity. OPF problems as they are formulated today comprise a very large number of variables derived from the size of the network. Fortunately, the resulting system of equations is very sparse. Hence, a suitable solution method has to consider sparsity and take advantage of the structure of the equations. Linear programming provides for sparsity in basic programs and there are possibilities to enhance the performance by suitably arranging and formulating the problem.

Quadratic programming [9], [10] seems to be ideally suitable for dealing with piece-wise quadratic functions which is true as long as there are not too many inequality constraints. When differentiating the Lagrangian one obtains immediately linear conditions for formulations as outlined above. The determination of active and inactive constraints, however, requires an additional effort. It is this problem area where further work will have to be done in order to arrive at an effective merge between the advantages of quadratic programming as they evolve naturally for quadratic problems and the efficiency of dealing with inequality constraints as exhibited by linear programming. In the following a method will be worked out which will achieve this aim to a certain extent. The requirement of handling sparsity in an efficient way will not be fulfilled in a satisfactory way.

4.2 LP advantages and disadvantages

Linear inequality constraints are treated by the Simplex method which is straight-forward in combination with a linear or linearized objective. Extensions or modifications of the basic method all turn around the base change whereby the tableau is adapted to the size of the problem or to a particular local area. The change of base can even be seen in a more general way as it is possible to start from infeasible solutions from where a feasible one can be reached by sequential changes. This aspect is quite important when it comes to treating a solution of the equality constraints which on the other hand does not fulfill the inequality constraints yet. New methods for solving the linear problem evolve as it is seen from [11]. The obvious disadvantage of LP lies in the necessity to model the objective by piece-wise linear functions which is reasonable for separable objectives only. Losses cannot be brought in a from which would lend itself to a separable formulation. The size of the problem, i.e the number of variables, equations, inequality constraints are no basic difficulty for LP. A special application of LP to minimize losses will be given in chapter 9.

4.3 QP advantages and disadvantages

In order to give a first idea of the obvious benefits of using a quadratic programming approach for a quadratic problem a simple problem is solved. The objective is assumed to be a quadratic form as given below

$$F = p^T \mathbf{x} + \frac{1}{2} \mathbf{x}^T Q \ \mathbf{x} \tag{27}$$

having also a linear term. This function should reach a minimum subject to a set of linear equations, i.e.

$$D\mathbf{x} = b_1 \tag{28}$$

There are no inequality constraints and thus a Lagrangian can be formulated which is to be differentiated and set to zero.

$$\mathcal{L} = p^T \mathbf{x} + \frac{1}{2} \mathbf{x}^T Q \ \mathbf{x} + \lambda^T (D\mathbf{x} - b_1) \Longrightarrow Min \tag{29}$$

$$\begin{bmatrix} \mathbf{Q} & \mathbf{D}^T \\ \mathbf{D} & \mathbf{0} \end{bmatrix} \begin{bmatrix} x \\ \lambda \end{bmatrix} = \begin{bmatrix} -\mathbf{p} \\ \mathbf{b}_1 \end{bmatrix} \tag{30}$$

In this case the optimality conditions are given by this system of linear equations and the solution is trivial. It is this basic approach of solving a set of linear equations which produces the solution in one step and which makes QP attractive. The moment inequality constraints are attached the solution takes on a quite different character. It will always be the case that a small number of inequality constraints is active only. If this active set would be known beforehand there would be a linear problem again. So the situation is comparable to the linear LP-problem where a strategy is needed for visiting feasible solutions in a sequence whereby the number of solutions to be visited should be as small as possible. There is a series of solution methods [9], [10], [12], [13] in the operations research literature which will provide a solution to a quadratic problem. However, the efficiency of these methods varies greatly. A typical example is Beale's methods which is quite easy to apply since the submission of data is similar to that of an LP method. For problem sizes in the order of about 100 variables the performance is acceptable but if the number of variables goes up to around 300 the computation time becomes prohibitive.

4.4 Classifying the approaches to the OPF-problem

The large size of the equality constraints, i.e the load flow part of the problem for which efficient solution methods exist suggests to take advantage of this experience in the way that non-optimum solutions are utilized in the various steps. This idea requires the solutions of load flows whose results are starting points for an optimization process. The results of the latter are set points for a next load flow and so on. The solved load flow allows the computation of an incremental load flow which could be beneficial for the subsequent optimization step. Based on this concept one class of OPF-solution methods may be defined which will be denominated **Class A**. In this approach the following computational steps are carried out:

- solution of a load flow for the given load with approximate control variables
- computation of an incremental power flow (coefficients)
- solution of the OPF with the incremental power flow as an equality constraint
- updating of the control variables
- solution of a new load flow if accuracy is not yet satisfactory

In this procedure the optimization method is not specified yet and could be any reasonable algorithm. The main point is that the solution of the load flow and the optimization are separated. The justification of this separation lies in the fact that the solution of the load flow is quite close to the final solution due to the fact that constraints confine the solution within a fairly narrow area. A consequence may be that the number of iterations is somewhat larger than in the other approaches. The second class named **Class B** is oriented towards the classical Lagrangian formulation where the optimality conditions are within one set of equations comprising both equality and inequality constraints. This set of equations has to be solved in one step or run which means that the load flow is solved together with the optimization problem, again however, iteratively. If the problem would be ideally quadratic there would be one computational step only using a QP-method. Due to the nonlinear nature iterations are necessary which is always true if the objective function is not ideally quadratic or if constraints are nonlinear. The class B formulation is well suited for the application of QP-methods if sparsity [14] can be exploited. Otherwise a Newton approach may be more advantageous. Class A and B have been introduced in the literature under name of compact and non-compact formulation [15]. The compact formulation corresponds to class A and the non-compact formulation to class B.

5 Economic dispatch - a sample problem

5.1 Problem definition

The dispatch of generated power with the objective of minimizing operating cost is a standard problem which normally is not conceived as an OPF-problem. For the sake of illustrating certain principles and approaches it is a useful example for applying an optimization method which later on can be extended to a true OPF-problem. The problem is defined based on the supply of a concentrated load as illustrated in Fig. 4 where several generators feed one single load. Each generator produces its output according to a quadratic cost function which is assumed to be valid over the complete operating range of the individual generator.

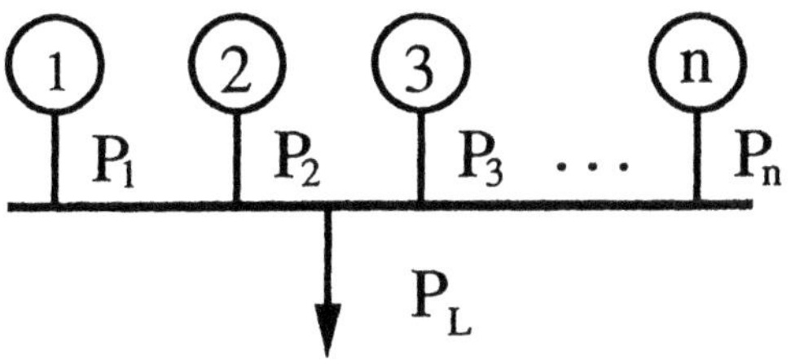

Figure 4: Schematic of a simplified network supplied by several generators

The cost function for each generator is given by

$$C_i = a_{i0} + a_{i1}P_i + a_{i2}P_i^2 \tag{31}$$

The objective function is the sum of these individual costs which is to be minimized

$$C = \sum_i C_i = \sum_i (a_{i0} + a_{i1}P_i + a_{i2}P_i^2) \tag{32}$$

There is an equality constraint which is simply the sum of all generated power equalling the load. This formulation is characterized by a separable cost function and by a single scalar equality constraint. The only nonlinearity is the cost function. At the moment no inequality constraints or limits are considered. In terms of dependent and independent variables there are n-1 independent variables and one single dependent variable both being identical to generated powers. In this simple problem inequality constraints would have to applied as upper and lower limits on the generated powers which is left to a later treatment. The optimality conditions and the corresponding solution are straight forward using a Lagrangian formulation, i.e.

$$\mathcal{L} = \sum_i (a_{i0} + a_{i1}P_i + a_{i2}P_i^2) + \lambda(\sum_i P_i - P_L) \tag{33}$$

The optimality conditions are linear equations

$$\begin{aligned}
\frac{\partial \mathcal{L}}{\partial P_1} &= a_{11} + 2a_{12}P_1 + \lambda &&= 0 \\
\frac{\partial \mathcal{L}}{\partial P_2} &= a_{21} + 2a_{22}P_2 + \lambda &&= 0 \\
&\vdots \\
\frac{\partial \mathcal{L}}{\partial P_n} &= a_{n1} + 2a_{n2}P_n + \lambda &&= 0 \\
\frac{\partial \mathcal{L}}{\partial \lambda} &= P_1 + P_2 + P_3 + \ldots + P_n - P_L &&= 0
\end{aligned} \tag{34}$$

to be solved by classical means in one single computational step.

5.2 The dispatch problem including a load flow

If the network is to be represented by its individual nodes the equality constraints become multi-dimensional, i.e the load flow. In a first step this load flow will be confined to active power only which can be done by setting up an ordinary load flow and extracting an incremental power flow which is limited to active power only. Thus it is assumed that the incremental form, i.e. its coefficients are available

$$
\begin{aligned}
\frac{\partial \mathcal{L}}{\partial \Delta P_1} &= a_{11} + 2a_{12}\Delta P_1 + \alpha_{P1}\lambda &&= -a_{11} - 2a_{12}P_{1o} \\
\frac{\partial \mathcal{L}}{\partial \Delta P_2} &= a_{21} + 2a_{22}\Delta P_2 + +\alpha_{P2}\lambda &&= -a_{21} - 2a_{22}P_{2o} \\
&\vdots \\
\frac{\partial \mathcal{L}}{\partial \Delta P_n} &= a_{n1} + 2a_{n2}\Delta P_n + (-1)\lambda &&= -a_{n1} - 2a_{n2}P_{no} \\
\frac{\partial \mathcal{L}}{\partial \lambda} &= \alpha_{P1}\Delta P_1 + \alpha_{P2}\Delta P_2 + \alpha_{P3}\Delta P_3 \ldots - \Delta P_n &&= 0
\end{aligned}
\qquad (35)
$$

with the consequence that the variables are increments. Since the real functions are not ideally quadratic it is reasonable to represent the objective in incremental form as well which can be easily done by inserting the dependent and independent variables in the cost function as already shown by (21). Again a Lagrangian can be formulated which is ideally quadratic in terms of the increments. The optimality conditions in classical form would be linear equations as above. If an LP-approach is used objective and constraints remain separated and will need some preprocessing.

6 Linear programming as a solution method

6.1 Representation of the cost function

In order to be able to accommodate a more general convex function for the representation of the cost relationships the concept of piece-wise quadratic functions was used above. If the same cost functions have to be modelled within an LP-program the concept of piece-wise linear functions has to be used whereby the size of the increments will be kept variable such that at the end a given accuracy can be achieved. For illustrative purposes a quadratic cost function is considered which is assumed to be valid over a given range. When representing this function by piece-wise linear sections of constant width the corresponding slopes of the approximation form a stair case function as shown in Fig. 5.

When the size of the segments is reduced the stair case function approaches the straight line in Fig. 5 (B).

6.2 LP-solution based on the incremental system model

For the following derivation a class A approach is assumed whereby a non-optimal solution of the load flow is taken as a starting point. Around this solution incremental forms are employed. In particular an incremental load

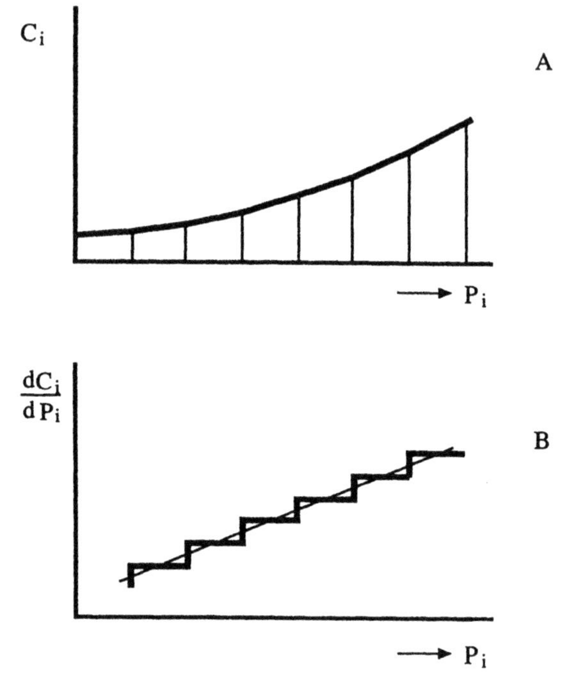

Figure 5: Quadratic cost function represented by a piece-wise linear function and corresponding derivatives (slope)

flow is assumed to be known. Since it is an active power dispatch for the moment coefficients α_{Pi} are retained only.

$$\alpha_{P_1}\Delta u_1 + \alpha_{P_2}\Delta u_2 + \alpha_{P_3}\Delta u_3 + ... + \Delta x = 0 \tag{36}$$

In this formula above control variables are distinguished from the increment of the slack power which is actually a dependent variable, i.e. this incremental power flow will be the equality constraint in the OPF problem at hand. As given by (23) the incremental contribution of each generator to the cost function around the starting point is represented by two non-negative variables

$$\begin{aligned} C_i &= C_{i0} + c_i\Delta u_i - \overline{c_i}\Delta\overline{u_i} \\ C &= \textstyle\sum_n C_i = \sum_n C_{i0} + \sum_{n-1}(c_i\Delta u_i - \overline{c_i}\Delta\overline{u_i}) + c_n\Delta x\overline{c_n}\Delta x \end{aligned} \tag{37}$$

Depending on the sign of Δx two different slopes c_n and $\overline{c_n}$ respectively have to be used.

Eliminating the dependent variable results in a cost function of control variables only.

$$C = \sum_{n} C_{i0} + \sum_{n-1} (c_i - \alpha_{P_i}\overline{c_n})\Delta u_i - \sum_{n-1} (\overline{c_i} - \alpha_{P_i} c_n)\Delta \overline{u_i} \tag{38}$$

which illustrates the influence of system losses via the coefficients α_{Pi} . Defining the size of the increments of the control variables completes the formulation for an LP-problem.

$$\begin{aligned} \Delta u_i &\leq h \\ \Delta \overline{u_i} &\leq h \end{aligned} \tag{39}$$

As seen from Fig. 5(A) a constant size can be taken making the LP-tableau quite simple. This tableau comprises the inequality constraints just defined and the coefficients of the cost functions as given below.

$$\begin{array}{cccccccc|c} \Delta u_1 & \Delta \overline{u}_1 & \Delta u_2 & \Delta \overline{u}_2 & \Delta u_3 & \Delta \overline{u}_3 & \dots & \Delta \overline{u}_{n-1} & \\ \hline 1 & & & & & & & & h \\ & 1 & & & & & & & h \\ & & 1 & & & & & & h \\ & & & 1 & & & & & h \\ & & & & 1 & & & & h \\ & & & & & 1 & & & h \\ & & & & & & \ddots & & \vdots \\ & & & & & & & 1 & h \\ \hline c_{1n} & -\overline{c}_{1n} & c_{2n} & -\overline{c}_{2n} & c_{3n} & -\overline{c}_{3n} & \dots & -\overline{c}_{n-1n} & \end{array} \tag{40}$$

$$c_{in} = c_i - \alpha_{P_i} c_n, \qquad \overline{c}_{in} = \overline{c}_i - \alpha_{P_i} \overline{c_n}$$

Since the matrix in this tableau is diagonal the base change for the step-by-step improvement of the solution is quite simple. According to the Simplex method a negative sign of a cost coefficient indicates a candidate for a change. In this special form no computations are necessary but sign changes and a change of designation from non-basic to basic variables only which means that all control variables showing negative signs have to change base, i.e all control variables having a negative sign move to the next limit. At this point the corresponding coefficients for the cost functions have to be updated. This is to be done including the coefficients of the incremental power flow. Thus further base changes will be necessary until all cost coefficients become positive. Finally up-to-date coefficients of the incremental power flow will have to be inserted and checked for optimality. Thus a first iteration for a given segmentation is completed.

6.3 Extension of the LP-solution to functional constraints

In real situations there are not only constraints of the control variables but also those on dependent variables, i.e. voltages, currents, line flows, etc. According to the concepts of modelling constraints these have to be given in incremental form expressed by control variables. A limit on a magnitude of a nodal voltage is formulated by a lengthy expression derived from the load flow (Jacobian), i.e.

$$
\begin{aligned}
\Delta V_i &\leq \Delta V_{imax} \\
\Delta V_i &= a\Delta E_i + b\Delta F_i \qquad\qquad (41) \\
\Delta V_i &= v_{i1}\Delta u_1 + v_{i2}\Delta u_2 + v_{i3}\Delta u_3 + \ldots + v_{in}\Delta u_n
\end{aligned}
$$

where the coefficients v_{i1} ... v_{in} are the result of a linearization. Strictly speaking there is a full size vector for these coefficients which is to be accommodated in the Simplex tableau. For each dependent variable which tends to exceed its limit such a row be inserted. Schematically the tableau will look as outlined below.

$$
\begin{array}{cccccccc|c}
\Delta u_1 & \Delta \overline{u}_1 & \Delta u_2 & \Delta \overline{u}_2 & \Delta u_3 & \Delta \overline{u}_3 & \ldots & \Delta \overline{u}_{n-1} & \\
\hline
1 & & & & & & & & h \\
 & 1 & & & & & & & h \\
 & & 1 & & & & & & h \\
 & & & 1 & & & & & h \\
 & & & & 1 & & & & h \\
 & & & & & 1 & & & h \\
 & & & & & & \ddots & & \vdots \\
 & & & & & & & 1 & h \\
* & * & * & * & * & * & \ldots & * & \Delta V_{jmax} \\
* & * & * & * & * & * & \ldots & * & \Delta V_{lmax} \\
\vdots & \vdots & \vdots & \vdots & \vdots & \vdots & \vdots & \vdots & \vdots \\
\hline
c_{1n} & -\overline{c}_{1n} & c_{2n} & -\overline{c}_{2n} & c_{3n} & -\overline{c}_{3n} & \ldots & -\overline{c}_{n-1n} &
\end{array}
\qquad (42)
$$

The LP-solution process has to consider these additional rows with the consequence of an extra effort which cannot be avoided. There are possibilities to include coefficients of the constraints from a certain magnitude on which will make the tableau more sparse. Since it is necessary to repeat the load flow anyway it will require a small extra effort to check the limits and work out a new constraint value which will improve the solution in the subsequent iteration.

6.4 Segment refinement

It should be made clear that at the point where the Simplex has terminated for a given load flow solution and a prescribed segmentation the optimal solution has not been reached yet or may not be satisfactory. Either the size of the segments is too large or the coefficients of the incremental power flow do not correspond to the solutions. Hence further iterations are necessary. In order to approach a more accurate solution in an efficient way the sizing of the segment has to be done in an intelligent way. From one-dimensional search processes it is known that a binary division or the use of Fibonacci numbers will be quite efficient. So by dividing the segments by two and working out the corresponding slopes a new LP-solution can be determined. The closer the solution gets to the final solution the smaller will the changes in the coefficents derived from the load flow be, i.e. α_{Pi} and coefficients of the functional constraints. Hence most of the work can be confined to the new segmentations. The strategy of choosing the segments is a matter of experience [8]. So it can be useful to prescribe the size of the segments as the iterations go on by a fixed schedule. In any case the iterations should terminate when the size of the segments have reached the magnitude of one MW or less. A binary segmentation will achieve this fairly fast when the process starts with segments in the order of one tenth of the maximum generator output being for example 300 MW, i.e. 30 MW. Subsequently five segmentations by two will reach the order of one MW. The substantial effort in this class A process will be the corresponding number of load flow solutions. Hence when comparing the performance of OPF methods it is quite practical to compare them on the basis of multiples of the time for the load flow solution.

6.5 Approximate reduction of losses in active power dispatch

It is to be noted that the value and the sign of the coefficients α_{Qi} contain substantial information of the influence of reactive control variables on the losses. It is gradient information and can be used in the iterative process in parallel. Since the size of the active power segments are reduced as the solution process goes on it is reasonable to change the reactive power by the same size if the sign of the corresponding α_{Qi} is large enough and of the right sign, i.e. indicating a reduction of losses. From a given value of α_{Qi} on the reactive influence can be ignored since it is known that in the unconstrained solution the coefficients tend to zero. Thus an approximate reduction of the losses will be achieved.

7 Use of a standard quadratic programming method

7.1 Solution of a sample problem - active power dispatch

Since there are quadratic cost relations it is reasonable to use a corresponding quadratic solution method. There are several quadratic programming methods (QP-methods) available which however are not always ideally suited for the solution of the OPF-problem. The formulations of the problem follow the ideal quadratic concept, i.e quadratic objective function and linear constraints.

$$
\begin{aligned}
F &= p^T\mathbf{x} + \tfrac{1}{2}\mathbf{x}^T Q\mathbf{x} \Longrightarrow Min \\
s.t. \quad & D\mathbf{x} - b_1 = 0 \\
& A\mathbf{x} - b_2 \leq 0
\end{aligned}
\tag{43}
$$

In the application to the foregoing dispatch problem all functions must be given in incremental form. In order to clearly express control and dependent variables the vector $\mathbf{x}$ is separated into $\Delta\mathbf{u}$ and $\Delta\mathbf{x}$.

$$
\begin{aligned}
F &= p^T \begin{bmatrix} \Delta\mathbf{u} \\ \Delta\mathbf{x} \end{bmatrix} + \tfrac{1}{2} \begin{bmatrix} \Delta\mathbf{u}^T & \Delta\mathbf{x}^T \end{bmatrix} Q \begin{bmatrix} \Delta\mathbf{u} \\ \Delta\mathbf{x} \end{bmatrix} \Longrightarrow Min \\
s.t. \quad & D \begin{bmatrix} \Delta\mathbf{u} \\ \Delta\mathbf{x} \end{bmatrix} - b_1 = 0 \\
& A \begin{bmatrix} \Delta\mathbf{u} \\ \Delta\mathbf{x} \end{bmatrix} - b_2 \leq 0
\end{aligned}
\tag{44}
$$

Thereby it is assumed that the problem is treated according to class A which seems to be the most effective way because of the sparsity of the quadratic form of the objective function. A standard QP-algorithm [13] accepts these functions, forms a Lagrangian as required or eliminates the equality constraints, etc. and sets up the matrices, as well as functionals.

The user has to program the loop around the QP-algorithm including the updating of the variables which in this case are control variables. According to the chosen class A approach a load flow is needed which is to be integrated in the loop. The load flow accepts the updated control variables and produces the incremental power flow, see appendix A1. The coefficients form the row vector which is submitted to the QP-algorithm as the equality constraint. The inequality constraints are formulated as a matrix either oriented towards control variables or towards dependent variables (functional constraints). The QP-algorithm could be Beale's method which processes the input data as described above. Another algorithm suitable for the dispatch of active power aimed at cost minimization will be introduced in a later chapter. The general experience with such a formulation is that the number of runs through the loops is surprisingly small. Only two to four runs which can be called iterations are necessary. It is the effort within the QP-method itself which counts. More details about the performance will be discussed in a later chapter.

7.2 Loss minimization

The treatment of the loss formula in chapter 3.2.2 has shown that losses in function of control variables, i.e. generator outputs are not separable but result in quadratic forms whereby the sparsity as originally observed is lost. When a standard QP-program is to be used then such a formulation cannot be avoided. Assuming that this formulation can be created a class A approach is possible which is quite similar to what has been said in the previous chapter. The difference lies in the formulation of the objective function which in this case comprises the equality constraint implicitly. The extended incremental power flow according to appendix A2 produces the linear incremental power flow and the quadratic form which augments the linear incremental power flow. In the loop controlling the iteration there has to be a load flow which produces the Jacobian from which the objective function is generated [5]. The objective function with a specified content of the vector of control variables is submitted to the QP-algorithm together with the inequality constraints which are identical to those necessary for the economic dispatch problem. The flow graph in Fig. 6 gives an idea of the sequence of steps necessary for the application of QP within the class A approach.

As already indicated above the convergence of the optimization process is excellent. This is shown by the decrease of the losses in function of the

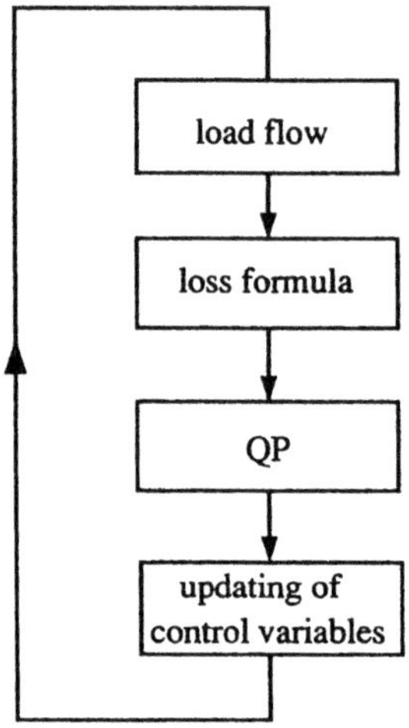

Figure 6: Flow graph for application of QP within class A

number of iterations as given in Fig. 7.

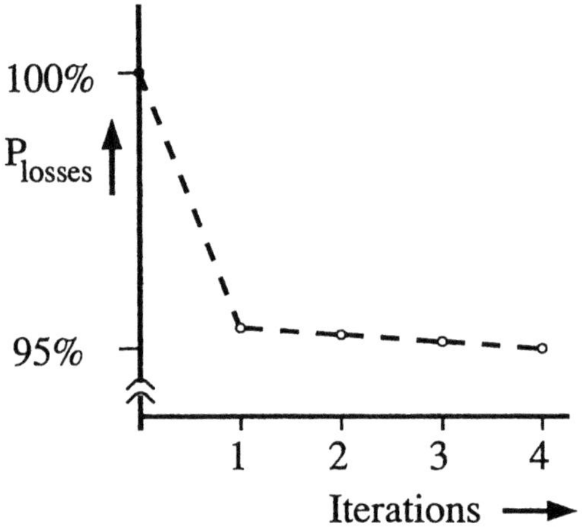

Figure 7: Decrease of losses in function of number of iterations

It can be said that the final result is reached in two iterations. Subsequent iterations improve the result insignificantly. This performance is due to the detailed modelling of the losses by a quadratic form. In order to give an idea of what type of results can be achieved by such a method the performance of a realtime application in the RWE-system in Germany [16] is outlined. The utility has applied the method as described above to their 380/220-kV-network comprising 300 nodes, 50 generators and 30 transformers equipped with tap changers. The online algorithm controls the voltage magnitude of the generators and the taps of transformers 380/220 kV. The objective is to minimize losses observing constraints on reactive inputs and line flows. Since the tap settings are discrete an approximation is necessary. Due to the relatively small size of the step it is justified to set the taps to the nearest analog value as it is calculated by the program. The algorithm is activated every half hour or following a major switching operation. The algorithm starts from a load flow which is verified by state estimation. The voltage profile of the optimized system is given in Fig. 8.

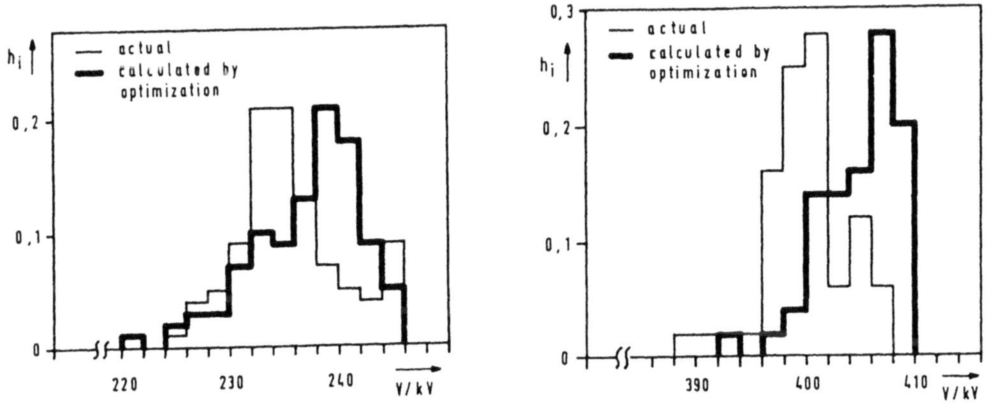

Figure 8: Voltage profile of RWE 380/220-kV-system

Relevant data for the RWE-system can be taken from Tab. 1. Losses range from 220 to 320 MW.

Load	22000 MW
Number of nodes	298
Number of branches	480
Generators with controllable reactive output	50
Transformers with controllable tap ratio	30

Table 1: Key data of the RWE system

Tests have shown that it is sufficient to run the program every half hour without sacrificing any substantial loss decrease which on the other hand allows enough time for the calculation of optimized control variables (new setpoints). The program is implemented on a VAX 780.

7.3 Discussion

It will be worth discussing the straightforward application of a QP-algorithm which has the obvious advantage that the user has to set up the objective function and the constraints as prescribed by the theory. As far as specifying the inequality constraints is concerned there is no difference with respect to other methods. The QP-algorithm has to be taken as it is and shows a performance which is due to general considerations which have been valid for a large class of problems and which are not necessarily applicable to problems of the power system. Sparsity might have been considered or not. The same is to be said for the repetitive application of the algorithm. In subsequent iterations active constraints do not vary considerably which is an important consideration when it comes to improving the performance. Beale's method is easy to apply but has the disadvantage that it manipulates the quadratic form after each base change. When the quadratic form is very large which is true for power system problems then the computing time for this part may become excessive. For this reason it is not feasible to apply the method for network sizes above 300 nodes and 50 control variables. In general Beale's method is suitable for the class A approach since the algorithm will start around a solved load flow and the solution will be near this point.

8 Dual method of quadratic programming

8.1 The dual algorithm of QP

For quadratic problem formulations as derived above a method of solution exists [17] which requires the solution of a set of linear equations and a manipulation of a Simplex tableau. The derivation is explained by using switching concepts in [18]. A short outline of the derivation is given in appendix A3. The system of linear equations represents the optimality conditions for the unconstrained problem

$$\begin{bmatrix} \mathbf{Q} & \mathbf{D}^T \\ \mathbf{D} & \end{bmatrix} \begin{bmatrix} \mathbf{u}_o \\ \mathbf{x}_o \\ \lambda \end{bmatrix} = \mathbf{Y} \begin{bmatrix} \mathbf{u}_o \\ \mathbf{x}_o \\ \lambda \end{bmatrix} = \begin{bmatrix} -\mathbf{p} \\ \mathbf{b}_1 \end{bmatrix} \tag{45}$$

As shown in the appendix the matrix specifying the inequality constraints is employed to generate a tableau which is to be made feasible. The variables in the tableau are the Kuhn-Tucker variables which must be non-negative.

$$[\mathbf{A}\ \ \mathbf{0}]\,\mathbf{Y}^{-1}\,[\mathbf{A}\ \ \mathbf{0}]^T\,\mu \geq \mathbf{A}\begin{matrix}\mathbf{u}_o\\ \mathbf{x}_o\end{matrix} - \mathbf{b}_2 \tag{46}$$

Feasiblity of these variables guarantees the optimality of the quadratic problem. Using the result of the feasible tableau the determination of the superimposed system variables is possible.

$$\Delta\mathbf{v} = -\mathbf{Y}^{-1}\,[\mathbf{A}\ \ \mathbf{0}]^T\,\mu \tag{47}$$

$$\mathbf{v} = \mathbf{v}_o + \Delta\mathbf{v} = \begin{bmatrix}\mathbf{u}_o + \Delta\mathbf{u}\\ \mathbf{x}_o + \Delta\mathbf{x}\\ \lambda_o + \Delta\lambda\end{bmatrix} \tag{48}$$

If all inequalities are satisfied, i.e. when no further limits are violated the final result is reached. It may be gathered immediately that the difficulty in the method lies in the formation of the tableau which may be quite large when all violations are considered as they appear from the unconstrained solution. The number of constraints in the final solution is substantially reduced. Therefore it is not practical to set up a tableau for all violated variables right from the beginning. There are techniques to generate the tableau for each violated constraint at a time and to reduce it when a limit becomes inactive. This way the tableau can be kept small, however, it is necessary to be able to exploit the sparsity of the matrices in (46). Examples will be given in the following chapter. The advantage in this method lies in exploitation of the sparsity of (45) which corresponds to the ordinary load flow. It is a method which is applicable to the non-compact form as well as to the compact form of the OPF-problem. When the linear system and the tableau have been established the solution techniques are straightforward.

8.2 Problem formulation

For the application of the dual QP-method the OPF problem has to be formulated using a Lagrangian with objective and constraints in incremental form. So the matrices Q, D, A and the vectors p, b_1, b_2 have to be known. The formulation can be done under class A or B. The advantage of class A is, of course, the single row in matrix D. The form of matrix Q will depend on the objective and affects the formulation of the tableau in a fundamental way.

8.3 Application to the dispatch problem - Class A

The dispatch problem as it was introduced above is used here again. Inequality constraints are applied right from the beginning. In the formulation below the method can be applied directly, however, there is the need for iterations as in any other application of optimization methods to the power flow problem.

$$\begin{array}{ll} F & = \sum_n (a_{ip} + a_{i1}P_{io} + a_{i2}P_{io}^2 + (a_{i1} + 2a_{i2}P_{io})\Delta P_i + a_{i2}\Delta P_i^2) \\ \Delta P_n & = \sum_{n-1} \alpha_{Pi} \Delta P_i \\ \mathbf{A}\Delta\mathbf{P} - \mathbf{b}_2 & \leq \mathbf{0} \end{array} \tag{49}$$

The dispatch problem offers a number of advantages which are worth to be treated in detail. First, it is the quadratic form which consists of a diagonal matrix

$$\mathbf{Q} = 2 \begin{bmatrix} a_{12} & & & & \\ & a_{22} & & & \\ & & a_{32} & & \\ & & & a_{42} & \\ & & & & a_{52} \end{bmatrix} \tag{50}$$

Second, the equality constraint is a scalar which is to be attached to the diagonal matrix thus forming the matrix which is to be inverted according to (46). This inversion can be done in terms of the triangular matrices of $Y = L \cdot R$ as given below for $n = 5$.

$$\mathbf{Y} = \begin{bmatrix} 2a_{12} & & & & & \alpha_{P1} \\ & 2a_{22} & & & & \alpha_{P2} \\ & & 2a_{32} & & & \alpha_{P3} \\ & & & 2a_{42} & & \alpha_{P4} \\ & & & & 2a_{52} & -1 \\ \alpha_{P1} & \alpha_{P2} & \alpha_{P3} & \alpha_{P4} & -1 & 0 \end{bmatrix} \tag{51}$$

$$Y^{-1} = (L \cdot R)^{-1} = R^{-1} \cdot L^{-1} \tag{52}$$

explicitly

$$\mathbf{L} = \begin{bmatrix} 1 & & & & & \\ & 1 & & & & \\ & & 1 & & & \\ & & & 1 & & \\ & & & & 1 & \\ \frac{\alpha_{P1}}{2a_{12}} & \frac{\alpha_{P2}}{2a_{22}} & \frac{\alpha_{P3}}{2a_{32}} & \frac{\alpha_{P4}}{2a_{42}} & \frac{-1}{2a_{52}} & 1 \end{bmatrix} \tag{53}$$

$$\mathbf{L}^{-1} = \begin{bmatrix} 1 & & & & & \\ & 1 & & & & \\ & & 1 & & & \\ & & & 1 & & \\ & & & & 1 & \\ -\frac{\alpha_{P1}}{2a_{12}} & -\frac{\alpha_{P2}}{2a_{22}} & -\frac{\alpha_{P3}}{2a_{32}} & -\frac{\alpha_{P4}}{2a_{42}} & \frac{1}{2a_{52}} & 1 \end{bmatrix} \tag{54}$$

$$\mathbf{R} = \begin{bmatrix} 2a_{12} & & & & & \alpha_{P1} \\ & 2a_{22} & & & & \alpha_{P2} \\ & & 2a_{32} & & & \alpha_{P3} \\ & & & 2a_{42} & & \alpha_{P4} \\ & & & & 2a_{52} & -1 \\ & & & & & b \end{bmatrix} \tag{55}$$

$$\mathbf{R}^{-1} = \begin{bmatrix} \frac{1}{2a_{12}} & & & & & -\frac{\alpha_{P1}}{2a_{12}b} \\ & \frac{1}{2a_{22}} & & & & -\frac{\alpha_{P2}}{2a_{22}b} \\ & & \frac{1}{2a_{32}} & & & -\frac{\alpha_{P3}}{2a_{32}b} \\ & & & \frac{1}{2a_{42}} & & -\frac{\alpha_{P4}}{2a_{42}b} \\ & & & & \frac{1}{2a_{52}} & -\frac{-1}{2a_{52}b} \\ & & & & & \frac{1}{b} \end{bmatrix} \tag{56}$$

where

$$b = -\sum_{i=1}^{5} \frac{\alpha_{Pi}^2}{2a_{i2}}$$

This is a very important point because realizing the sparsity of the matrices will allow to generate the coefficients of the tableau $[A \;\; 0]\, Y^{-1} [A \;\; 0]^T$ adhoc whenever a nonfeasible solution shows up. As long as limits of the control variables are to be considered the single coefficients $+1$ or -1 appear in the matrix A only, e.g. take the upper limits of ΔP_1, ΔP_3, ΔP_5 and lower limits of ΔP_2 and ΔP_4. The form of the matrix A is then

$$\mathbf{A} = \begin{bmatrix} 1 & & & & \\ & -1 & & & \\ & & 1 & & \\ & & & -1 & \\ & & & & 1 \end{bmatrix} \tag{57}$$

Assuming that the unconstrained solution indicates that ΔP_1 and ΔP_4 are exceeding the limits it is necessary to simply compute the coefficients of the tableau $[A \;\; 0]\, Y^{-1} [A \;\; 0]^T$ by the following operations where z_{ij} are the coefficients of the tableau.

$$\begin{array}{ll} z_{11} = \frac{1}{2a_{12}} + \frac{\alpha_{P1}^2}{4a_{12}^2 b}; & z_{12} = -\frac{\alpha_{P1}\alpha_{P4}}{4a_{12}a_{42}b} \\ z_{21} = z_{12} & z_{22} = \frac{1}{2a_{42}} + \frac{\alpha_{P4}^2}{4a_{42}^2 b} \end{array} \tag{58}$$

The sign of z_{12} would be positive if ΔP_4 exceeds the upper limit. In order to find out which of the variables has to change base the diagonal coefficients give the necessary information. Hence diagonals should be calculated first. The largest positive ratio of the right hand side over z_{ii} indicates the pivot i around which the base change has to be performed. When this is known the other elements z_{ij} of the column have to be calculated and the base change is completed. Further base changes may be necessary.

When all values of μ in (46) have become non-negative the optimum solution is reached and the variables to be superimposed on the unconstrained solution can be determined according to (48). In a next iteration a load flow for an improved set of control variables is worked out and the corresponding incremental power flow is calculated. The dual method is then applied as before. The acquired knowledge about active constraints is utilized and will make the subsequent processes more efficient.

8.4 Application to the dispatch problem - Class B

The advantage of using the class B approach is the fact that the sparsity of the original problem formulation is retained and that the load flow is solved as an optimal load flow. In applying the dual QP, however, there are disadvantages which are particularly obvious in the dispatch problem. Control variables and dependent variables appear in the system, i.e. corresponding coefficients have to be accommodated in the matrix which enlarge the dimension of the latter. In each iteration the system would have to be solved like a load flow and subsequently the coefficients for the tableau have to be worked out. Thereby sparsity cannot be exploited as in the case of the class A approach. Hence the approach according to class B for the dispatch problem is not advisable.

8.5 Loss minimization

Losses can be minimized by applying a general QP-method as outlined in one of the previous chapters. The necessity associated with a general method is the need to generate the extended incremental power flow according to appendix A2. The disadvantage, however, of the EIPF lies in the fact that the effort to generate the quadratic form increases substantially with the number of control variables i.e. with the network size. So for larger networks it does not seem to be a promising method. To avoid this drawback an effort has been made to obtain a performance for the minimization of losses which is comparable to that of the load flow solution as far as the dependence of the solution time on the network size is concerned. This can be achieved by expressing losses in terms of nodal voltages and nodal currents [19] and by using a load flow relation in terms of voltages and currents. Thereby PQ- and PV- nodes have to be modelled in a special way. The system of equations expressing the optimality conditions are sparse and lends itself to an exact decoupling, as a matter of fact into two identical matrices. This system can be subjected to the dual QP-algorithm in such a way that the LP-tableau is generated in an adaptive fashion which means that rows are generated for a few variables which exceed their limit by the largest amount. After handling these variables by performing base change operations the next variables are treated which still exceed their limits and so on. By this strategy the size of the LP-tableau can be kept small. Variables which move away from their limit are taken out of the tableau for the same reason. Thus the handling of constraints is fully systematic. Due to the formulation of the problem in terms of voltages and currents the matrices are constant with the consequence that one factorization of the one decoupled matrix of the unconstrained problem is necessary only. The major effort lies in finding active constraints. The two curves in Fig. 9 give an idea of the performance of this method as compared to forming the EIPF and applying Beale's method. Solving an OPF (loss minimization) on a PC 486 (33 MHz) for a 712 node system having 40 active constraints takes 21 seconds whereby the final mismatch reaches 10^{-7} p.u. Thereby preparatory matrix operations (ordering) are not counted.

Although this method may not be the final answer to achieving the best performance for the loss minimization problem it shows the way in which efforts should be made. The main points to this end are summarized as follows.

- express losses in terms of dependent variables, i.e. voltages
- exploit the similarity between the quadratic form of the loss formula and the matrix of the load flow
- apply the dual QP-method by exploiting the sparsity of the matrices whereby introducing functional constraints results in about the same

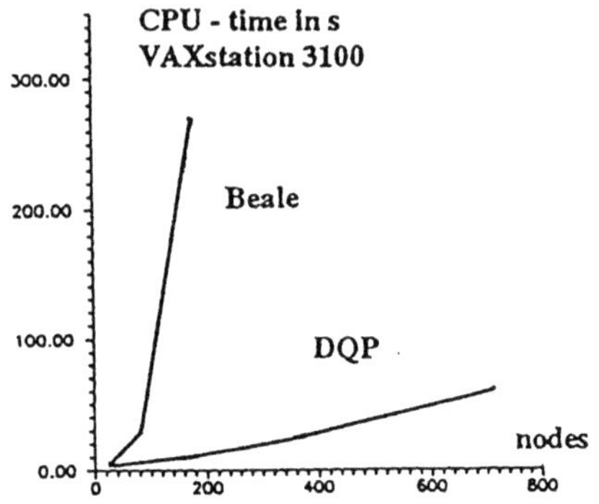

Figure 9: Computing time in function of network size - standard QP (Beale) and dual QP

complexity as experienced in the class A approach.

9 Loss minimization by reactive optimization based on LP

In the power industry a method of loss minimization [8] is in use which is based on linear programming whereby reactive injections and tap ratios are the control variables. In practice this approach responds to a quite useful objective since active generation frequently is given by generating units which cannot be changed from minute to minute whereas the terminal voltage of a generating unit can be adjusted easily thereby affecting the losses. The incentive for using an LP-approach for this problem lies in the performance of the algorithm, in handling constraints and is based on the success of segment refinement for active power dispatch. There are quite a few similarities to the active power dispatch problem such as class A approach, formulation of constraints both for variables and for dependent quantities (functional) and use of the incremental power flow. The central issue in this approach is the use of the incremental power flow, i.e. the α_{Qi}'s for the creation of a hyperplane (gradient) along which the control variables are allowed to move. Thereby the crucial point is the control of the step length of ΔQ_i for which a logic has been developed which considers the size of the components of the gradient and the success of the foregoing iteration. The control of the step length corresponds to segment refinement and all the techniques developed for that can be used here. The success of this approach is somewhat unexpected since gradient techniques since the introduction of the method [20] have lost their attractiveness and have not proven effective for large systems and many constraints. The performance as documented in [8] is not as good as for active power dispatch and is inferior to the Newton method for loss minimization. The advantage on the other hand lies in the possibility to have one single program package for active dispatch and loss minimization. The important aspect certainly is the fact that loss minimization can be done by an LP-approach.

10 The Newton optimal power flow

The publication of [21] at the time was considered a breakthrough in the area of power flow optimization since it allowed to solve the problem according to a clearly formulated Lagrangian and based on a proven solution technique. The optimality conditions are solved using the Newton method which amounts to treating a system which was known from the load flow. This latter aspect had a great appeal to the analysts. The drawback of the Newton method which still exists is the treatment of constraints by penalty terms which are still reasonable to handle for the limitation of variables but become more complicated for functional relations. There is no systematic procedure for selecting active and non-active constraints. Heuristic methods have been developed for setting constraints and removing them. Interesting enough the Newton OPF is 2 to 3 times faster than an LP-based method for loss minimization when a large number of controls is considered [8]. A closer look at the method described under 8.5 [19] will indicate that the dual QP method applied to the formulation as used for the Newton method merges with the latter. It turns out that the formulations are the same irrespective of the use of the form of the load flow and the choice of the variables. Both approaches use class B and the main difference lies in the treatment of constraints. With the dual QP there is a systematic method available which could lead to a substantial improvement of the loss minimization problem.

11 Concluding remarks

It was the objective to show in this treatment of the OPF that all solutions and methods are based on incremental models and formulations from where standard methods or methods adapted to the load flow problem can be applied. The load flow problem itself and the type of objectives determine the choice of method to a large degree. The classification into the two basic classes A and B are typical for the problem area. The separability or non-separability of the objective function is determinant for the suitability of linear programming as the basic tool which has evolved as a preferred method because of its effectiveness for handling constraints. As it was shown the dual QP-method also relies on LP and is able to treat a quadratic problem in an elegant way. Active power dispatch is best solved by a class A approach where a load flow is obtained first and from where the actual optimization starts. It seems that LP- and QP-methods are equally efficient in solving this problem where functional constraints lead to about the same difficulties. Loss minimization does not show a similarly clear picture. For large systems with many control variables the Lagrangian formulation (class B) is the most effective solved by Newton or otherwise. For less demanding problems the LP-method which was derived in similarity to the segment refinement method for active power dispatch is suitable. The number of iterations is considerably larger and the solution time is longer than that for active power dispatch for the same size problem. Hence there is obviously a need for further development in the area of loss minimization.

A APPENDIX

A.1 Method for determining the incremental power flow

From a solved load flow the Jacobian is available

$$\begin{bmatrix} \Delta \mathbf{P}_{\mathbf{n-1}} \\ \Delta \mathbf{Q}_{\mathbf{n-1}} \\ \Delta P_n \end{bmatrix} = J \begin{bmatrix} \Delta \mathbf{E}_{\mathbf{n-1}} \\ \Delta \mathbf{F}_{\mathbf{n-1}} \\ \Delta E_n \end{bmatrix} = L \cdot R \begin{bmatrix} \Delta \mathbf{E}_{\mathbf{n-1}} \\ \Delta \mathbf{F}_{\mathbf{n-1}} \\ \Delta E_n \end{bmatrix} \tag{A1}$$

Multiplying both sides by L^{-1}

$$\begin{bmatrix} L^{-1} & & \\ \alpha_P^T & \alpha_Q^T & 1 \end{bmatrix} \begin{bmatrix} \Delta \mathbf{P}_{\mathbf{n-1}} \\ \Delta \mathbf{Q}_{\mathbf{n-1}} \\ \Delta P_n \end{bmatrix} = R \begin{bmatrix} \Delta \mathbf{E}_{\mathbf{n-1}} \\ \Delta \mathbf{F}_{\mathbf{n-1}} \\ \Delta E_n \end{bmatrix} \tag{A2}$$

The last row of L^{-1} is of interest only.
For the computation of α_P, α_Q a fictitious system of equations

$$\mathbf{y} = \mathbf{L}^T\mathbf{x} \tag{A3}$$

is considered which can be solved by

$$\left\{\mathbf{L}^T\right\}^{-1}\mathbf{y} = \mathbf{x} \tag{A4}$$

Solving for a vector **y**

$$\mathbf{y} = \begin{bmatrix} 0 \\ 0 \\ 0 \\ 1 \end{bmatrix} \tag{A5}$$

results in **x** which has components exactly equal to α_{P_i}, α_{Q_i}.
The solution requires a backward substitution of (A4) only.

A.2 Derivation of the extended incremental power flow (EIPF)

A.2.1 Starting point

It is assumed that for a large number of variables of the power system a quadratic relation of the form is possible

$$u_i = \mathbf{x}^T J_{u_i}\mathbf{x} \tag{A6}$$

whereby

$\mathbf{x}$	vector of dependent variables
u_i	control variable, e.g. P_i, Q_i, $\lvert V_i\rvert$
J_{u_i}	quadratic matrix

A.2.2 Aim of the derivation

An incremental change of a variable y is to be represented by a $\Delta\mathbf{u}$ in the form

$$\Delta y = \frac{\partial y}{\partial \mathbf{u}}\Delta\mathbf{u} + \frac{1}{2}\,\Delta\mathbf{u}\,\frac{\partial^2 y}{\partial \mathbf{u}^2}\,\Delta\mathbf{u} \tag{A7}$$

A.2.3 Procedure

Determination of the first and second derivatives of y with respect to **u** whereby it is also assumed that y can be represented by a quadratic form in **x**

$$y = \mathbf{x}^T R\,\mathbf{x} \tag{A8}$$

R is a matrix with constant coefficients.

A.2.4 Determination of the first derivative

$$\frac{\partial y}{\partial \mathbf{u}} = \frac{\partial y}{\partial \mathbf{x}} \cdot \frac{\partial \mathbf{x}}{\partial \mathbf{u}} \tag{A9}$$

$$\frac{\partial y}{\partial \mathbf{x}} = 2\ \mathbf{x}^T R \tag{A10}$$

For $\partial \mathbf{x}/\partial \mathbf{u}$ the following steps are useful. The vector $\mathbf{u}$ is given by

$$\mathbf{u} = \begin{bmatrix} u_1 \\ u_2 \\ u_3 \end{bmatrix} \tag{A11}$$

and

$$u_i = \mathbf{x}^T J_{u_i} \mathbf{x} \tag{A12}$$

$$\mathbf{u} = L(\mathbf{x}) \cdot \mathbf{x}, \quad \text{where} \quad L(\mathbf{x}) = \begin{bmatrix} \mathbf{x}^T J_{u_1} \\ \mathbf{x}^T J_{u_2} \\ \vdots \\ \mathbf{x}^T J_{u_m} \end{bmatrix} \tag{A13}$$

$L(\mathbf{x})$ is identical to the Jacobian of the load flow besides the factor 1/2 by choosing $\mathbf{x}$ appropriately.

$$L(\mathbf{x}) = \frac{1}{2}\ J(\mathbf{x}) \tag{A14}$$

$\mathbf{u} = L(\mathbf{x}) \cdot \mathbf{x}$ has the following property

$$\alpha^T \mathbf{u} = \sum_1^n \alpha_i u_i = \sum_1^n \alpha_i \mathbf{x}^T J_{u_i} \mathbf{x} = \sum_1^n \mathbf{x}^T \alpha_i J_{u_i} \mathbf{x} = \mathbf{x}^T J(\alpha)\ \mathbf{x} \tag{A15}$$

$$J(\alpha) = \sum_1^n \alpha_i J_{u_i} \tag{A16}$$

which results in

$$\mathbf{x}^T J(\alpha) = \alpha^T L(\mathbf{x}) \tag{A17}$$

This can be applied to the derivation of the first derivative

$$\alpha^T \frac{\partial}{\partial \mathbf{x}} (L(\mathbf{x})) = \frac{\partial}{\partial \mathbf{x}} \left(\alpha^T L(\mathbf{x})\right) = \frac{\partial}{\partial \mathbf{x}} \left(\mathbf{x}^T \cdot J(\alpha)\right) = J(\alpha) \tag{A18}$$

A change in the inverse of $L(\mathbf{x})$ is then

$$d \cdot \{L^{-1}(\mathbf{x})\} = -L^{-1}(\mathbf{x})\, dL(\mathbf{x})\, L^{-1}(\mathbf{x}) \tag{A19}$$

$L(\mathbf{x}) \cdot \mathbf{x}$ is composed of quadratic forms and thus

$$d\mathbf{u} = dL(\mathbf{x})\, \mathbf{x} + L(\mathbf{x}) \cdot d\mathbf{x} \tag{A20}$$

The first term above can be represented by

$$dL(\mathbf{x})\, \mathbf{x} = \mathbf{x}^T\, dL(\mathbf{x}) = L(\mathbf{x}) \cdot d\mathbf{x} \tag{A21}$$

Hence

$$\frac{\partial \mathbf{x}}{\partial \mathbf{u}} = \frac{1}{2}\, L^{-1}(\mathbf{x}) \tag{A22}$$

Finally the first derivative is given by

$$\frac{\partial y}{\partial \mathbf{u}} = \mathbf{x}^T R\, L^{-1}(\mathbf{x}) = \left(L^{-1}(\mathbf{x})\right)^T R\, \mathbf{x} = \alpha^T(\mathbf{x}) \tag{A23}$$

A.2.5 Determination of the second derivative

The second derivative is worked out by differentiating the first derivative

$$\frac{\partial^2 y}{\partial \mathbf{u}} = \frac{\partial}{\partial \mathbf{u}} \alpha^T(\mathbf{x}) \tag{A24}$$

Using (A23)

$$d\alpha^T(\mathbf{x}) = d\mathbf{x}^T R\, L^{-1}(\mathbf{x}) + \mathbf{x}^T R\, d\{L^{-1}(\mathbf{x})\} \tag{A25}$$

For the second term (A19) is useful

$$\begin{aligned} \mathbf{x}^T R\, dL^{-1}(\mathbf{x}) &= -\mathbf{x}^T R\, L^{-1}(\mathbf{x})\, dL(\mathbf{x})\, L^{-1}(\mathbf{x}) \\ &= -\alpha^T(\mathbf{x}) \cdot dL(\mathbf{x})\, L^{-1}(\mathbf{x}) \\ &= -d\mathbf{x}^T J(\alpha) L^{-1}(\mathbf{x}) \end{aligned} \tag{A26}$$

Thereby one obtains for $d\alpha^T(\mathbf{x})$

$$\begin{aligned} d\alpha^T(\mathbf{x}) &= d\mathbf{x}^T \left(R\, L^{-1}(\mathbf{x}) - J(\alpha)\, L^{-1}(\mathbf{x})\right) \\ &= d\mathbf{x}^T \left(R - J(\alpha) L^{-1}(\mathbf{x})\right) \end{aligned} \tag{A27}$$

The derivative of $\alpha(\mathbf{x})$ with respect to $\mathbf{x}$

$$\frac{\partial \alpha}{\partial \mathbf{x}} = \left[(R - J(\alpha))\, L^{-1}(\mathbf{x})\right]^T = \left(L^{-1}(\mathbf{x})\right)^T (R - J(\alpha))^T \tag{A28}$$

The derivative of $\alpha(\mathbf{x})$ with respect to $\mathbf{u}$ is obtained by applying the chain rule

$$\frac{\partial \alpha}{\partial \mathbf{u}} = \frac{\partial \alpha}{\partial \mathbf{x}} \cdot \frac{\partial \mathbf{x}}{\partial \mathbf{u}} = \frac{1}{2} \left(L^{-1}(\mathbf{x})\right)^T (R - J(\alpha))^T L^{-1}(\mathbf{x}) = \frac{\partial^2 y}{\partial \mathbf{u}^2} \tag{A29}$$

which is the final result for the second derivative.

A.3 Dual quadratic programming (DQP)

Based on the given quadratic programming problem

$$F(\mathbf{x}) = \mathbf{p}^T\mathbf{x} + \frac{1}{2}\,\mathbf{x}^T Q\,\mathbf{x} \Longrightarrow Min \tag{A30}$$

subject to

$$\begin{aligned} D\mathbf{x} - \mathbf{b}_1 &= 0 \\ A\mathbf{x} - \mathbf{b}_2 &\leq 0 \end{aligned} \tag{A31}$$

where Q, D, A are matrixes
$\mathbf{p}$, $\mathbf{b}_1$, $\mathbf{b}_2$ are vectors
and $\mathbf{x}$ is the unknown vector
a Lagrangian can be formulated

$$\begin{aligned} \mathcal{L} &= \mathbf{p}^T\mathbf{x} + \tfrac{1}{2}\,\mathbf{x}^T Q\,\mathbf{x} + \lambda(D\mathbf{x} - \mathbf{b}_1) + \mu^T(A\mathbf{x} - \mathbf{b}_2) \Longrightarrow Min \\ \mu &\geq 0 \end{aligned} \tag{A32}$$

In matrix form the optimality conditions are

$$\begin{aligned} Q\,\mathbf{x} + D^T\lambda + A^T\mu &= -\mathbf{p} \\ D\,\mathbf{x} &= \mathbf{b}_1 \\ A\,\mathbf{x} &\leq \mathbf{b}_2 \end{aligned} \tag{A33}$$

For the following derivations $\mathbf{x}$ and λ are treated as one vector $\mathbf{v}$

$$\mathbf{v} = \begin{bmatrix} \mathbf{x} \\ \lambda \end{bmatrix} \tag{A34}$$

The solution is considered as the superposition of two components consisting of an unconstrained part $\mathbf{v}_0$ and a superimposed part $\Delta\mathbf{v}$ due to constraints.

$$\mathbf{v} = \mathbf{v}_0 + \Delta\mathbf{v} \tag{A35}$$

The unconstrained solution results from

$$\begin{bmatrix} Q & D^T \\ D & \end{bmatrix} \cdot \mathbf{v}_0 = \begin{bmatrix} -\mathbf{p} \\ \mathbf{b}_1 \end{bmatrix} \tag{A36}$$

where the matrix can be designated by a single symbol Y and the vector by $\mathbf{b}_0$

$$\begin{bmatrix} Q & D^T \\ D & \end{bmatrix} = Y; \quad \begin{bmatrix} -\mathbf{p} \\ \mathbf{b}_1 \end{bmatrix} = \mathbf{b}_0 \tag{A37}$$

Hence

$$Y \cdot \mathbf{v}_0 = \mathbf{b}_0 \tag{A38}$$

Assuming that the unconstrained solution is available the unknown vector $\mathbf{v}$ is inserted in

$$\begin{array}{llll} Y & (\mathbf{v}_0 + \Delta\mathbf{v}) & +[A\ \ 0]^T \mu & = \mathbf{b}_0 \\ [A\ \ 0] & (\mathbf{v}_0 + \Delta\mathbf{v}) & & \leq \mathbf{b}_2 \end{array} \tag{A39}$$

Since $\mathbf{v}_0$ fulfills (A36) the latter can be subtracted leaving

$$\begin{array}{llll} Y & \Delta\mathbf{v} & +[A\ \ 0]^T \mu & = \mathbf{0} \\ [A\ \ 0] & (\mathbf{v}_0 + \Delta\mathbf{v}) & & \leq \mathbf{b}_2 \end{array} \tag{A40}$$

Here it is seen that $\Delta\mathbf{v}$ is only due to inequality constraints and thus due to μ.
$\Delta\mathbf{v}$ can be eliminated

$$\Delta\mathbf{v} = -Y^{-1}[A\ \ 0]^T \mu \tag{A41}$$

and inserted in the inequality

$$-[A\ \ 0]Y^{-1}[A\ \ 0]^T \mu \leq \mathbf{b}_2 - [A\ \ 0]\,\mathbf{v}_0 = \mathbf{b}_2 - A\mathbf{x}_0 \tag{A42}$$

or

$$[A\ \ 0]Y^{-1}[A\ \ 0]^T \mu \geq A\mathbf{x}_0 - \mathbf{b}_2 \tag{A43}$$

The latter is a tableau as known from LP without the row with a relative cost factor. According to [17] this is the dual quadratic programming problem.
It is solved by changing the base in consecutive steps aiming at a feasible μ (non-negative). Thereby diagonal elements are pivots only [19]. A feasible μ will guarantee the optimal solution of (A30), (A31).

References

[1] R. Bacher; *Power System Models, Objectives and Constraints in Optimal Power Flow Calculations*, SVOR/ASRO Tutorial on: Optimization in Planning and Operation of Electric Power Systems, October 15-16, 1992, Thun (Switzerland)

[2] H. Glavitsch, R. Bacher; *Optimal Power Flow Algorithms*, Advances in Electric Power and Energy Conversion Systems, Editor C.T. Leondes, Academic Press, 1991

[3] L.K.Kirchmayer; *Economic Operation of Power Systems*, Wiley, New York (1958)

[4] B. Stott, O. Alsac, A. Monticelli; *Security and Optimization*, Proceedings of the IEEE, Vol. 75, No. 12, Dec. 1987

[5] M. Spoerry, H. Glavitsch; *Quadratic Loss Formula for Reactive Dispatch*, IEEE PICA Proceedings, 17-20 May 1983 Houston USA

[6] H.W.Kuhn, A.W.Tucker; *Non-linear programming, Proc. 2nd Berkeley Symposium on Mathematics, Statistics and Probability*, University of California Press, Berkeley, California (1951)

[7] B.Stott, E.Hobson; *Power System Security Control Using Linear Programming*, Part I+II, IEEE Trans. PAS, Vol. 97. (1978), No. 5

[8] O. Alsac, J. Bright, M. Prais, B. Stott; *Further Developments in LP-Based Optimal Power Flow*, IEEE Transactions on Power Systems, Aug. 1990

[9] H.H.Happ; *Optimal Power Dispatch. A Comprehensive Survey*, IEEE Trans. PAS, Vol. 96. (1977), pp. 841-853

[10] H.H.Happ; *Optimal Power Dispatch*, IEEE Trans. PAS, Vol. 93. (1974), pp. 820-830

[11] J.-P. Vial; *Interior Point Methodology*, SVOR/ASRO Tutorial on: Optimization in Planning and Operation of Electric Power Systems, October 15-16, 1992, Thun (Switzerland)

[12] R.C. Burchett, H.H. Happ, D.R. Vierath; *Quadratically Convergent Optimal Power Flow*, IEEE Trans. PAS-103, No.11, 1984, pp. 3267-3275

[13] A.H. Land, S. Powell; *Fortran Codes for Mathematical Programming*, Pitman Publ. 1968

[14] W.F. Tinney, V.Brandwajn, S.M. Chan; *Sparse Vector Methods*, IEEE Transactions on Power Apparatus and Systems, Vol. PAS-104, No.2 pp. 295-301, February, 1985

[15] J. Carpentier; *Optimal Power Flows*, Electrical Power & Energy Systems, Butterworths; Vol 1 No.1, April 1979

[16] D. Denzel, K.W. Edwin, F.R. Graf, H. Glavitsch; *Optimal Power Flow and its Real-Time Application at the RWE Energy Control Centre*, CIGRE Session 1988 Report 39-19

[17] D.P. Bertsekas, J.N. Tsitsiklis; *Parallel and Distributed Computation: Numerical Methods*, Prentice Hall, Englewood Cliffs, N.J. 1989

[18] W. Hollenstein, H. Glavitsch; *Constraints in Quadratic Programming Treated by Switching Concepts*, Proceedings of the 10th Power Systems Computation Conference PSCC, Graz, Austria, August 19-24, 1990

[19] W. Hollenstein, H. Glavitsch; *Linear Programming as a Tool for Treating Constraints in a Newton OPF*, Proceedings of the 10th Power Systems Computation Conference PSCC, Graz, Austria, August 19-24, 1990

[20] H.W. Dommel, W.F. Tinney; *Optimal Power Flow Solutions*, IEEE Transactions on Power Apparatus and Systems, Vol. PAS-87, pp. 1866-1867, Oct. 1968

[21] D.I. Sun, B. Ashley, B. Brewer, A. Hughes, W.F. Tinney; *Optimal Power Flow by Newton Method*, IEEE Transactions on Power Apparatus and Systems, Vol. PAS 103, No. 10, pp. 2864-2880, Oct. 1984

OPTIMAL POWER FLOW PACKAGES REQUIREMENTS AND EXPERIENCES

A. Papalexopoulos

Pacific Gas & Electric Company
San Francisco, California, U.S.A.

Abstract. The purpose of an Optimal Power Flow (OPF) function is to schedule the power system controls to optimize an objective function while satisfying a set of nonlinear equality and inequality constraints. The equality constraints are the conventional power flow equations; the inequality constraints are the limits on the control and operating variables of the system. Mathematically the OPF can be formulated as a constrained nonlinear optimization problem.

Practical, constrained active and reactive OPF problems have complicated non-analytical, non-static and partially discrete formulations. At the same time, however, most OPF development efforts have centered on the mathematical optimization of simple classical OPF formulations, expressed in smooth nonlinear programming form. As they stand, these formulations are far too approximate and incomplete descriptions of the real life problems to be adequate for on-line use. Furthermore, at the present time they cover only a limited area of system operations.

This paper will address specific operational requirements that need to be met for a successful implementation and use of an on-line OPF package. These include a) response time requirements, b) robustness with respect to starting point, c) expansion of the scope of the OPF problem to be able to solve realistically posed problems, d) infeasibility detection and handling, e) ineffective "optimal" rescheduling, f) discrete modeling, g) development of techniques/guidelines for selecting an "optimal trajectory" that steers the power system as reliably and as far as possible in the direction of the optimum, h) modeling of contingency constraints, i) consistency of OPF with other on-line functions, j) data quality and other practical requirements, k) maintenance and MMI and l) on-line OPF based external modeling.

Over the last two decades several approaches have been proposed to solve the constrained nonlinear OPF problem. A straightforward solution could be to use quadratic programming. Other approaches that have been implemented with various degrees of success include, Generalized Reduced Gradient, Newton's method, Linear Programming and Interior Point techniques. Based on the above requirements, experience and conclusions reached from the development and/or use in a practical environment of these techniques will also be discussed.

1. Introduction

Thirty years ago, Carpentier [1] introduced a generalized, nonlinear programming formulation of the economic dispatch problem including voltage and other operating constraints. This formulation was later named the Optimal Power Flow problem (OPF) [2]. It has since been generalized to include many other problems. Today any problem that involves the determination of the instantaneous "optimal" steady state of an electric power system is an Optimal Power flow problem [3,4,5]. The optimal steady state is achieved by adjusting the available controls to minimize an objective function subject to specified operating and security requirements. Different classes of OPF problems, tailored towards special-purpose applications, are defined by selecting different functions to be minimized, different sets of controls and different sets of constraints. All these classes of the OPF problem are subsets of the general problem. Historically different solution approaches have been developed to solve these different classes of the OPF problem.

It would be very difficult to accurately classify all the approaches that have appeared in the literature, since many employ a mix of specific methodologies. However, it seems that the most promising ones that have been developed over the last ten years are based on real and reactive power decoupling [6], successive sparse quadratic programming [7], successive nonsparse quadratic programming [8], successive nonsparse separable programming [9]-[12], Newton's method [13],[14] and Interior Point Method [15][16].

Some of these techniques have resulted in production OPF programs that have obtained a fair level of maturity and have overcome some of their earlier limitations in terms of flexibility, reliability and performance requirements. However, many times they are plagued by weak convergence, unrealistic assumptions, poor input data and inadequate models of the power system to be able to solve real life problems. These limitations are especially valid in an on-line environment where operational problems impose the most onerous requirements on the OPF technology.

Although these deficiencies may potentially limit the practical value and scope of OPF applications they have not received adequate attention in the industry. The deficiencies exist because significant aspects of the OPF problem in an on-line environment have been overlooked or ignored. In most cases there is a good reason for that; realistic treatment of the problems leads to very complicated and intractable formulations that are not amenable to existing classical mathematical optimization techniques. Simplistic formulations and solutions that may be sufficient for other applications, i.e., planning studies, are usually unrealistic and produce unacceptable results.

Another reason is that there is not enough interaction between users, researchers and software developers to address specific practical requirements due to lack of successful OPF applications already in use in an on-line environment. Utility involvement in developing production grade on-line OPF tools that can be useful to operators is of critical importance. Only then, the necessary approximations of OPF formulations and modeling to make the problems more solvable with available optimization techniques can be made without undue degradation of the quality of the recommended solutions. This observation usually runs contrary to existing practices, where OPF functions are being specified and installed in most new Energy Management Systems (EMSs) as part of the whole package of the EMS network applications, without concrete prior knowledge about how they will be defined and formulated, what they will be used for, how valuable they will be, etc.

PG&E as part of its efforts to improve system operations has invested several man-years in defining, developing, implementing, testing and quantifying the benefits of an OPF function. Our EMS optimization capability presents one of the first serious attempts to move away from simplistic formulations for the on-line optimal scheduling problem and focus on implementations that meet specific operational requirements. Although it falls short of fulfilling all realistic requirements it is suitable for getting started with wide industrial on-line applications. The experience from using different OPF packages for planning or operations planning studies have been of critical importance for the success of this effort.

This paper reports on the experience and the results gained from implementing an on-line OPF function. It also attempts to present some of the conclusions reached from the extensive use of other OPF packages. The paper is organized as follows. Section II presents a brief overview of the OPF problem with an operationally oriented audience in mind. Section III describes the main operational requirements along with modeling issues that need to be addressed for a successful on-line implementation. Section IV briefly presents some of the major properties and limitations of OPF algorithms that have been used in a practical environment. Section V summarizes the conclusions.

2. Overview of the OPF problem

The optimal power flow problem can be formulated as:

$$\begin{aligned} \text{minimize } & f(u,x) \\ \text{s. t. } & g(u,x) = 0 \\ & h(u,x) \leq 0. \end{aligned} \tag{1}$$

where u is the control variable and x is the decision variable; $f(u,x)$ is a real-valued objective function; $g(u,x)$ represents the power flow constraints, occasionally augmented by a few special equality constraints; $h(u,x)$ consists of the limits of the control variables and the operating limits of the power system.

As part of an Energy Management Systems (EMS), the OPF function is designed: to operate in real time or study mode; to schedule active- or reactive-power controls or both; and to optimize a defined operational objective function. In the following the basic attributes of the OPF, i.e., the controls, and the constraints as they relate to the active-and reactive-power subproblems, the dependent variables and the objective function will be discussed in more detail.

Most of the controls are separated in active and reactive based on their impact on the active- or reactive-power subproblem conditions. This separation is consistent with the principle that the active- and reactive-power subproblems are weakly coupled to each other. This is reflected in all power system operational practices that treat the scheduling for the two subproblems separately from each other. The active power OPF consists of determining the values of the active power controls which minimize an objective which is a function of active power variables while satisfying the active power constraints. During this optimization, the reactive power control variables are kept constant, except for possibly some generator bus voltage controls that may be released due to lack of sufficient reactive support. The reactive power OPF consists of determining the value of the reactive power controls which minimize an objective which is a function of reactive power variables while satisfying the reactive power constraints. Control and constraints (in addition to power balance) of the decoupled subproblems may be summarized:

Controls for Active Power Subproblems : MW Generations, Economy Interchanges, Phase Shifter Positions, HVDC Line MW Flows, Load Shedding.

Controls for Reactive Power Subproblems : Generator Voltages of MVARs, SVC MVARs, Capacity / Reactor Status, LTC Tap Positions.

Constraints for Active Power Subproblems : Control Variable Limits, Voltage Angles between Specific Buses, MW Branch Flows, MW Reserve Margins, Area MW Interchanges, Transmission Corridor MW Transfers, Net Area MW Generations.

Constraints for Reactive Power Subproblems : Control Variable Limits, Bus Voltage Magnitudes, MVAR Generations, MVAR Branch Flows, MVAR Reserve Margins, Area MVAR Interchanges, Transmission Corridor MVAR Transfers.

The advantages of a decoupled approach (control/constraint decoupling) are several.

1. Decoupling greatly improves computational efficiency.
2. Decoupling makes it possible to use different optimization techniques to solve the active- and the reactive-power OPF subproblems. This has become a necessity given that there is no single basic solution approach today that offers the right combination of flexibility, speed, reliability and other desirable properties for all types of on-line OPF problems. Recent developments [17] aimed at establishing the LP based approach into a truly general-purpose OPF solver may bear fruits but more experience and field testing is required before it proves its validity for a wide range of complex applications in the OPF field.
3. Decoupling makes it possible to have a different optimization cycle for each subproblem. In general, active power controls are scheduled frequently to satisfy economic requirements, while the reactive power controls are optimized less frequently to provide a secure post contingency voltage level or a voltage/VAR dispatch which minimizes transmission losses. (It has been reported that a reactive power control periodicity of about one hour could produce most of the savings that might be achieved with transmission loss reduction [18].)

Simultaneous scheduling of all system active and reactive power controls to minimize a global objective subject to all constraints is referred to as "full OPF." Under some conditions that require excessive rescheduling such as during abnormal operations or heavily stressed scenarios, cross coupling features (for example use of active power rescheduling for voltage violation elimination) or a full OPF may be necessary. For system specific bottlenecks where the traditional separation of active/reactive power scheduling breaks down, a full OPF may also be necessary. However, extensive testing [19] has strongly indicated that for normal operating conditions, decoupled formulations produce solutions that are close to the full OPF solutions. Given this mass of experimental evidence it would be unwise to sacrifice decoupling just to accommodate a few exceptional cases. For those cases, implementation of special cross coupling features that provide the needed coupling is straightforward and conceptually simple without altering the sparse matrix structures that make decoupling possible [20].

In addition to decoupling some controls are continuous (for example, real and reactive generation rescheduling) and some are discrete (for example, transformer taps, capacitor banks, branch switching etc.). Different classes of discrete controls have different step sizes. Even within one class, the step sizes may be different. Changing one control by one step may cause another control to change by several steps. The effects are very nonlinear and usually most pronounced close to a solution.

Rigorous solution of problems with continuous and discrete variables requires the solution of a nonlinear mixed-integer problem for the computation of the optimum settings. Optimization problems with integer decision variables are difficult to solve because brute force enumeration procedures of the discrete variables require computations that increase exponentially with the number of discrete variables. Standard integer programming methods such as branch and bound and cutting plane algorithms are slow and intractable for large-scale problems like the OPF.

All variables in an OPF problem that are not controls are classified as dependent variables. The main dependent variables are the bus voltage angles and magnitudes. Other variables could also be viewed as dependent variables. For example, if a transformer in the OPF is modeled as a local rather than a global control, its tap ratio could be classified as a dependent variable. Most of the dependent variables are continuous, but a few of them, such as locally controlled transformers are discrete. The state of the power system is completely determined by the values of the control and dependent variables.

The power system model often does not retain this clear distinction between control and dependent variables during the solution process. In optimization methods, such as Sequential Quadratic Programming no differentiation is made between control and dependent variables by the solution process since both are represented within the comprehensive model that is used.

The inequality constraints, in terms of physical realizability, can be divided into two groups:

1. *Physical Limits of Controls.* Physical limits on the control variables cannot be violated. For example, a transformer tap cannot go beyond its upper and lower bounds. A solution in which these limits are violated would be meaningless because it would not be physically realizable.
2. *Operating Limits.* These limits are imposed to enhance security and do not represent physical bounds. They can be relaxed temporarily, if necessary, to obtain feasible solutions. Infeasible problems of this kind are encountered in some on-line applications and a good OPF program must be able to cope with them.

In addition to the constraints, there are also the active and reactive power bus injections that are related to the power flow equations. Each scheduled injection is an equality constraint that must be satisfied exactly and unconditionally. Violation of the equality constraints that correspond to scheduled injections could yield a physically realizable solution, but this would alter the characteristics of the problem. There are other quantities, aside from the bus injections, that have to be enforced exactly. For example, the scheduled exchange of power between different control areas is an equality constraint.

Among the least well-developed aspects of the OPF problem is the formulation and modeling of the objective function. At the present time it covers only a limited area of system operations. Classical OPF formulations utilize single scalar objectives. In most cases this is not sufficient. Composite objectives are required but if they are not properly modeled they can lead to unrealistic solutions from an operations view point. The four most common objectives are very briefly discussed next.

1) *Minimization of Production Cost:* This is the most used basic objective that reflects current economic dispatch practices. The objective consists of the sum of the costs of the controlled generations of thermal plants, plus the costs of controlled interchange transactions. All system control variables are eligible to participate in minimizing the objective and enforcing operational constraints such as transmission line flow constraints, reserve margin constraints, etc. If only active power controls participate in the optimization, the OPF calculation is referred to as Security Constrained Economic Dispatch (SCED). If controls with no direct costs

participate in the optimization, artificial costs based on practical experience and engineering judgment are used. The modeling of the generator I/O cost curves are the most critical factor in this application. Cost curves of thermal and hydro units are usually approximated by piecewise linear or quadratic segmentation, or smooth polynomials. Compatibility of these curves with the ones used in AGC or the conventional Economic Dispatch is of paramount importance for consistent results.

2) *Minimization of Active Power Transmission Lines Losses:* The controls that are used with this objective are usually the ones without direct costs, such as transformer taps, generator voltages, switching capacitors, etc. The optimization process tends to reduce circulating VARs and also maintain acceptable voltage profiles. A very similar but potentially competitive objective is to minimize reactive power losses. In general, it near-minimizes active power losses and near-maximizes generator VAR reserves. If only part of an interconnected system is optimized, the OPF calculation may produce lower losses for the area of interest at the expense of the external portions of the network.

3) *Optimization of the Active/Reactive Power Security:* The goal of the active / reactive power security is to reschedule active/reactive power controls the minimum amount necessary to relieve all constraint violations. This objective is often referred to as the minimum shift or minimum deviation objective. If the initial power flow solution is free of constraint violations, then no rescheduling occurs. The desired values correspond to the initial or some other specified operating point. This objective appears to be simple, however its application especially with different control types require extreme care in selecting the cost curve weighting factors. Practical experience with this objective strongly indicates that the weighting factors depend on the operating conditions and the types and locations of the violations. Consequently they should be updated accordingly for acceptable performance.

4) *OPF Sensitivity Calculation:* While knowledge of the optimal solution for a given set of conditions (e.g., network configuration, loads and equipment settings) is important, for some applications knowledge of the optimal solution changes as conditions change may be even more important [21]. Such changes, which are usually small, include the evolution of bus real and reactive loads over time, control variables which move to their limits over time, and changes in other power system operating constraints,

and constraint parameters, such as line flow limits and bus voltage limits. Sensitivities of the OPF solutions with respect to above changes can be directly used in many practical applications. For example, Bus Incremental Costs (BICs), i.e., the sensitivities of the production cost of generation with respect to changes in the bus active power injections, provide an insight in the economic dispatch mechanism and the ability to estimate the influence of active power injection variations on the optimal solution. Therefore, BICs can enhance a utility's ability to maximize services that involve economically beneficial transactions. In the planning environment, BICs and other related sensitivities can be used in studies regarding incremental generation and transmission additions. The need for such studies is becoming increasingly important as economy energy transactions, reliance on external sources of capacity, and competition for transmission resources have resulted in higher loading of the transmission system.

The above objectives can be combined in many different ways to produce composite objective problems that are intended to optimize different aspects of the power system operations.

In addition to the above objectives other non-traditional composite objectives may become important as utilities evolve toward more open transmission access regimes. For example, identification of expensive system constraints, maximization of power transfers with other entities and network-type of transmission services will be become increasingly important in the future.

3. Operational requirements for an on-line OPF implementation

On-line implementations pose the most onerous requirements on the OPF technology. As it currently stands classical OPF formulations expressed in smooth nonlinear programming form are far too approximate descriptions of the real life problems to lead to successful on-line implementations. This is mainly true because current OPF formulations do not have the capability to incorporate all operational considerations into the solutions. Furthermore, occasionally, utility operating practices are incompatible with the existing OPF problem formulations.

In both cases the proposed "optimal", from an algorithmic perspective, solutions are of little value, and under some conditions worse than useless, to the control center dispatchers who almost constantly are confronted with concurrent events (especially in times of emergency) that fall outside the scope of the OPF problem definition. These limitations, if properly addressed, do not have to prevent OPF programs from being used in control centers, especially when the operational optimal solution may also not be known. However it is important to keep in mind that the OPF technology will become a standard tool in an EMS environment only when it will be able to provide solutions to the dispatchers that are optimal from an operational, rather than an algorithmic perspective. In the following, some of the requirements, that need to be met so that OPF applications are useful to and usable by the dispatchers, are presented and discussed.

3.1 Response time requirements

The power system state is changing continuously through time, sometimes slowly and at other times rapidly or even abruptly during emergency conditions. Such changes, include evolution of bus real and reactive generations and loads over time, control variables which move to and off their limits over time, changes in other power system operating constraints, and topology changes due to switching and other planned or forced outages. Consequently, the security status of the system is continuously changing. These changes put a high premium on the solution speed of an OPF program designed for on-line calculations. This is especially true when excessive amount of calculations due to modeling of contingency constraints or repeated OPF runs is involved. As it is seen later, contingency constraints are important in many applications. On the other hand, the need for iterative OPF runs is critical a) when the dispatcher is trying to adapt the OPF problem definition to the current state of the power system and input changes are required and b) when modeling of various types of discreteness and decision making such as network switching, control/constraint prioritization and unit startup/shutdown is a necessity. As a general rule an on-line OPF calculation should be completed before the state of the power ystem has drifted to another state that is substantially different from the previous one. Furthermore, it would be desirable to be able to run an on-line OPF several times per hour. The determination of the optimal execution frequency that maximizes the benefits of the calculation depends on the specific application and almost invariably is limited by finite computing resources. Given the nature of the unreachable moving target of optimality of the power system, it is then preferable to focus on developing algorithms that are incrementally correct and flexible to provide very fast and frequent scheduling. Consequently, powerful algorithms with quadratic convergence characteristics based on classical formulations that provide very accurate and "mathematically optimal" solutions that neglect operational realities may not be suitable for on-line implementations.

Fast and frequent scheduling calls for the development of "hot start" OPF capabilities that take advantage of the optimal status of previously optimized operating points. The hot start option is mostly beneficial when the rate of change of the power system state is small and previously optimized points are still "relevant" to the current operating conditions. Criteria that determine this "relevance" are usually system specific, time consuming to develop and require extensive testing. If properly designed, a hot start OPF capability can improve performance for on-line applications by 30 to 50 percent. Utilities are encouraged to develop this capability (and the associated applicability criteria) that is conceptually straightforward but complicated and time consuming from an implementation view point.

3.2 Robustness of the OPF with respect to starting point

An OPF program must produce consistent solutions if it is to be used to guide the decision-making of power system operators. This requires that the OPF solution not be sensitive to the (arbitrarily or randomly) selected starting point used by the OPF program, and that changes in the OPF solution point be consistent with the changes in the power system operating constraints. Such changes include the evolution of bus loads over time, control variables which move to their limits over time, and changes in topology due to disturbances. Due to the iterative nature of the solution process, the solutions, when starting from different starting points, will never be exactly the same. But any differences should be within the tolerances associated with the convergence criteria, and of a magnitude which would be considered insignificant to a power system operator. One of the major reasons that first-order OPF methods were not well received was that noticeably different solutions could be obtained when the OPF was simply initialized from different starting points, with only one (or even none) of the solutions actually constituting a local optimum.

There are several possible reasons for potential sensitivity of an OPF solution to initial starting points. Among them:

a) There are multiple local minima in the feasible region, and it is these different local minima that are reached from different initial points.
b) The solution method is unable to reach a true optimal solution.

If (a) is the reason, then the problem may lie in the nature of the actual power system. In some cases, lack of a unique solution could be due to an under-specified problem. Most of the time this problem is due to inadequate modeling of the power system. Usually adding the proper models and the operating information needed to fully specify the problem would eliminate the non-uniqueness. Typical examples that require special attention include: a) parallel controlled capacitors or transformers, b) a radial transformer used as control variable with taps on the radial bus side, c) parallel generators connected through step-up transformers with no resistance to a high voltage bus, etc. In all these cases proper modeling and information of what constitutes a good solution from operational view point would eliminate the arbitrariness.

Theoretically, if one can show that the objective function and the feasible region are convex, the optimal solution will be unique [22]. But the complexity of the nonlinear equations and inequality constraints involved in OPF problems has thus far defeated attempts to establish convexity. If multiple local minima truly do exist, then additional computational or heuristic methods must be used to either select one of the minima as an acceptable solution or to more exhaustively search for the multiple minima and then select one.

If, as in (b), the solution method is unable to reach a true local or global minimum, then the problem may lie with the OPF program used, or may even be due to an inherent limitation of the solution methodology chosen, and is thus more serious.

Numerous robustness tests on the PG&E system strongly suggest that OPF solutions are insensitive to starting operating points [19]. On the basis of the mass of this experimental evidence, and lacking any evidence to the contrary, it seems reasonable to accept the premise that in the normally feasible region the OPF solution is unique, i.e., the OPF solution space is convex. We acknowledge that this can only be an empirical, rather than a rigorous theoretical conclusion. Yet it is a conclusion of great practical significance, for such convexity is a necessary condition for the robustness. We consider this to be of vital importance to the ultimate acceptability of the OPF in the EMS environment.

The two most important factors that may prevent the normally feasible OPF solution space from being convex (thus leading to multiple OPF solutions) are a) the discontinuous techniques being used to model specific operating practices and preferences and b) the modeling of local controls. The extent of the problem in a) has not been fully analyzed yet and is probably system specific. On the other hand, the conventional power flow problem with local control capability, whose implicit objective is feasibility with respect to a limited set of inequalities, does not have a unique solution; nevertheless, solutions of the same problem from different starting conditions usually match quite closely. Occasionally, however, different starting conditions can lead to different solutions. This occurs when the nonlinear loads can be satisfied by two or more feasible voltage levels. OPF applications, however, should be able to overcome this kind of ambiguity.

Extensive testing with the PG&E system has strongly indicated that local controls, in general, do not cause any path dependence problems. However, if for a specific system and/or case, a local control, especially on the reactive power side, causes some sort of path dependence it is preferable, if all other corrective measures have failed, to reassign that control as an optimized variable with a narrow range on the target value. This reassignment may slightly restrict the feasible region, however, at the same time it may help reduce the inherent lack of coordination between global and local controls, and it will eliminate or mitigate the path dependence problem.

Utilities are encouraged to include robustness requirements in their EMS Work Statement Specifications to protect themselves against the possibility of acquiring/developing an OPF capability that fails to produce consistent solutions for continuously changing conditions.

3.3 Expansion of the scope of the OPF problem

The main role of on-line OPF is seen as a tool helping to maintain viable system operation (i.e., no limit violations), while tracking the unreachable moving target of optimality. The OPF function has to calculate practically implementable control moves that steer the power system as reliably and as far as possible in the direction of the optimum, while avoiding and alleviating limit violations. The main emphasis should be devoted to the operating-constraint and implementability issues. All relevant constraints (more than it is now typical) must be included in each OPF formulation: "optimal" solutions respecting many, but not all important constraints could be worse than useless. For example, we have observed that under a wide range of operating conditions for the PG&E system, when transmission corridor and net area generation constraints are not included in the formulation, the OPF proposed MW schedules are completely unrealistic. These additional requirements include: more flexible control and constraint priority strategies; incorporation of control and load dynamics; inclusion of start-up/shut-down constraints of certain controls and other operating constraints that meet specific practical requirements; voltage stability and other dynamic constraints; modeling of the prohibited operating zones on the cost curves; time restrictions on control/constraint violations; cost penalty thresholds on an individual basis for constraint enforcement (this is becoming increasingly important as utilities move toward more competition and security under some conditions will be traded off for economy); other system specific operational requirements that each utility should identify and develop for successful OPF implementations.

An important control action that utility dispatchers have at their disposal for use under some conditions is branch switching. Incorporation of branch switching in OPF formulations would greatly expand the scope of OPF applications. Modeling of branch switching should be consistent with utility operating practices. Many utilities do not use branch switching for economy improvements through the changing in losses caused by the switching. Others use branch switching extensively only as a constraint enforcement tool of last resort. In any case, branch switching drastically changes the network topology (and consequently the security) of the system. It introduces large discontinuities and non-linearities in the model that cannot be easily handled with smooth analytical OPF formulations. However, branch switching actions can be easily modeled in a rule-based expert system approach that incorporates utility guidelines involving switching. In this case, emphasis should be placed on including in the rule base as many predetermined situations as possible to cover a wide variety of power system states.

Efforts to incorporate some of the above enhancements in OPF applications have produced some benefits but more extensive testing and experience is needed for large scale industrial applications.

3.4 Infeasibility detection and handling

As the requirements for satisfactory system operation proliferate, the region of feasible solutions, satisfying all constraints simultaneously, may shrink to non-existence, and a new problem emerges, for developing criteria to establish the relative precedence among the constraints. For OPF applications, this means that when a feasible solution cannot be found, it is still very important for the algorithm to recommend the "best" engineering solution that is optimal in some sense even though it is infeasible. This is even more critical for OPF applications that include contingency constraints in their formulation.

There are several approaches to this problem. One approach calls for a least squares violation solution in which a) all power flow equations are satisfied and b) only the soft constraints that truly cause the bottlenecks are violated in a least squares sense. With the LP approach, this is achieved by introducing a weighted slack variable for each binding constraint. If a constraint can be enforced, the slack variable will be reduced to zero and the constraint will be observed. Those constraints causing infeasibility will have non-zero slack variables whose magnitudes are proportional to the amounts they need to be relaxed to achieve feasibility. Usually with this approach all binding constraints of a particular type are modeled as if they have identical infeasibility characteristics. That is, all slack variables corresponding to these binding constraints share the same cost curve, and their sensitivities (a column of zeros, with +/-1 in the row of the corresponding constraint) are scaled by a weighting factor associated with the type of the corresponding constraint. After extensive testing, it comes to realize that this type of modeling, even though typical in the industry, many times leads to inaccurate results. Certain individual constraints may have exceptional properties. For example, if a certain violation can be solved by reconfiguring the network (branch switching), a very low weighting factor will be needed for this constraint. Otherwise, the OPF will reschedule an undue amount of MWs to alleviate this violation. Thus, certain slack variables will need to be costed by exception for their unique infeasibility characteristics. Utilities are encouraged to identify similar specific cases and modify the infeasibility algorithm appropriately for better results. With the Newton method, if the OPF does not converge in the first set of iterations specified by the user, the constraint weighting factors, corresponding to the penalty functions associated with the load bus voltage limits and the branch flow limits, will be reduced successively until a solution is reached. This approach normally results in all constraints being met except for those load bus voltage limits and branch flow limits that are preventing feasibility. Special care should be taken in selecting the proper weighting factors to avoid numerical problems and produce acceptable solutions. Field experience with this infeasibility approach indicates that in many instances the recommended optimal, but infeasible, solutions are not realistic and consistent with operating practices.

Another approach calls for the development of hierarchical rules that operate on the controls and constraints of the OPF problem. The rules introduce discontinuous changes in the original OPF formulation. These changes include use of different sets of control/constraint limits, expansion of the control set by class or individually, branch switching, load shedding, etc. They are usually implemented in a predefined priority sequence to be consistent with utility practices. The decision when to proceed to the next priority level of modifications to achieve feasibility is critical, especially when it involves radial overloads, normally overloaded constraints and constraints known to have "soft" limits. As different OPF applications come on-line and field experience is gained, the sophistication of the rules will increase covering a wide range of conditions. It is expected that this approach will gain broad acceptance and credibility. Note that both approaches are not mutually exclusive and can be applied in combination. However, the order of application results in different OPF modifications and, consequently, in different solutions. Both approaches can be viewed as a form of a multi-criteria optimization problem. The solution to this problem leads to the identification of "non-inferior" sets of solutions within which no one criterion can be improved without subsequent degradation of the others. The selection of a final optimal solution among all the others in the set is achieved with the implementation of a "preference index." An application of the preference index approach that minimizes postcontingency line overloads due to generator outages is found in [23].

One of the most important attributes of any infeasibility algorithm is the speed of infeasibility detection. There are algorithms, especially NLP algorithms, which fail to detect the infeasibility status of the problem early enough in the optimization process. This is a serious drawback that deserves special attention. Rapid detection of infeasibility is of paramount importance for on-line applications.

Finally, in many instances an OPF problem is infeasible because it is badly posed. For example, constraint enforcement using only reactive power controls often leads to infeasible solutions. Also simple data errors may cause infeasibilities. OPF algorithms equipped with some level of intelligence, perhaps a rule-based expert system, should be able to handle these problems.

3.5 Ineffective "optimal" rescheduling

Existing OPF algorithms use all available control actions to obtain an OPF solution, but for many applications it is not practical to execute more than a limited number of control actions. For example, in on-line applications, the OPF is used to find changes in control settings with respect to an existing setting. For various reasons it is desirable or necessary to limit the number of such actions. Especially for the reactive power dispatch problem, the rescheduling of large numbers of reactive controls is unmanageable in real-time. The OPF problem then becomes one of selecting the best set of actions of a limited size out of a much larger set of possible actions. The problem was identified in [24] but no concrete solutions were offered.

It is not possible to select the best and most effective set of a given size from existing OPF solutions that use all controls to solve each problem. The control actions cannot be ranked and the effectiveness of an action is not related to its magnitude. Each control facility participates in both, minimization of the objective function and enforcement of the constraints. Separation of the two effects for evaluation purposes is not possible. The problem is difficult to be defined analytically and existing formulations are not adequate. One possible definition could be to include the total number of control actions, or the limits on the number of control action for each control class in the formulation [24]. Another definition could be to assign an initiation cost for each control action whose number needs to be limited. In this case, the minimum cost solution, including the starting costs of control actions, would produce the correct number of actions. Neither definition, however, can directly lead to acceptable solutions. The compromise in the short run should be to rely on near-optimal solutions that incorporate sound engineering rules that are fast enough for practical applications. In the long run a solid methodology to address this problem is very much needed.

The problem of ineffective rescheduling is related to but is not identical with the "minimum number of controls" objective. It is also closely linked to the problem of discrete control variables, since methods that recognize the discrete nature of some control facilities tend to decrease the number of control actions by keeping inefficient discrete controls at their initial setting. Researchers are encouraged to work in this open area recognizing the imperative to progress beyond the rigorous "optimal" solutions of smooth, simplistic and classical OPF formulations, and to concentrate on obtaining "near-optimal" solutions to more realistically stated OPF problems that minimize ineffective rescheduling.

3.6 Discrete modeling

The OPF problem is discrete in nature. Discrete OPF elements include: discrete control facilities; branch switching; prohibitive zones of generator I/O cost curves; and priority sequence levels for infeasibility handling. OPF algorithms designed for on-line applications should be able to handle the discrete aspects of the problem sufficiently. In the following we'll concentrate on the modeling of the discrete control variables.

Discrete controls are widely used by the utility industry. For example, transformers are used for voltage control, shunt capacitors and reactors are switched on or off in order to correct the voltage profile and reduce active power transmission losses, and phase shifters are used to regulate the MW flows of transmission lines. An efficient and effective OPF discretization procedure is needed to help the operators utilize these discrete controls in a realistic and optimal or near optimal manner.

In general, LP based OPF algorithms permit substantial recognition of control discreteness by setting the cost curve segment break points at discrete control steps. However, most methods that solve for a non-separable objective function by nonlinear programing methods do not properly model discrete controls.

The use of both discrete and continuous controls converts the OPF into a mixed discrete-continuous optimization problem. A possible rigorous solution using a method such as mixed integer-nonlinear programming, would be orders of magnitude slower than the ordinary nonlinear programming methods [22].

Presently, most OPF algorithms treat all controls as continuous variables during the initial solution process. Once the continuous solution is found, each discrete variable is moved to its nearest discrete setting. This procedure gives acceptable solutions provided the step sizes for the discrete controls are sufficiently small, which is usually the case for transformer taps and phase shifter angles [19]. However, shunt capacitors and reactors with larger bank sizes usually have greater impact on the optimization. Currently, two different approximation approaches are used after the rounding off: One is to execute a conventional power flow solution with all the discrete variables fixed on their steps. The other is to solve the optimization problem again with respect to the remaining continuous variables using the first continuous solution as the initial point. The former approach is widely used because of its computational efficiency. The latter approach gives a better solution in terms of feasibility and minimization, but the second optimization significantly increases the total time for an OPF execution. The final solution is still not guaranteed to be optimal because incorrect values for the discrete control steps may have been selected.

Given the intractability of rigorous solution methods, approximate solutions that can produce near optimal results appear to be a reasonable alternative. The use of penalty functions for discrete controls is one such scheme [25]. The aim is to penalize the continuous approximations of discrete control variables for movements away from their discrete steps. An attractive feature of this scheme is that it merges well with existing Newton OPF algorithms. The scheme consists of a set of rules which determine the timing of introduction and criteria of updating the penalties in the optimization process. The algorithm has been extensively tested under a wide variety of conditions on two large scale power systems. Test results indicate that the solutions are close to the optimum and the extra computational time required is definitely acceptable for on-line applications.

The above heuristic algorithm is compatible only with the Newton method and consequently is of limited scope. Substantial more work is needed to effectively resolve all problem associated with the discrete nature of controls and other discrete elements of the OPF problem.

3.7 Selection of an "optimal trajectory"

No serious attempts have been made yet in implementing a "trajectory" of the OPF control shifts that does not exacerbate existing violations or cause additional ones. The sequence in which the different control settings are altered may inadvertently create new problems. In general, the trajectory is probably less important for thermal violations than for voltage problems.

The limited amount of time to correct constraint violations is itself a security concern but it is further complicated by the fact that controls cannot move instantaneously. For some controls, the time required for movement is not trivial. This is an important consideration to be taken into account in designing a trajectory for the OPF solutions. For example, generator ramp rates can significantly restrict the speed with which active power is rerouted in the network. Delay times for switching capacitors and reactors, and transformer tap changing mechanisms can preclude the immediate correction of serious voltage violations. The time-urgency of the violations and the time-constraints on control movement can together determine the character of an OPF solution. If the violation is severe enough, slow controls that would otherwise be preferred may be rejected in favor of fast, less preferred controls.

Comprehensive guidelines and procedures need to be developed to resolve the trajectory problem in a satisfactory way. Utilities are encouraged not to overlook the significance of this problem in their efforts to develop on-line OPF applications.

3.8 Contingency constraints

Typical on-line OPF applications at the present time produce secure and optimal solutions with respect to the "base case" security and operating constraints. However, serious erosion of the power system's steady-state security in case of a contingency is possible. The inclusion of contingency-constraints is a major challenge but it is expected that eventually contingency-constrained OPF will become a standard tool in the industry. The need for modeling contingency constraints in OPF formulations for many applications is well understood. As utilities move toward a more open and competitive environment more and more third party generation, such as qualifying facilities and cogeneration, will seek access to their markets. Evaluation of requests for transmission access in the context of the system security will put a high premium on utilities to respond rapidly to an increasing number of energy players that undoubtedly will stress the power system networks even further. The increasing number of transactions that need to be evaluated calls for the development of new tools among which the contingency constrained OPF will play a central role.

The implementation of a contingency constrained OPF for on-line applications should be able to handle any number of contingency cases and any number of the following contingencies in each case: branch and generator outages; bus outages and bus splits; load outages; shunt and DC link outages and network islands [17]. A contingency may create new buses, different types of buses, new electrical islands, local control configurations whose elements belong to different islands, MW imbalances, substantial loss changes, etc. A contingency constrained OPF implementation should be able to model all of the above changes and should be flexible enough to produce higher risk, lower cost plans if necessary. The determination of the proper balance of security and economy is of critical importance as more and more utilities are willing to trade off security for economy. Furthermore, the proper balance depends on the current security state of system which is continuously changing. For example, if the base case state has serious constraint violations, the dispatchers may prefer to concentrate on corrective actions alone, neglecting the risk of contingencies. On the other side of the spectrum, when a utility is not willing to rely on post-contingency control action, then all contingencies must be addressed with preventive action.

All the operational requirements, few of which were mentioned above, pose major challenges for contingency constrained OPF implementations. Some programs already exist in the market [17], [26], but more experience and field testing are needed to prove their value in terms of performance, reliability and reasonableness. As utilities' operational needs evolve, it is expected that the need for contingency constrained OPF programs will increase. Ultimately, their success depends on the acceptance, on behalf of the dispatchers, of the base case on-line OPF applications.

3.9 Consistency of OPF with other on-line functions

On-line OPF programs are implemented in either study or closed loop mode. In the study mode, the OPF solutions are presented as recommendations to the dispatcher. In the closed loop mode, the control actions are implemented in the power system, typically via the SCADA system of the EMS [27]. In the closed loop mode, the execution of the OPF is triggered by wall clock time, operator request, execution of the Real-Time Sequence and Security Analysis, structural change, large load change, etc. A major problem of an OPF in closed loop mode is the design of its interface with the other on-line functions which are executed with different periodicities. Some of these functions are: Unit Commitment, classical Economic Dispatch (ED), Real-Time Sequence, Security Analysis, Automatic Generation Control (AGC), etc. To reduce the discrepancy between idealized and realistic OPF problems, emphasis should be focused in establishing consistency between these functions and static optimal solutions produced by the OPF. Consistency requires proper interfacing and integration of the OPF with these functions. The integration design should be flexible enough to allow OPF formulation modifications consistent with the ever dynamic and sometimes ill-defined security problem definition.

A central aspect of consistency between the OPF and other on-line applications is the coordination of the OPF-ED-AGC control hierarchy. The overall objective here is how to impose the security constrained MW schedules produced by an OPF to AGC through the ED. The industry is investigating a range of different possibilities. One of the simplest approaches, that has been traditionally proposed, is to retain the classical ED and periodically modify the limits (upper or lower) on the generation controls to observe current security constraints. This approach is straightforward to implement but its success is limited. Another approach calls for the substitution of the classical ED with a parametric OPF that updates previous OPF solutions as the system load is changing, observing a fixed set of binding constraints [28], [29]. This approach also needs substantial more development and testing before it is used for large scale applications.

A promising approach is to install a security constrained economic dispatch (SCED) which plays the combined role of OPF and ED for the active power subproblem. The basic SCED concept is presented in [30]. When there are no violations the ED is executing and is passing base points and participation factors to AGC. (The SCED, in this case, should give identical results with ED.) If a violation is identified by the Real-Time Sequence, the SCED is substituting the ED. The SCED determines the most effective remedial actions which can be implemented through economic dispatch for eliminating violations of active power operating constraints. The SCED function also calculates optimum generating unit base points and constrained economic participation factors for use by the AGC, in order to optimally dispatch generation while enforcing binding active power-related constraints. Substantial effort will be required to implement SCED in "closed loop" mode where base points and participation factors are passed automatically to AGC. This will be a very important step toward implementing a secure automatic generation control.

PG&E has embarked in a major project to implement the SCED capability. Development has been completed, however, extensive field testing is needed to establish the SCED as a viable approach to the OPF-ED-AGC hierarchical control problem. A similar concept can be developed for the reactive power problem, where target setpoints produced by a transmission loss objective are passed to a real-time reactive dispatch that schedules voltages at key buses in the system.

3.10 Data quality and other practical requirements

The accuracy of an OPF program is only as good as its input data. Data uncertainties sometimes cast doubt on the usefulness of the results. The increase of non-utility generation and customer load management further aggravates this problem. This is especially true for the real-time portion of the data that drive an OPF. The need for reliable real-time databases puts a high premium on the performance and accuracy of the front end Real-Time Sequence functions. Also the need for accurate generator cost curve modeling, line and other equipment ratings, network parameters, etc., calls for the development of a series of tools such as on-line dynamic thermal rating, parameter estimation, etc., that will ensure the accuracy of the OPF solutions.

Usually the data needed for an OPF run is metered, estimated or forecasted and consequently is subject to large errors. These uncertainties have yet to be analyzed and may result in such a large degree of arbitrariness as to make the goal of optimal operation illusory. The modeling of a generator I/O cost curve, for example, is extremely important for accurate OPF solutions. Costs for thermal units are derived from the heat-rate curves which generally are far from being smooth or convex. Usually cost curves are approximated by a convex polynomial or exponential or they are segmented with quadratic or linear segments while maintaining overall convexity. Traditionally, for LP-based methods, a separate LP variable has to be used for each segment of a curve. For large scale problems this modeling is impractical, because of the need to model each curve with a large number of segments, thus reducing the segment size. If the segment size is not small enough, several problems, depending on the application, will hinder the optimization process. These problems include incompatibility with AGC/ED, large circulating VARS and severe oscillations. The best approach that has been successfully tested at the present time, involves the successive refinement of the generator cost curves, during the OPF solution, into smaller segments around the generator MW level, thus improving the solution accuracy. At each OPF iteration each cost curve is modeled with a smaller segment size, until the final number of segments has been reached. This modeling allows cost refinement to any desired practical accuracy [17]. The idea is conceptually straightforward, yet very complicated to implement. Extensive field testing strongly indicates that LP-based OPF programs without the cost curve successive refinement capability are severely limited as the technology of choice for large scale on-line

OPF applications.

Another important issue related to input data is load modeling. In OPF applications, as in other EMS advanced applications, the loads typically represent aggregate demands at power system substations. Their characteristics are quite complex, yet they are usually represented as constant MVA loads or as a linear function of polynomials of the voltage. Due to lack of data the determination of parameters is seldom done for particular loads. On the other hand, load modeling has a major impact on the optimization and the OPF solutions. If a MW load is predominantly modeled as decreasing with voltage in a production cost minimization objective, which includes the use of reactive power controls, the resulting OPF solution will produce an unacceptable voltage profile. The reason is that the bus voltages, being global controls, will tend to become as low as possible in order to reduce the production cost of generation by reducing the electric demand. Utilities are encouraged to do substantial more work for physically based load models compatible with OPF applications.

For practically useful results for operations planning OPF applications the following issues are also important.

- If generators are taken off AGC to alleviate a violation, how will their outputs be managed as the power system state changes through time? When can a generator that has been manually interrupted return to automatic control?

- How often will the user be required to interact with the OPF? For example, if the increasing system load, for example during peak load hours, is detrimental to specific branch flow violations, constraints on these branch flows enforced by a prior OPF run could be violated again. Frequent execution of the OPF will keep these constraints enforced, however the nuisance to the operator from frequent manual intervention may be significant. Infrequent OPF execution is also possible and will tend to alleviate this operator nuisance. For example, the OPF can correct violations based upon forecasted system states rather than correcting the estimated state. This may have the desired effect, but with an economic penalty. Also, the OPF may need to be re-run if a power system event invalidates the forecast. Similar observations can be made when the system load is decreasing. A related question is then, how often should the user move these previously rescheduled units for economic reasons? When can a unit be restored to its initial control status?

The answer to the above and many other questions may be system specific, yet it is critical in implementing on-line OPF applications that are useful and usable.

3.11 Maintenance and MMI

There is a host of practical implementation problems that each utility should overcome in establishing a fully operational Optimal Power Flow. Some of these problems include; initial start up, tuning, maintenance and performance requirements; case setup and interpretation of the OPF results; user friendly MMI and user training. The investment in money, time and manpower to successfully address these problems should not be overlooked. On-line OPF applications constitute a quantum leap technology in existing control centers and the effort involved in tuning, maintenance, training, etc. is substantial, especially during the first few years of operations. OPF applications are not a requirement for dispatchers to perform their duties. A dispatcher does not necessarily need to use OPF applications to "get by." Therefore training dispatchers to use OPF applications takes on special requirements whereby one is not only interested in teaching the basics of these tools, but providing incentives and motivation for their usage.

The design of the MMI is also a crucial aspect of OPF applications. Care should be taken to layer the output so that top level information is available at a quick glance, yet more detailed information is available if desired. Artificial intelligence technology could assist in the MMI design and could make the OPF a more useful decision support tool. The proper interpretation of OPF results (which may consist of thousands of numerical values of voltages, flows, control settings, etc.) and their subsequent usage in decision making is a complex process that very often entails a judicious judgment and/or many combinatorial searches. Considering the typical demands on the dispatchers' time and attention, fast interpretation of the OPF results so that they can accurately assess various possibilities or solution strategies before proceeding with the implementation of the preferred solution is of critical importance. For example, implementation of an optimal control "trajectory" or suppression of ineffective re-scheduling may require additional executions of the OPF and further analysis until the dispatcher is satisfied with a solution to the original problem. Artificial intelligence technology may be able to provide such decision support capability.

Artificial Intelligence technology can also assist in the OPF input case setup. One of the most critical and complex tasks in the execution of an OPF program is the preparation of meaningful, error free input cases. This process normally requires the participation of highly skilled and experienced personnel. Artificial Intelligence technology can provide a tool that will facilitate this process by allowing rapid and meaningful case setup.

3.12 On-line OPF based external modeling

Power systems are interconnected. Each control area with its control center(s) is responsible for the control of only a portion of the interconnected system. The control center receives telemetered data of real-time measurements. The monitored part of the power system covered by these measurements is called internal system. This system usually consists of one's own system and a portion of other systems which is "electrically close" to that system. The rest of the interconnected system is called external system, and is usually unmetered and "electrically more distant" and has less effect on the internal system. Any unmonitored portions of the internal system such as lower voltage networks or unmonitored substations, must also be incorporated in the "external" system.

The internal system is solved by the State Estimator. A portion of the external system is usually retained intact and the rest is equivalenced. The extent of the equivalencing is usually dictated by memory limitations. In real-time mode the external network is solved by the External Estimator, which ensures consistency with the SE-based solution of the internal network. In the study mode the entire network (internal and external) is solved by the Dispatcher Power Flow. The equivalent network should be accurate enough to ensure the accuracy of network applications such as Dispatcher Power Flow, Security Analysis, Optimal Power Flow and Dispatcher Training Simulator.

Existing power flow based reduced models have been very effective in Security Analysis and DTS. However, if applied blindly for security control (in OPF applications for example) the results of the optimization would be unacceptable even for the base case, let alone the contingency cases. The problem has been identified in [24] but no solid solutions were offered. The errors result from poor accuracy of loss modelling in the equivalents and inability to monitor or enforce inequality constraints of the external system represented by the equivalent. If the losses in an equivalent are too low, the OPF algorithm will route too much flow through the equivalent. If the losses are too high, the effect will be opposite. If an equivalent is unconstrained, the OPF algorithm will circumvent binding inequalities by routing flows through the equivalent. To the extent that these effects occur, the solution of the network optimization is in error. Differences between the parameters of the equivalent and those of the external system will cause derivatives of the Lagrangian with respect to variables associated with the boundary buses to become nonzero. Thus a power flow based equivalent will change, sometimes substantially, an OPF problem.

We have been investigating different approaches in our efforts to develop reduced equivalents which are suitable for security analysis studies as well as for security control. One approach calls for the determination of an area in the system, called buffer zone, that contains the OPF controls. The buffer zone should be accurately modeled, correctly constrained, and large enough to make the effects of the equivalent on the base case OPF solution small enough to be acceptable. The drawback of this approach, of course, is that it can make OPF problems very large. Good criteria for determining the extent of the buffer zone are presently unknown. After some tedious testing, we elected not to pursue this approach any further. Another approach calls for the development of OPF based equivalents by reducing to the boundary buses the composite Hessian/Jacobian matrix of the Newton OPF. The resulting equivalent is a mathematical incremental model of the external system linearized at a selected base case solution point.

We have pursued this approach over the last two years and have developed an OPF based equivalent model that we are in the process of testing with large scale system data. Preliminary test results are very promising. A simplified outline of the method is as follows:

1. Create the entire (internal plus external) network.
2. Determine the internal, boundary and external buses.
3. Compute all of the Lagrange multipliers for the OPF solution.
4. Compute the Hessian / Jacobian matrix $\nabla_{yy}L$ where

$$L(y) = f(x) + \lambda^t \cdot g(x),$$

   $f(x)$ is the objective function, $g(x)$ is the binding set and $y = (x, \lambda)$.

5. Partition the Hessian / Jacobian by ordering the variables in the order of external (e), boundary (b) and internal (i) blocks.

$$\nabla_{yy}L = \begin{vmatrix} \nabla_{yy}L_{ee} & \nabla_{yy}L_{eb} & 0 \\ \nabla_{yy}L_{be} & \nabla_{yy}L_{bb} & \nabla_{yy}L_{bi} \\ 0 & \nabla_{yy}L_{ib} & \nabla_{yy}L_{ii} \end{vmatrix}$$

6. Perform factorization of the external block of the Hessian / Jacobian

$$\nabla_{yy}L = \begin{vmatrix} \nabla_{yy}L_{ee} & \nabla_{yy}L_{eb} & 0 \\ 0 & \nabla_{yy}L_{bb1} & \nabla_{yy}L_{bi} \\ 0 & \nabla_{yy}L_{ib} & \nabla_{yy}L_{ii} \end{vmatrix}$$

with

$$\nabla_{yy}L_{bb1} = \nabla_{yy}L_{bb} - \nabla_{yy}L_{be} \cdot \nabla_{yy}L_{ee}^{-1} \cdot \nabla_{yy}L_{bb1} \tag{2}$$

7. Disconnect the existing branches between boundary and external buses.

8. Insert new fictitious branches between boundary buses. The branch parameters are computed based on the fill-in values of the off-diagonal blocks of matrix (2) using a least squares technique. A least squares approach was utilized because the number of parameters to be determined (two for each fictitious line) is much less than the number of equations to be satisfied. The number of equations for each off-diagonal block element of the matrix (2) is equal to the number of fill-ins in that block.

9. Compute the modifications of the diagonal blocks of the matrix (2) due to the newly-added branches.

10. Add shunt and constant current loads to each of the boundary buses. The shunts and constant current loads are computed based on the fill-in values of the diagonal blocks of matrix (2) and the first order Kuhn-Tucker conditions, using a constrained least squares technique. Constant current loads are used in order to satisfy the first order Kuhn-Tucker conditions associated with the voltage angles of the boundary buses.

11. Perform conventional boundary MVA load matching. The above methodology has been tested using small scale systems. The results are very promising. Tables 1 and 2 present the comparison of the performance of an OPF reduced model (column 2) and a power flow based reduced model (column 3). The power flow reduced model is based on the Extended Ward method. The performance of each reduced model is evaluated based on its accuracy with respect to the full system OPF solutions. The difference of the objective function values (production cost in this example) and the largest difference in voltages, MW/MVAR generations and MW/MVAR flows were used to establish the accuracy of each reduced model. Table 1 presents the results for the base case and Table 2 presents the results for a generator outage (320 MW). As can be seen the performance of the power flow based reduced model is very poor even

for the base case (note the 107.3 MVAR largest flow difference between full and reduced system as compared to the 0.01 MVAR largest flow difference for the OPF based reduced model). The system that was used for this test consists of 20 buses (out of which 14 belong to the internal system), 31 branches and six generators. Substantial more testing is required to assess if this approach can be developed into a completely satisfactory OPF equivalencing scheme suitable for large scale OPF applications. Nevertheless the results so far are very promising.

In the short run, without a solid OPF based equivalencing scheme, utilities should keep using power flow based equivalents for security/control with extreme caution and should not trust the results blindly.

Largest Difference	OPF-based Reduced Model	PF-based Reduced Model
Objective	0.0	0.3
Voltage %	0.0	-1.5
MW Generation	-0.1	0.8
MVAR Generation	0.1	-58.8
MW Flow	0.7	3.8
MVAR Flow	0.01	107.3

Table 1: Performance Evaluation of the OPF-based and PF-based Reduced Models (Base Case)

Largest Difference	OPF-based Reduced Model	PF-based Reduced Model
Objective	0.0	0.5
Voltage %	0.33	-1.33
MW Generation	-0.5	0.93
MVAR Generation	27.6	-49.85
MW Flow	0.56	4.33
MVAR Flow	27.6	106.6

Table 2: Performance Evaluation of the OPF-based and PF-based Reduced Models (Generator Outage)

4. Experience with OPF packages in a practical environment

In this section some of the major properties and limitations of different OPF methods that have been developed/used for planning and operations planning studies will be briefly presented. The experience and knowledge gained from the development/use of these methods are heavily influenced by the specific code implementations. Therefore the information presented here should be used judiciously. It should also be recognized that modeling requirements and specific implementation aspects of a particular approach often have the greatest impact on its ultimate success.

4.1 Generalized Reduced Gradient (GRC) approach

This approach is essentially the same with the Dommel-Tinney method presented in [2]. At each iteration, the objective is approximated by a quadratic function of the changes in the system control variables about the current point. The inequality constraints are also expressed by linear functions of the changes. The reduced problem is then solved by a general-purpose quadratic programming package. The main difference with the Dommel-Tinney method is the handling of the functional inequality constraints. In the GRC approach, when a constraint is violated it enters the control set and a control (independent) variable is selected to become a dependent variable in its place. The selection for this variable swap is made based on the limit conditions of all of the original control variables and their sensitivity to the limiting dependent variable. In general the optimization process consists of finding a feasible solution and then while maintaining feasibility it attempts to minimize the objective function. Despite the mathematical vigor of this approach and its flexibility there are practical difficulties associated with large scale applications. Some of these difficulties that have been identified in practice are:

- The method requires a load flow solution at every iteration.
- The method exhibits slow convergence (linear convergence) because it is a first order method.
- The variable swap destroys the favorable properties of the system resulting in sparse matrix factorizations that are inefficient and slow. Furthermore, it does not work as expected in large scale systems.

- The method cannot handle efficiently the functional inequality constraints.
- The overall efficiency of the method depends very much on the problem size.
- The penalty function and gradient step mechanism require careful tuning. If an optimal gradient move is made, some inequality constraints may become violated to the extent that recovery may be oscillatory, even to the point of divergence.
- Detection of infeasibility is difficult and slow. This is especially important for real-time applications where the OPF should give a clear indication that the solution is not feasible as soon as possible, and then provide the "best" solution it can.
- There is an unattractive speed / reliability trade off especially for heavily stressed and ill-conditioned systems.

4.2 Quadratic Programming (QP)

This approach is straightforward. It can be implemented in a sparse [7] or nonsparse (compact) form [8]. In a sparse form the problem is approximated by a sparse quadratic objective and sparse linearized constraints. The approximated problem is solved by a QP package. In nonsparse form the objective is approximated as a nonsparse quadratic function of the control variable changes. The inequality constraints are also expressed as linear nonsparse functions of the control variable changes. The reduced problem is solved by a QP package. The dimensionality of this approach is low but the loss of sparsity can be a severe problem.

The code that has been used for planning studies utilizes a quasi-Newton formulation to compute the Newton descent direction for the super-basic control variables. A positive definite approximation of the reduced Hessian is computed at each iteration by systematically updating a starting matrix with a series of vector transformations (BFGS update) [22]. Thus, curvature information is built as the iterations of the descent method proceed, using the observed behavior of the objective function and the gradient. Therefore, by construction, the projection of the Hessian of the Lagrangian onto the subspace of the super basics is always positive definite and the second order optimality condition that guarantees minimality of the stationary point is satisfied.

Extensive experience with this method strongly indicates the following:

- The method is very robust and reliable if the QP optimizer is efficient.
- For well posed problems the method normally converges in a few iterations but is consistently slower than LP-based methods for most applications.
- Computing time tends to rise very rapidly with the number of controls/ constraints. For difficult problems it can be very slow.
- The method can handle efficiently linearized constraints. For marginally feasible problems the method may fail to converge.
- Detection and handling of infeasibilty is a major problem.
- The method is not flexible enough to handle nonclassical and nonsmooth OPF formulations. For real-time applications this can be a major drawback.
- The method is not well suited to suppress ineffective rescheduling, or minimize the number of control actions.

4.3 Newton approach

The Newton-based OPF algorithm is a major breakthrough [13],[14]. Equality constraints are imposed by the Lagrange multiplier method. The Lagrangian function is minimized and the zero-gradient equations are solved iteratively by Newton's method. If the correct binding inequality constraints are known, the Newton solution will converge quadratically. However, the binding set is not known a priori. Therefore, heuristic trial techniques have to be used between Newton iterations.

Preliminary field experience with this approach indicates the following:

- The success of this approach depends more than in any other approach on the specific implementation.
- Computational effort depends linearly on the problem size. Only for difficult problems this method can be slower than expected.
- Decoupled formulations are sufficient most of the time.
- Detection and handling of infeasibility may be a problem.

- The method is not well suited to suppress ineffective rescheduling or minimize the number of control actions.
- The method is efficient for nonseparable objective functions.
- The method effectively exploits the sparsity of the network.
- Depending on the implementation, detection of the binding set may be a problem.

4.4 Linear Programming

Linear Programming based optimization methods are very efficient when the objective is separable. A special dual LP solution with upper and lower bounding, and handling of the separable piecewise-linear convex cost curves has been especially successful. Computationally, the approach is very favorable for the production cost minimization. However, transmission loss minimization is not handled efficiently because its objective function is strongly non-separable. It appears that the method is currently the choice for real-time applications.

Extensive experience with this method strongly indicates the following:

- The method is reliable, robust and consistently faster than any nonlinear programming method.
- It can handle efficiently functional inequality constraints.
- It provides unequivocal and rapid detection of infeasibilities.
- The method is probably more suitable than other NLP methods for suppression of ineffective rescheduling.
- The method is flexible to model discontinuous modifications, control/ constraint priority strategies, priority sequence schemes and other non-analytical engineering rules that reflect operational practices.
- It can accurately model generator cost curves.
- It is probably better suited than other methods to accurately model contingency constraints for preventive rescheduling.

4.5 Interior Point method

The field of Interior Point (IP) methods has experienced a remarkable surge of activity following publication of Karmarkar's projective algorithm in May 1984 [31]. Different approaches of IP methods have been considered for LP problems, that were later extended to convex quadratic programming and other problems. It appears that the wide attention that the IP methods are receiving recently is not only due to their theoretical properties but also to their computational efficiency. Software packages based on IP methods are not only competitive with other packages based on simplex method but also they can be even more attractive when the problem size is very large [32].

The IP-based OPF package that has been used for planning studies in PG&E over the last two years is based on an algorithm that linearizes the constraints, like the familiar Newton-Raphson power flow solver, and optimizes a sequence of constrained sub-problems. In each iteration the OPF solver computes an optimal correction of the variables that minimize a quadratic model of the objective function. The sequence of the optimal corrections converge onto the true unique optimal solution of the nonlinear problem.

The optimization solver, which is based on an Interior Point method, that utilizes a primal-dual logarithmic barrier technique, eliminates all inequality constraints (limits on voltages, etc.) from the problem by adding a term to the objective function. These terms are called "barrier functions," because they form an infinite barrier, or "wall," around the boundaries of the feasible region. Unlike penalty functions used by other methods, the barrier function treats all limits as hard constraints, permitting no violation whatsoever unless directed to do so by the user in the event of infeasibility. By replacing the inequality constraints with barrier functions, the algorithm is able to converge onto the correct set of binding inequality constraints while at the same time converging to a zero mismatch solution.

The "smooth" non-iterative strategy for determining the binding constraints, the main weakness of QP based methods, greatly speeds the optimization process. The barrier functions added to the nonlinear objective, apparently eliminate numerical problems associated with indefiniteness of the Lagrangian Hessian matrix (which produces negative curvatures in the QP) and consequently aid the reliable computing of the optimum. The smaller the barrier parameter specified by the user (i.e., the coefficient in the objective function that multiplies the barrier function), the closer the voltages, taps and other variables can approach their bounds.

Reasonable field experience from the use of this method strongly indicates the following:

- The method is consistently faster (10 to 20 times, depending on the problem) than the QP-based method available to PG&E.
- An increase of the final value of the barrier parameter from 10^{-6} to 10^{-10} (to produce more accurate solutions) increases the execution time by as such as 20 to 30 percent.
- The method is very robust and reliable even for large disturbances. Singularity or near singularity problems reported in [33] that are inherent to Newton OPF have never been observed.
- The method produces solutions that are as accurate and reasonable as the solutions produced by other mature and well accepted methods.
- For marginally feasible cases the method usually manages to compute reasonable solutions.
- Infeasibility detection and handling is consistently better than the one observed with the QP-based approach. However, the reasonableness of the least squares violation solutions many times is questionable. It appears that LP-based methods are better suited to handle infeasibility.
- For on-line applications, it may be necessary to resort to discontinuous modifications of the original OPF problem. Also engineering rules that reflect operating practices may be necessary. It appears that IP methods are not well suited to handle these on-line requirements.
- The method may also not be suited to handle other on-line modeling requirements, such as modeling of local control facilities.
- It appears at the present time, that IP methods may not well be suited to develop an effective "hot start" capability. LP-based techniques seem to have a clear advantage in meeting this requirement. LP techniques using the optimal basis of the previous LP need only a few rank 1 updates to solve the current LP. For starting points that are close to the optimal solution the method is losing its advantage even over QP-based methods. This method may also encounter numerical problems for starting points that are close to the optimum because the condition number of the Hessian goes to infinity as the barrier variable goes to zero, if the number of active constraints is less than the number of control variables.

- It is not clear, at this point, how successful IP methods can be in modeling contingency constraints, an essential requirement for many on-line applications.

The ultimate success of IP methods in on-line environment will eventually depend on their ability to solve not just classical and smooth nonlinear programming OPF problems, but also to provide acceptable solutions to realistically posed OPF applications.

5. Conclusions

On-line implementations pose the most onerous requirements on the OPF technology. As it currently stands classical OPF formulations expressed in smooth nonlinear programming form are far too approximate descriptions of the real-life problems to lead to successful on-line implementations. This is mainly true because current OPF formulations do not have the capability to incorporate all operational considerations into the solutions.

In this paper some of the requirements, that need to be met so that OPF applications are useful to and usable by the dispatchers, were presented and discussed. These include:

a) response time requirements,
b) robustness with respect to starting point,
c) expansion of the scope of the OPF problem to be able to solve realistically posed problems,
d) infeasibility detection and handling,
e) ineffective "optimal" rescheduling,
f) discrete modeling,
g) development of techniques/guidelines for selecting an "optimal trajectory" that steers the power system as reliably and as far as possible in the direction of the optimum,
h) modeling of contingency constraints,
i) consistency of OPF and other on-line functions,
j) data quality and other practical requirements,
k) maintenance and MMI and
l) on-line OPF based external modeling.

Over the last two decades several approaches have been proposed to solve the constrained nonlinear OPF problem. Some of these approaches have been implemented with various degrees of success into production grade OPF programs. These approaches include: Generalized Reduced Gradient, Quadratic Programming, Newton's method, Linear Programming and Interior Point techniques. Based on the above requirements, experience from the development and/or use in a practical environment of these techniques was also discussed.

References

1. J. Carpentier: Contribution a.' l'etude du Dispatching Economique; Bulletin de la Societe Francaise des Electriciens, Vol. 3, pp. 431-447. Aug. 1962.
2. H.W. Dommel and W.F. Tinney: Optimal Power Flow Solutions; IEEE Transactions on Power Apparatus and Systems, Vol. PAS-87, pp. 1866-1876, Oct. 1968.
3. H.H. Happ: Optimal Power Dispatch - A Comprehensive survey; IEEE PAS, Vol. PAS-96, pp. 841-854, May/June 1977.
4. B. Stott, O. Alsac, and A. Monticelli: Security Analysis and Optimization; Proceedings of IEEE, December 1987.
5. J. Carpentier: Towards a Secure and Optimal Automatic Operation of Power Systems; PICA 87, pp. 2-37.
6. R.R. Shoults, D.T. Sun: Optimal Power flow Based Upon P-Q Decomposition; IEEE Transactions on Power Apparatus and Systems, Vol. PAS-101, No. 2, pp. 397-405. February 1982.
7. R.C. Burchett, H.H. Happ, D.R. Vierath: A Quadratically Convergent Optimal Power flow; IEEE Transactions on Power Apparatus and Systems, Vol. PAS-103, pp. 3267-3276, November 1984.
8. H. Glavitsch and M. Spoerry: Quadratic loss formula for reactive dispatch; IEEE Transactions on Power Apparatus and Systems, vol. PAS-102, pp. 3850-3858, Dec. 1983.
9. B. Stott and E. Hobson: Power System Security Control Calculations Using Linear Programming, Parts I and II; IEEE Transactions on Power Apparatus and Systems, vol. PAS-97, pp. 1713-1731, Sept./Oct. 1978.
10. B. Stott and J.L. Marinho: Linear Programming For Power System Network Security Applications; IEEE Transactions on Power Apparatus and Systems, vol. PAS-98, pp. 837-848, May/June 1979.
11. K.R.C. Mamandur and R.D. Chenoweth: Optimal Control Of Reactive Power Flow For Improvements In Voltage Profiles and For Real Power Loss Minimization; IEEE Transactions on Power Apparatus and Systems, vol. PAS-100, pp. 3185-3193, July 1981.

12. K.R.C. Mamandur: Emergency Adjustments To VAR Control Variables To Alleviate Over-Voltages, Under Voltages and Generator VAR Limit Violations; IEEE Transactions on Power Apparatus and Systems, vol. PAS-101, pp. 1040-1047, May 1982.

13. D. I. Sun, B. Ashley, B. Brewer, A. Hughes, and W.F. Tinney: Optimal Power Flow By Newton Approach; IEEE Transactions on Power Apparatus and Systems, vol. PAS-103, pp. 2864-2880, Oct. 1984.

14. G.A. Maria and J.A. Findlay: A Newton Optimal Power Flow Program For Ontario Hydro EMS; IEEE Transactions on Power Apparatus and Systems, vol. PWRS-2, pp. 576-584, Aug. 1987.

15. K.A. Clements, P.W. Davis, K.D. Frey: Treatment of Inequality Constraints in Power System State Estimation; IEEE Winter Meeting 1992, Paper 92WM 111-5 PWRS.

16. L.S. Vargas, U.H. Quintana, A. Vannelli: A Tutorial Description of an Interior Point Method and its Applications to Security-Constrained Economic dispatch; IEEE Summer Power Meeting 1PP2, Paper 92SM 416-8 PWRS.

17. O. Alsac, J. Bright, M. Prais, B. Stott: Further Developments in LP-Based Optimal Power Flow; IEEE Transactions on Power Apparatus and Systems, Vol. 5, No. 3, pp. 697-711, August 1990.

18. M.A. El-Kady, B.D. Bell, V.F. Carvalho, R.C. Burchett, J.J. Happ and D.R. Vierath: Assessment Of Real-Time Optimal Voltage Control; IEEE Transactions on Power Apparatus and Systems, Vol. PAS-1, No. 2, pp. 98-107, May 1986.

19. A.D. Papalexopoulos, C.F. Imparato and F.F. Wu: Large-Scale Optimal Power Flow: Effects of Initialization, Decoupling and Discretization; IEEE Transactions on Power Apparatus and Systems, Vol. PWRS-4, pp. 748-759, May 1989.

20. D.S. Kirschen and H.P. Van Meeteren: MW/Voltage Control in a Linear Programming Based Optimal Power Flow; IEEE Transaction on Power Apparatus and Systems, Vol., PWRS-3, pp. 481-489, May 1988.

21. S.V. Venkatesh, E. Liu, A.D. Papalexopoulos: A Least Squares Solution For Optimal Power Flow Sensitivity Calculations; IEEE Transactions on Power Apparatus and Systems, Vol. 7, No. 3, pp. 1394-1401, August 1992.

22. P.E. Gill, W. Murray, M.H. Wright: *Practical Optimization*; Academic Press, 1981.

23. R. Yokoyama, S.H. Bae, T. Morita, H. Sasaki: Multiobjective Optimal Generation Dispatch Based On Probability Security Criteria; IEEE Transactions on Power Apparatus and Systems, vol. PWRS-3, n. 1, pp. 317-324, 1988.

24. W.F. Tinney, J.M. Bright, K.D. Demaree and B.A. Hughes: Some Deficiencies in Optimal Power Flow; IEEE PICA Conference Proceedings, pp. 164-169, Montreal, May 1987.

25. E. Liu, A.D. Papalexopoulos, W.F. Tinney: Discrete Shunt Controls in A Newton Optimal Power Flow; IEEE Winter Power Meeting 1991, Paper 91WM 041-4 PWRS.

26. W.C. Merritt, C.H. Saylor, R.C. Burchett and H.H. Happ: Security Constrained Optimization - A Case Study; IEEE Transactions on Power Apparatus and Sytems, Vol. PWRS-3, pp. 970-977, Aug. 1988.

27. IEEE Current Operating Problems Working Group Report: On-Line Load Flows From a System Operator's Viewpoint; IEEE Transactions on Power Apparatus and Systems, vol. PWRS-102, pp. 1818-1822, June 1983.

28. M. Innorta and P. Marannino: Very Short Term Active Power Dispatch With Security Constraints; in Proceedings IFAC Symposium on Planning and Operation of Electric Energy Systems (Rio de Janeiro, Brasil, July 1985), pp. 379-386.

29. J. Carpentier: Optimal Power Flows: Uses, Methods and Developments; in Proceedings IFAC Symposium on Planning and Operation of Electric Energy Systems (Rio de Janeiro, Brazil, July 1985), pp. 11-21.

30. R. Bacher and H.P. Van Meeteren: Real-Time Optimal Power Flow in Automatic Generation Control; IEEE Transactions on Power Apparatus and Systems, Vol. PWRS-3, pp. 1518-1529, Nov. 1988.

31. N. Karmarkar: A New Polynomial Time Algorithm for Linear Programming; Combinatorica, May 1984.

32. P.E. Gill, W. Murray, M.A. Saunders: Interior-PointMethods for Linear Programming: A Challenge to the Simplex Method; Technical Report SQL 88-14, Department of Operations Research, Stanford University, 1988.

33. A. Monticelli, E. Liu: Adaptive Movement Penalty Method for the Newton Optimal Power Flow; presented at the IEEE/PES 1990/Winter Meeting, Atlanta, Georgia, February 4-8, 1990.

COST/BENEFITS ANALYSIS OF THE OPTIMAL POWER FLOW

K. Kato

ECC, Inc.
Garden Grove, California U.S.A.

Abstract. In order to acquire an optimal power flow (OPF) for a utility control center, it is usually necessary to justify the OPF to the utility management, especially during these difficult financial times. An example is presented of an accurate simulation of the cost savings obtained by the OPF for a large utility. The optimization problem that was originally investigated was restricted to the minimization of production cost while eliminating thermal overloads using real power controls, such as generation MW, interchange MW, and load shedding. Thermal overloads usually occur during outage conditions in the power system. Several actual outage conditions were replicated to determine the cost savings of an optimal dispatch over an economic dispatch. The optimization problem that was studied was to operate the above real power controls in a preventive manner such that in the event of contingency outages, the post-contingency operating state does not have any thermal overloads.

The methodology of the simulation approach was to simulate an optimal dispatch (using the OPF) and a conventional economic dispatch of the utility power system for a number of actual operating conditions obtained from historical records. The difference in production cost between the optimal dispatch and economic dispatch represented the cost savings; i.e., the benefits of the OPF.

The results of the simulation study showed that, for the given conditions of the particular utility, the OPF can produce annual savings of 900.000 ECU in 1992 and increase each year to 1.6 million ECU in 2001. The total savings over a ten year time interval is 11.6 million ECU. Furthermore, the simulations showed that the OPF can reduce significantly the amount of load shedding over other methods, such as manual action.

Consequently, the benefits analysis was able to provide accurate quantitative cost savings and was able to convince the utility management to proceed with the optimal power flow.

1. Introduction

1.1 Motivation

In order for a utility control center to acquire an optimal power flow (OPF), it is usually necessary to justify the OPF to the utility management due to the large expense and significant manpower effort to support it. This is especially necessary during these difficult financial times. The type of questions that utility managements are likely to ask regarding justification of the OPF are: What are the benefits of the OPF for the utility? How much money can the OPF save for the utility?

Benefits analysis can answer these types of questions and thus is an important part of the management justification process. The benefits of the OPF can be expressed in quantitative and qualitative terms. The quantitative benefits of the OPF can be described in a variety of ways, such as fuel cost savings, productions cost savings, real power loss reduction, etc. Qualitative benefits are typically reduction of load shedding or improvement in security of the power system.

1.2 Types of cost/benefits analysis

The quantitative benefits can be expressed either in approximate terms using results from published results and scaled to a utility or more accurately by means of a simulation of the OPF benefits for a utility. An example is presented of the approximate method using published results scaled to a utility. An example is presented of an accurate simulation of the cost savings obtained by the OPF for a large utility.

2. Approximate analysis

The methodology of the approximate analysis approach is to scale the results of published studies to the particular utility [3,4]. The approximate analysis method has the advantage of obtaining results quickly with relatively small amounts of data. The disadvantage is that the results are only approximate and hence may not be believed by the utility management.

As an example, one objective function of the OPF is the minimization of real power losses in the network. The minimization of losses is achieved by using the optimal settings of reactive controls such as generator unit vars, load tap changing transformers, synchronous condensor Mvars, and switchable shunt capacitors. The constraints to be observed are typically the voltage limits on network busses and the finite range of the controls, such as unit and synchronous condensor vars, and LTC tap limits.

The annual cost savings due to the loss minimization mode of operation of the OPF can be approximately calculated as (see [3,4]) :

$$\text{Annual Cost Savings} = \text{Loss Reduction} \times \text{Network Losses} \times \text{Annual Fuel Cost}$$

It has been shown that the loss minimization objective function of the OPF can reduce the losses in transmission networks by 4 to 10% [1,2]. The losses in a transmission network are typically in the range of 2 to 3%. As an example, assume that a utility has network losses of 3%, and the loss reduction due to the OPF is 5%. Then, the annual cost savings can be calculated as:

$$\text{Annual Cost Savings} = 0.15\% \times \text{Annual Fuel Cost.}$$

Typically, the percentage savings of the OPF are quite small. However, since the annual fuel costs are quite large, the annual cost savings, which is the product of the two numbers, is usually quite significant.

The annual fuel cost of a large utility can easily be 1 Billion ECU (European Currency Units). The annual cost savings due to loss minimization for a utility of this size is:

$$0.15\% \times 1 \text{ Billion ECU} = 1.5 \text{ Million ECU/year}\ .$$

3. Accurate analysis

The methodology of the accurate analysis is to accurately simulate the particular utility. The advantage of this method is the accuracy of the results. The disadvantages are that it is very time consuming, requires a large amount of historical data and other information, and requires an OPF and perhaps other computational tools.

The following material is an example of a cost/benefit analysis involving detailed simulation of the OPF in an Energy Management System (EMS) of a large utility.

3.1 Background

The utility system is presently dispatched by means of an economic dispatch (ED) technique. Economic dispatch does not handle network constraints such as thermal overloads in transmission lines and transformers. Consequently, manual adjustments to the economic dispatch solutions are necessary when overloads occur in the network. Thermal overloads are modelled in the OPF as transmission line/transformer flows that violate their thermal overload limits. Thermal overload limits are limits, or inequality constraints, beyond which the lines/transformers overheat and can eventually fail.

3.2 Objective

The objective of this particular study was to quantify the cost/benefits of the OPF to minimize production costs while eliminating thermal overloads. In addition, some non-quantitative benefits of OPF use in network optimization are discussed. The purpose of this cost/benefits analysis was to determine the cost savings from the use of the OPF function. This, among other incentives, helped the utility management decide the level of funding for future development in the network optimization area.

3.3 Scope of study

This particular study was restricted to the enhancement of the OPF in the first application. The first application of the OPF minimizes the production cost of the utility power system while eliminating thermal overloads by using only active power controls, such as generator MWs, phase shift transformers, interchange schedules, and load shedding.

Many other applications of the EMS OPF had been identified through interviews with system operations staff. Some of them are outlined below:

- Minimize production cost while eliminating thermal overloads
- Assist system operations engineers in contingency planning
- Maximize power transfers with other companies
- Identify limits for power transfers between the utility and other companies
- Evaluate generation clearances
- Define more realistic voltage schedules
- Recommend tap settings for fixed tap transformers

- Minimize active power transmission losses

This OPF application was chosen over the others because it added value to the company. Furthermore, it was the least difficult among the various OPF applications to implement and had minimal risk. Some of the other OPF applications may produce more benefits for the utility but they may also have significantly more risk. It is anticipated that experience gained from the use of the first OPF application will pave the way for the successful implementation of additional OPF applications.

3.4 Quantifiable cost/benefits

Presently, the utility uses economic dispatch to determine the least cost settings of the generators. Since economic dispatch does not recognize and observe line/transformer limits, manual adjustments are necessary to handle these network constraints.

The quantifiable benefits of the study were in terms of the reduction of the production cost between the optimal dispatch solution produced by the OPF and the present economic dispatch method. The quantification was in terms of the production cost savings reduction between the economic dispatch and the optimal dispatch methods, rather than the absolute production costs of the economic dispatch and the optimal dispatch methods.

This study considered the following two aspects of the production cost of operating the utility power system.

Corrective operation to eliminate presently existing thermal overloads

The first aspect of production cost was associated with the elimination of thermal overloads occurring in the power system. In the current operating condition, the utility system is dispatched so that thermal overloads do not occur. The only time that thermal overloads existed were when outages occurred. The number of transmission network outages per year was in the range of a hundred. The average duration of the transmission outages was approximately one hour. For an outage that resulted in a thermal overload, the production cost savings was essentially the product of the difference between the optimal and economic dispatches and the time duration of the outage. As a result of the short durations of the outages and the corresponding thermal overloads, it was expected and the study results confirmed that the production cost savings of the optimal dispatch over the economic dispatch were rather small.

Preventive operation to eliminate contingency overloads

The other aspect of production cost was related to the preventive nature of operation of the utility power system in order to eliminate thermal overloads due to the occurrence of contingency outages. Typically, the utility operates in the preventive mode for long time periods in order to eliminate the thermal overloads caused by contingency outages and to mitigate the effects of contingency outages, e.g. minimize the duration of the overloads. One example is the preventive operation during heavy transaction flows on the utility 500 kV network which may occur during the winter months. Consequently, it was not unexpected that the bulk of the production cost savings of the optimal dispatch as opposed to the economic dispatch came from the preventive operating mode.

More efficient utilization of the utility network resources

Another aspect of the quantifiable benefits of the OPF in the Control Center Energy Management System (EMS) is the capability to develop optimal operating strategies for current power system problems using real-time data. Presently, the systems operation group of the utility develops operating guidelines based upon many studies run in the off-line mode. These operating guidelines must be very broad in scope in order to handle a wide range of operating conditions. Consequently, in order to prevent jeopardizing the safe and reliable operation of the power system, the operating instructions must be conservative. As a result, they may not allow dispatchers to operate the system as efficiently or as economically as possible, given the actual operating conditions. Whereas, the EMS OPF, because it uses actual real-time data, can develop optimal operating strategies to obtain maximum efficiency and economy. The EMS OPF can be used by the system operations group to develop more efficient operating guidelines. It can also be used by the dispatchers to determine optimal control settings to handle real-time operating problems.

Lost revenue due to load shedding

Another quantitative benefit was the reduction of lost revenue due to reducing the amount of load shedding used as a last resort control in order to eliminate thermal overloads. While it may be possible to determine the amount of lost revenue due to load shedding, it would be very time-consuming to obtain such results. Consequently, due to the time limitations in conducting this study, the revenue loss was not quantified. However, the media impact of load shedding, or customers dropped, is discussed below.

3.5 Non-quantifiable benefits

There were several non-quantifiable benefits of the OPF. The first benefit was the potential reduction in the amount of load shedding. The amount of load shed is an indirect indication of the number of customers dropped. This benefit was very important because of the adverse public reaction and media attention on events of this nature. These factors may have a significant effect on the utility due to the possibility of indirectly affecting the approval or denial of rate increases and other utility activities with the regulatory agencies.

The second benefit was the reduction in the over-stressing of the equipment which can shorten the operating life of the equipment. This benefit also was not quantified but it is certain that OPF dispatches can provide some relief in over-stressed equipment and in doing so, prolong their operating life.

The third benefit was that the OPF can formulate corrective strategies that the dispatchers may not be aware of. This last benefit can arise in the event that the operating guidelines do not cover the operating conditions that the system is presently in. While it is the case that system operations does run many contingency studies that cover many hypothesized operating conditions, it is also conceivable that certain unusual conditions may not have been studied. For instance, due to forced outages, the power system may not be in the normal base conditions for the various seasons of the year. Then, when additional outages occur, the power system is in a second or third order contingency condition, for which operating studies may not be available and thus there may be no operating guidelines.

3.6 Cost/benefits assessment methodology

The basic methodology of the study was to compare the production costs of the optimal dispatch and economic dispatch for a variety of operating conditions, all of which require some re-dispatching to eliminate thermal overloads. The annual production cost differential between the optimal and economic dispatch was determined. A cost/benefits evaluation program was used to analyze the costs and benefits [5].

Total life cycle benefits

The benefits of the OPF accrue over each year of its operating life, consequently the benefits will be accumulated over its operating life. For the purposes of this study the life span of the OPF was arbitrarily defined to be 10 years.

Simulation of optimal dispatch

Simulation of the optimal dispatch was performed by running the EMS OPF program using models of the utility power system for the conditions under study.

Simulation of the economic dispatch

The simulation of the economic dispatch was performed by the system operations group of the utility. The economic dispatch was primarily based on economics with some manual adjustment of the units in accordance with the operating guidelines and control sensitivities for the various operating conditions (e.g. thermal overloads) in the utility system. The system operations group made manual adjustments to the economic dispatch solutions to eliminate any overloads.

3.7 Test data description

Power system

In order to obtain results that were realistic for the utility, real-time data from the utility EMS was used. The power system models used in this study were operations models provided by the system operations group of the utility.

An extensive review of recent past conditions of the utility was performed. The network base case selected for this study was based on the 1990 summer peak case. The rationale for selecting this case was that it represented the most recent complete year and the system operations group had a network model for peak load conditions of that year. It exhibited thermal overloads due to the heavy loading during the peak load condition.

Historical records were studied to understand the recent past outage conditions of the utility power system. A number of actual or credible contingency cases applied to the base case were determined. These contingency cases consisted of outages that caused thermal overloads. The following four outage cases were established for this study.

Outage case 1 involved the outage of a 500 kV line during the summer peak base case with a heavy interchange flow with a neighboring company.

Outage case 2 involved the outage of a 500/230 kV transformer bank during the summer peak base case. This outage case required extensive corrective redispatch. There were two 500/230 kV transformer banks at the substation involved. With one bank out and normal peak load conditions, the second bank overloaded to 122% without any corrective control action.

Outage case 3 involved the outage of a 230/115 kV transformer bank during the summer peak base case. It required network switching and load shedding as corrective actions. With one 230/115 kV bank out at the station involved, the other 230/115 kV bank overloaded to 150% without any corrective action.

Outage case 4 involved the outage of a 500 kV line at an off peak winter system load with heavy interchange flows. This outage was different than the four previous outages in that preventive action was required instead of post-outage corrective action. This preventive action was necessary to prevent severe overloads of 230 kV lines after the outage. One of the main factors in this outage case was the limitation on the power transfers with an external company, which occur primarily during the winter and occasionally during the late autumn and early spring.

The hydro and steam generation data for the test cases were obtained from historical records. The gamma values (worth of water) of the hydro units were calculated by a scheduling program for the summer peak load conditions for the given load levels, thermal generator loadings, hydro unit loadings, and hydro conditions.

The unit incremental cost curves, fuel cost of the thermal units, and the equivalent fuel cost of the hydro units were obtained from the EMS database for the summer peak load day.

The utility long term fuel price forecast was provided by the utility fuel department. The average annual fuel price increase for 10 years, from 1992 to 2001, was 5.40% per year, according to the long term fuel price forecast.

Operating guidelines for economic dispatch

The utility operating guidelines were used in the economic dispatch of the power system.

Cost information

The procurement cost of the OPF was not considered because the OPF was a part of the EMS already delivered to the utility. The utility manpower to enhance this OPF application was estimated to be two man-years. The utility manpower estimate to maintain it was three man-months per year for each year of operation. The annual salary cost per programmer for the enhancement/maintenance of the OPF was 70.000 ECU. The salary rate increases were assumed to be 6% for each year of the OPF enhancement/maintenance period.

3.8 Simulation results

It was noticed that thermal overloads occurred in most of the economic dispatch cases for the Base Case and Outage Cases 1 through 4. The utility does not operate continuously with thermal overloads. It may be possible that the overloads in the economic dispatches could have been eliminated by gas turbines (GTs) being turned on. However, these cases were not re-run due to time and budget limitations of the study. In spite of these conditions, the results of the study were still valid for the following reason. If the economic dispatch cases and optimal dispatch cases had used gas turbines, the thermal overloads in all likelihood would be eliminated. The production costs of the economic dispatches and optimal dispatches for the GTs turned on in both cases would be higher than that with the GTs off. It is highly likely that the increase in the production costs of the economic and optimal dispatches would be very close. Hence, the cost savings (difference between the two production costs) would be the same (or possibly even slightly better) than the situation without the GTs being on. Since the primary quantitative output of this study was the cost savings, or difference between the two dispatches, the study results were still valid.

For the cases in which the economic dispatch runs had overloads, two OPF dispatch runs were made. One optimal dispatch was to minimize the production cost and eliminate the thermal overloads in the power system. Eliminating the overloads in the OPF run was referred to as enforcing the constraints. The other OPF dispatch run was to determine the effect on production cost if the overloads were permitted to exist as they did in the economic dispatch. This was necessary in order to replicate the same conditions to permit a valid comparison of production cost between the economic and optimal dispatches. Not requiring the overloads to be eliminated in the OPF run was referred to as relaxing the constraints. It was noted that negative cost savings implied an increased cost in the optimal dispatch over the economic dispatch, whereas a positive cost savings implied a reduced cost.

Controllable generation was the generation that can be controlled by hydro and conventional steam/fossil units on AGC. The production cost was for only the controllable generation. Non-controllable generation consisted of base loaded generation, such as nuclear plant, cogeneration, or non-utility generation (NUG). The generation of the NUGs and the cogenerators were treated as negative loads.

Base case

The economic dispatch of the summer peak load base case had several minor overloads. In all likelihood these overloads would not be permitted to exist in the actual operating power system. The production costs of the utility system were calculated using the fuel costs, incremental cost curves, no-load generation cost, etc. as supplied by the system operations group.

There were two optimal dispatch runs for this test case as shown in Table 3.8.1. One optimal dispatch was to eliminate the thermal overloads in the power system. For this optimal dispatch, the production cost savings was -0.00218% per hour. The negative sign implied that there was a small penalty imposed in order to eliminate the thermal overloads in the base case. The other optimal dispatch run was to determine the effect on production cost if the overloads were permitted to exist as they did in the economic dispatch. This was necessary in order to replicate the same conditions to permit a valid comparison between the manual and optimal dispatch. In this case, the production cost savings was +0.127% per hour. This implied that the optimal dispatch that reduced production cost while allowing the same thermal overloads as in the economic dispatch can save 0.127% of the hourly production cost. It should be noted that at the peak load, the units were almost fully utilized. Consequently, there was rather limited maneuvering room for the units to be optimized by the OPF. It is expected that, at lower load levels, the OPF would produce even larger cost savings.

Summer peak case	OPF/enforced	OPF/relaxed
Production Cost Savings, %/H	-0.00218	+0.127

Table 3.8.1: Summer Peak Base Case Production Cost Savings

Outage case 1

The economic dispatch for the Outage Case 1 500 kV line post-outage case on the summer peak base case resulted in several overloads. As noted above, in the actual system, the economic dispatch would be further adjusted to eliminate the overloads.

As above, two optimal dispatches were run for this post-outage case as shown in Table 3.8.2. One optimal dispatch eliminated the thermal overloads at a production cost savings of -0.351% per hour, the negative sign indicating increased cost. The second optimal dispatch relaxed the constraints; i.e., allowed the same level of overloads to exist but minimized the production cost. Again, in the real world, this condition would not be allowed to happen. However, for this study, in order to obtain a valid comparison, it was necessary to replicate the same overload conditions to get a true measure of the OPF's ability to minimize production cost. The production cost savings for this optimal dispatch was 0.287% per hour.

Outage Case 1	OPF/enforced	OPF/relaxed
Production Cost Savings (%/H)	-0.351	+0.287

Table 3.8.2: Outage Case 1 Production Cost Savings

Outage case 2

The 500/230 kV bank outage was taken against the summer peak base case. The economic dispatch had a number of overloads.

It was noted that the OPF run with enforced constraints was infeasible; i.e., using only the conventional generation sources, no OPF solution was found to eliminate the thermal overloads. Consequently, only the results of the OPF dispatch with relaxed constraints were provided. The production cost savings for the relaxed constraint OPF run with the same constraint relaxation as in the economic dispatch run was 0.100% per hour.

Outage case 1	OPF/enforced
Production Cost Savings (%/H)	-0.100

Table 3.8.3 Outage Case 2 Production Cost Savings

Outage case 3

The test cases for the 230/115 kV bank outage require some explanation. The first test case for the bank outage was an economic dispatch in which load was already shed, a line was switched out, and several gas turbines were started. The next two test cases were OPF runs in which the same amount of load was shed. Consequently, in these runs the OPF did not try to optimize the load shedding since the load was already shed. These two OPF runs optimized the generator units to minimize production cost with the constraints enforced and with the constraints relaxed. The last test case was an OPF run in which the load shedding control was actually optimized.

The 230/115 kV bank outage was taken against the summer peak base case. The economic dispatch of the post-outage case had a number of thermal overloads. The economic dispatch of the post-outage case included line switching to transfer load to other busses, shedding load, and using gas turbines.

Outage case 3	OPF enforced/no L.S.	OPF relaxed/no L.S.	OPF enforced L.S. Cntl
Savings (%/H)	-0.536	+1.83	-5.71

L.S.: Load Shedding

Table 3.8.4 Outage Case 3 Production Cost Savings

There were two optimal dispatches that corresponded to the economic dispatch, both involved line switching with load shedding already accomplished. One optimal dispatch totally eliminated the thermal overloads for the same amount of load shedding. The production cost savings for this optimal dispatch was -0.536% per hour. The other optimal dispatch relaxed the constraints to correspond to the overloads in the economic dispatch run. The production cost savings for this optimal dispatch was 1.83% per hour.

A third OPF dispatch eliminated the overloads and reduced the amount of load shedding by 16.4%. The production cost savings for this optimal dispatch was -5.71% per hour. The OPF modelled load shedding as very costly generation (about 10 times more expensive than the most costly generator). Consequently, the cost savings was negative and implied an increasing cost. Whereas, in the first two OPF runs, the load had already been shed and hence had no costs associated with these shed loads.

Outage case 4

a) Interchange flow pre-outage condition

The economic dispatch had no overloads. According to the utility dispatching instructions, the interchange flow with another company was restricted. The optimal dispatch had a production cost savings of 1.50% per hour and no overloads. In the OPF run of the pre-outage condition, the optimal dispatch was able to increase the power transfer capability of the interchange flow by 5.69% and yet was able to meet the post-outage line/transformer constraints with the optimal post-outage corrective control actions. The annual cost savings for optimal dispatch for 12 hours per day for six months (mid-October through mid-March) was 2190 hours $\times$ 1.5% $\times$ Hourly Production Cost = 0.375% of Annual Fuel Cost.

b) Post-outage condition

The economic dispatch of the post-outage case did not result in any overloads. The optimal dispatch likewise did not have any overloads. The production cost savings of the optimal dispatch was 1.18% per hour.

Outage case 4	OPF/enforced
Production Cost Savings (%/H)	1.5

Table 3.8.5.1 Outage Case 4 Pre-Outage Production Cost Savings

Outage case 4	OPF/enforced
Production Cost Savings (%/H)	1.18

Table 3.8.5.2 Outage Case 4 Post-Outage Production Cost Savings

3.9 Cost/benefits quantification analysis

The potential annual cost savings of this OPF application were obtained by the preventive operating mode, as exemplified by the pre-outage condition of Outage Case 4. There were additional savings due to eliminating the thermal overloads in the corrective operation mode, but these savings were negligible in size compared to the savings due to preventive operations. Thus, the savings due to corrective operation were not included in the analysis. The one preventive operation example of Outage Case 4 accounted for potential savings of 0.375% of the annual fuel cost. Using this example only, the annual and total cumulative 10 year cost savings (in 1992 ECU) of the OPF in the first application, taking into account the annual fuel price increases, were as shown below.

YEAR	BENEFITS	COST	BENEFITS-COST
1992	1.035.952	140.000	895.952
1993	1.091.893	18.550	1.073.343
1994	1.150.856	19.663	1.131.193
1995	1.213.002	20.843	1.192.159
1996	1.278.504	22.093	1.256.411
1997	1.347.543	23.419	1.324.124
1998	1.420.311	24.824	1.395.487
1999	1.497.007	26.314	1.470.693
2000	1.577.846	27.892	1.549.954
2001	1.663.049	29.566	1.633.483
TOTAL	13.275.963	353.164	11.598.675

Table 3.9.6: Annual Cost/Benefits (ECU) of OPF

The annual and total life span cost to enhance and maintain this OPF application are shown above. The first year cost includes two man-years of

OPF enhancement. The succeeding years show cost associated with the three man-months of OPF maintenance effort per year and take into account salary increases. All cost figures are expressed in 1992 ECU.

3.10 Lessons learned

Need for appropriate hydro unit gamma values

During the course of this study, it became very obvious that the gamma values can change the OPF results significantly.

This fact was demonstrated in several OPF runs, in which the hydro unit gamma values for a typical autumn day were used. The resulting OPF and economic dispatch solutions were observed to be unrealistic. The gamma values themselves can change significantly during a week and even during a day from hour to hour. Consequently, in real-time there is a need for the OPF to access the most recently updated hourly gamma values of the hydro units as computed by the scheduling programs. Similarly, for OPF studies, the scheduling programs should be run in order to produce gamma values for the appropriate conditions.

Need for constraint relaxation for short duration overloads

Another observation that was made during the study is that there may be short durations, e.g. load peaks, when minor overloads may be tolerated without jeopardizing the power system. Consequently, there may be a need for the OPF to be used in a constraint relaxation mode.

Another reason to have the relaxation mode is that it may be very expensive or very difficult to observe the constraints. In these conditions, depending upon the discretion of the users, relaxation may be tolerable, even desirable.

4. Conclusions

The benefit/cost analysis showed that the OPF can yield significant savings for utilities. Therefore, the OPF can be justified to utility managements for inclusion in the control center energy management system.

Using an approximate approach, it was shown that 0.15% of the annual fuel cost of a utility can be saved using the OPF in the loss minimization mode. The simulation study showed, under the assumptions used in the study, that one OPF application can potentially achieve savings (benefits-cost) of 0.375% of the annual fuel cost. For the large utility, the savings amounted to approximately 900.000 ECU in 1992 and increasing each year to 1.6 million ECU in 2001. The total savings over a 10 year time span were estimated to be 11.6 million ECU (in 1992 ECU). It was also shown that, under conditions in the study, the OPF can reduce the amount of load shedding by 16.4% and can increase power transfer capability by 5.69%. The cost associated with the OPF enhancement is 140.000 ECU in 1992. In the succeeding years, the maintenance costs range from ECU 18.550 in 1993 to ECU 29.566 in 2001. The total cost for the enhancement and maintenance for the 10-year time interval is ECU 353.164 (expressed in 1992 ECU).

References

1. M. A. El-Kady, B. D. Bell, V. F. Carvalho, R. C. Burchett, H. H. Happ, D. R. Vierath: Assessment of Real-Time Optimal Voltage Control; IEEE/PES 1985 Summer Meeting, Paper 85 SM 489-0.
2. D. Denzel, K.W. Edwin, F. R. Graf, H. Glavitsch: Optimal Power Flow and Its Real-Time Application at the RWE Control Center; CIGRE Session, Study Group 39, Paper #39-19, Paris, 1988.
3. K. Kato, J. T. Robinson: Economic Justification for an Energy Management System; Pennsylvania Electric Association, Allentown, Pennsylvania, January 10-11,1991.
4. K. Kato, J.T. Robinson; Justification of Power Application Functions; Edison Electric Institute, Engineering & Operating Computer Forum, St. Louis, Missouri, September 9-12, 1990.
5. ECC Inc. : BCE, Benefit/Cost Evaluator; San Jose, California, 1990.

GPSR Compliance

The European Union's (EU) General Product Safety Regulation (GPSR) is a set of rules that requires consumer products to be safe and our obligations to ensure this.

If you have any concerns about our products, you can contact us on ProductSafety@springernature.com

In case Publisher is established outside the EU, the EU authorized representative is:

Springer Nature Customer Service Center GmbH
Europaplatz 3
69115 Heidelberg, Germany

Batch number: 10726298

Printed by Printforce, the Netherlands